# African Wildlife

## A Photographic Journal

*Photography By Ronn Maratea, Writing by Pamela Barcita*

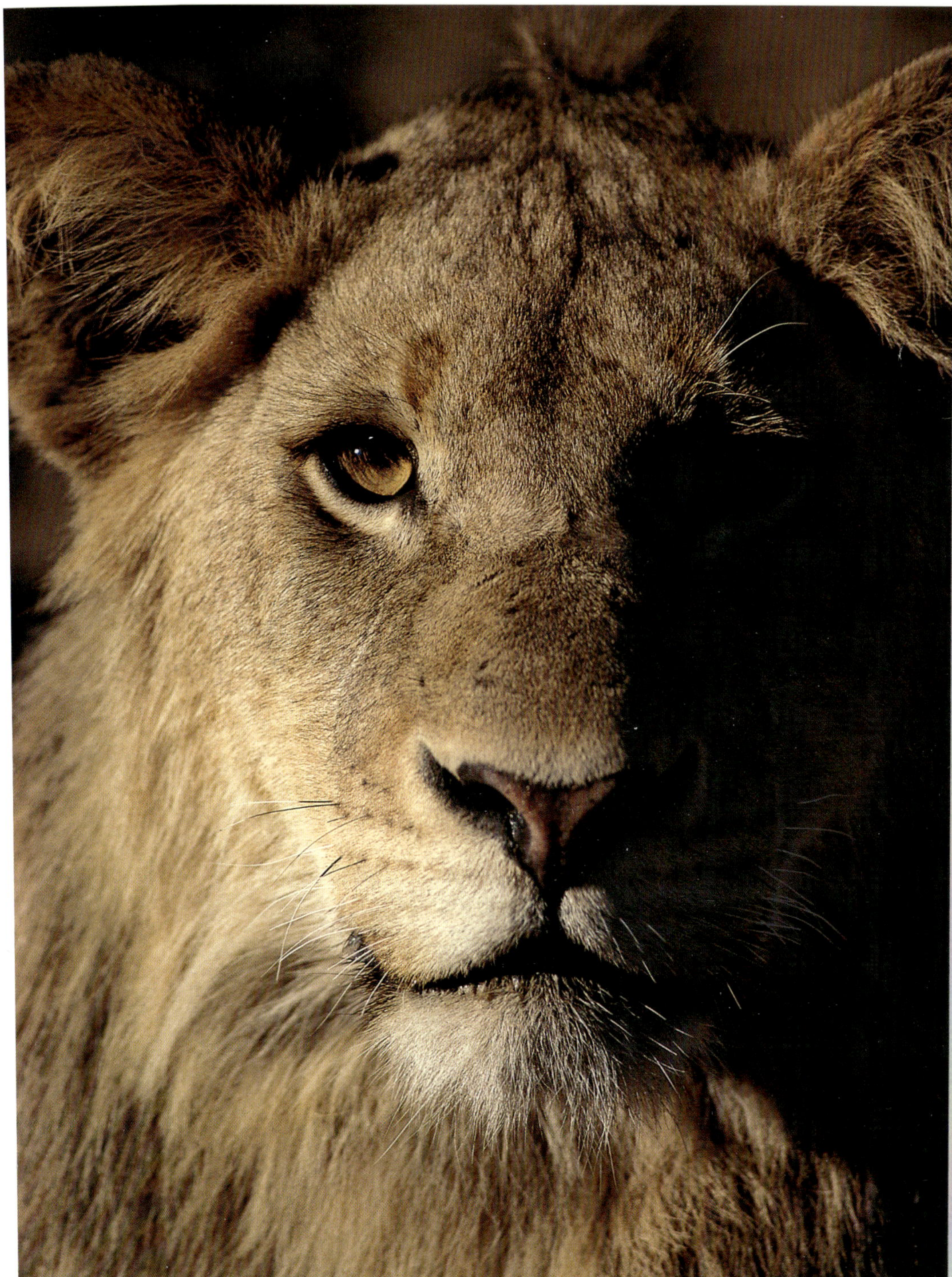

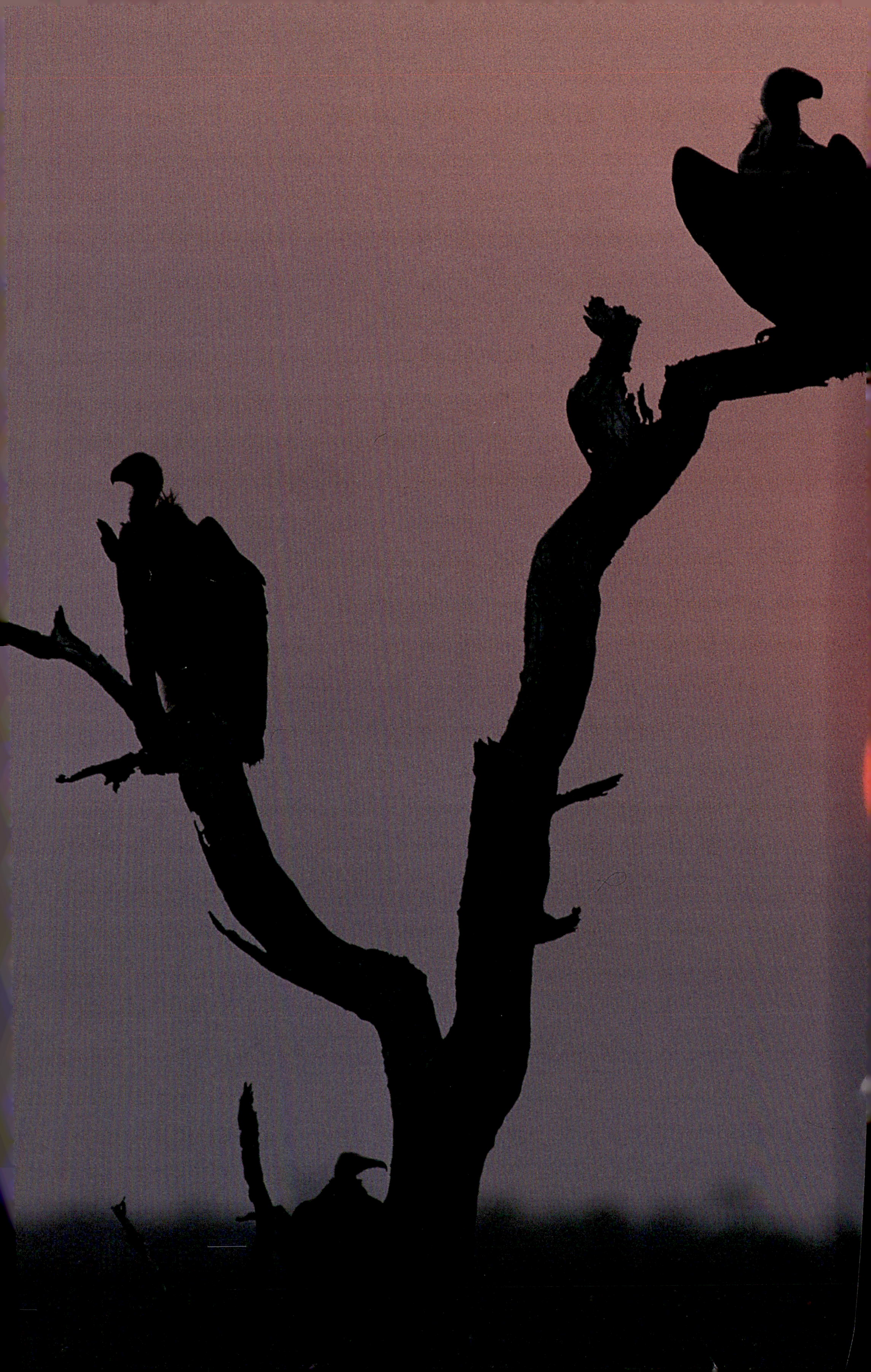

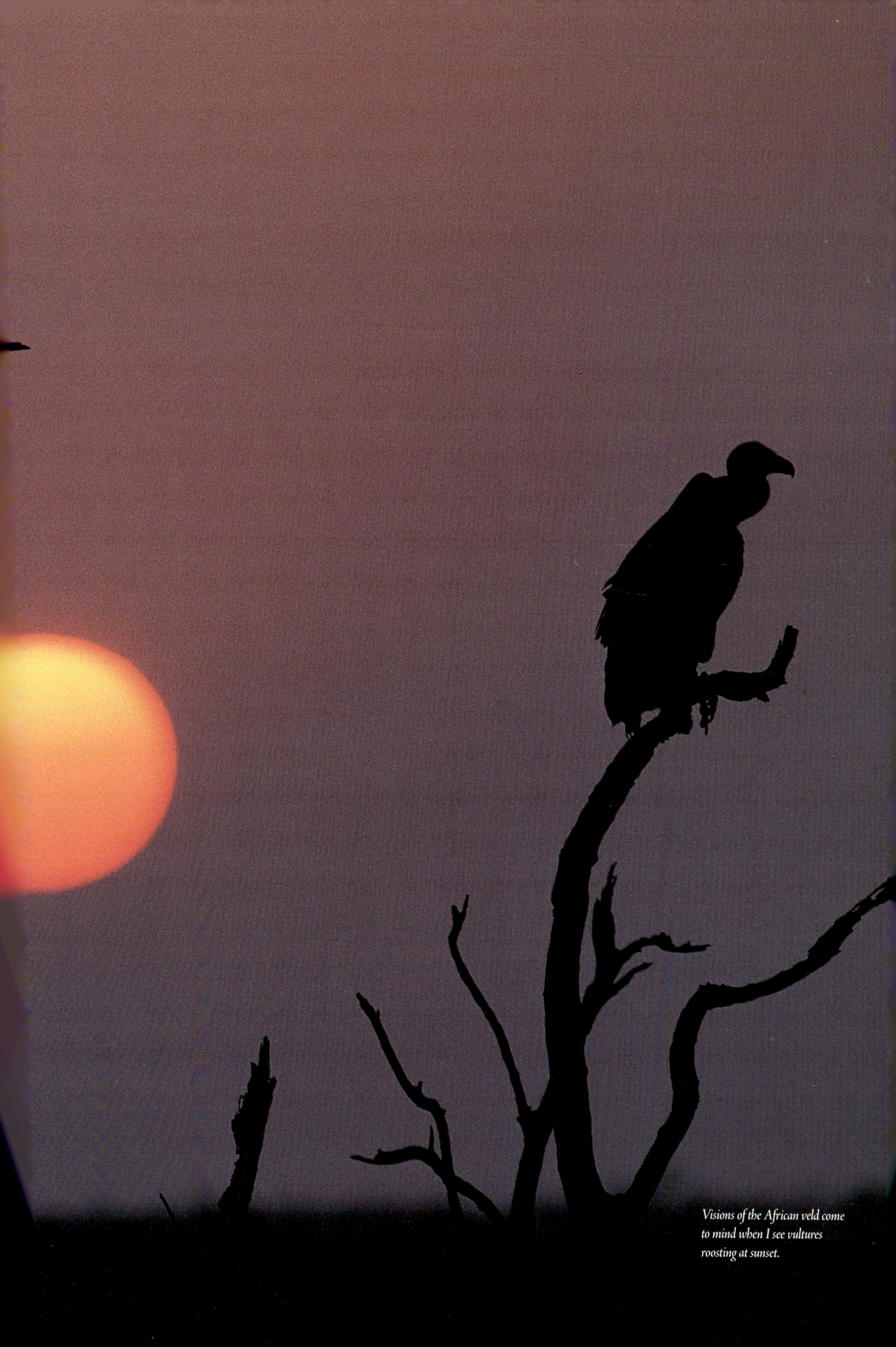

*Visions of the African veld come to mind when I see vultures roosting at sunset.*

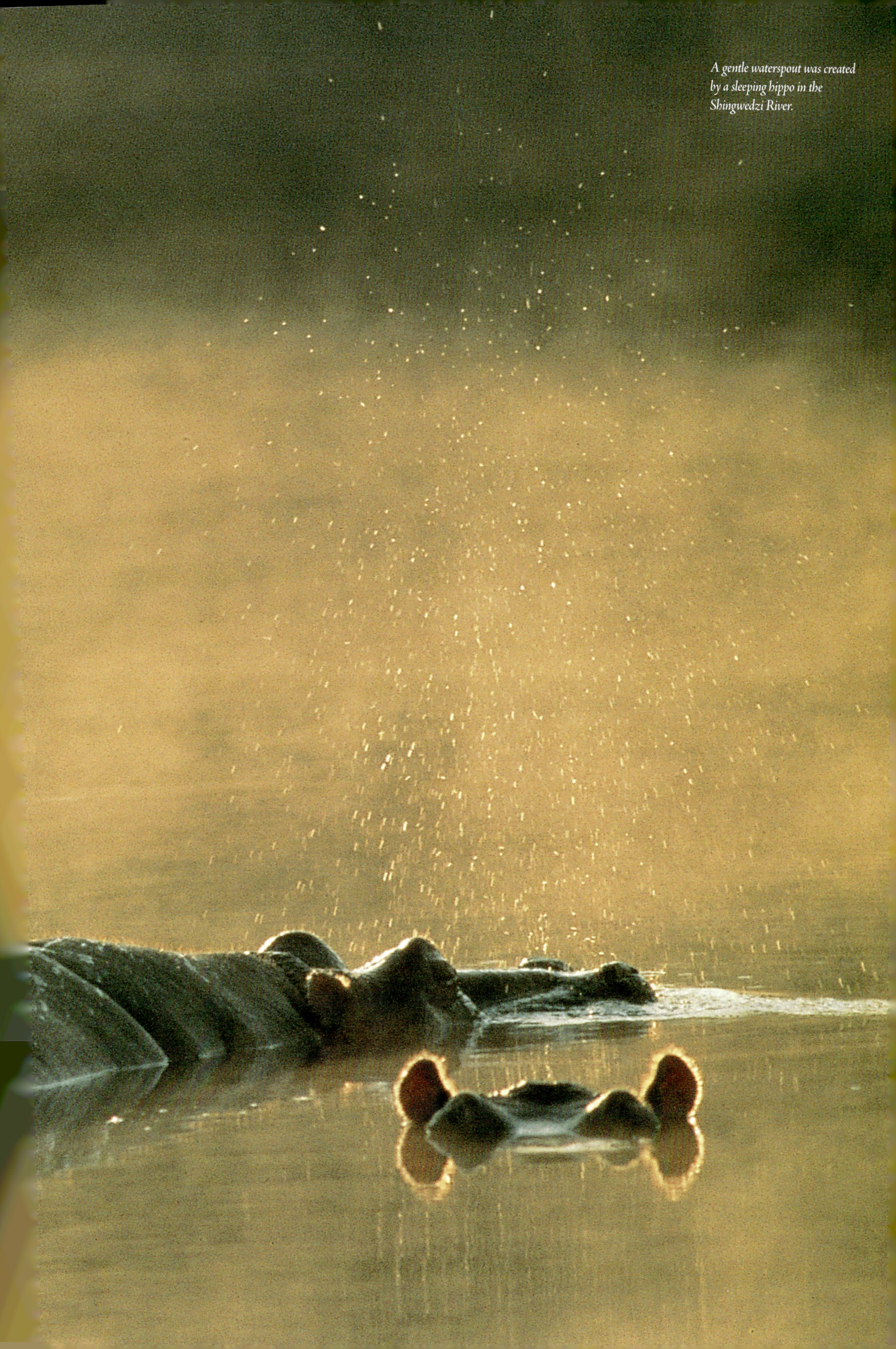

*A gentle waterspout was created by a sleeping hippo in the Shingwedzi River.*

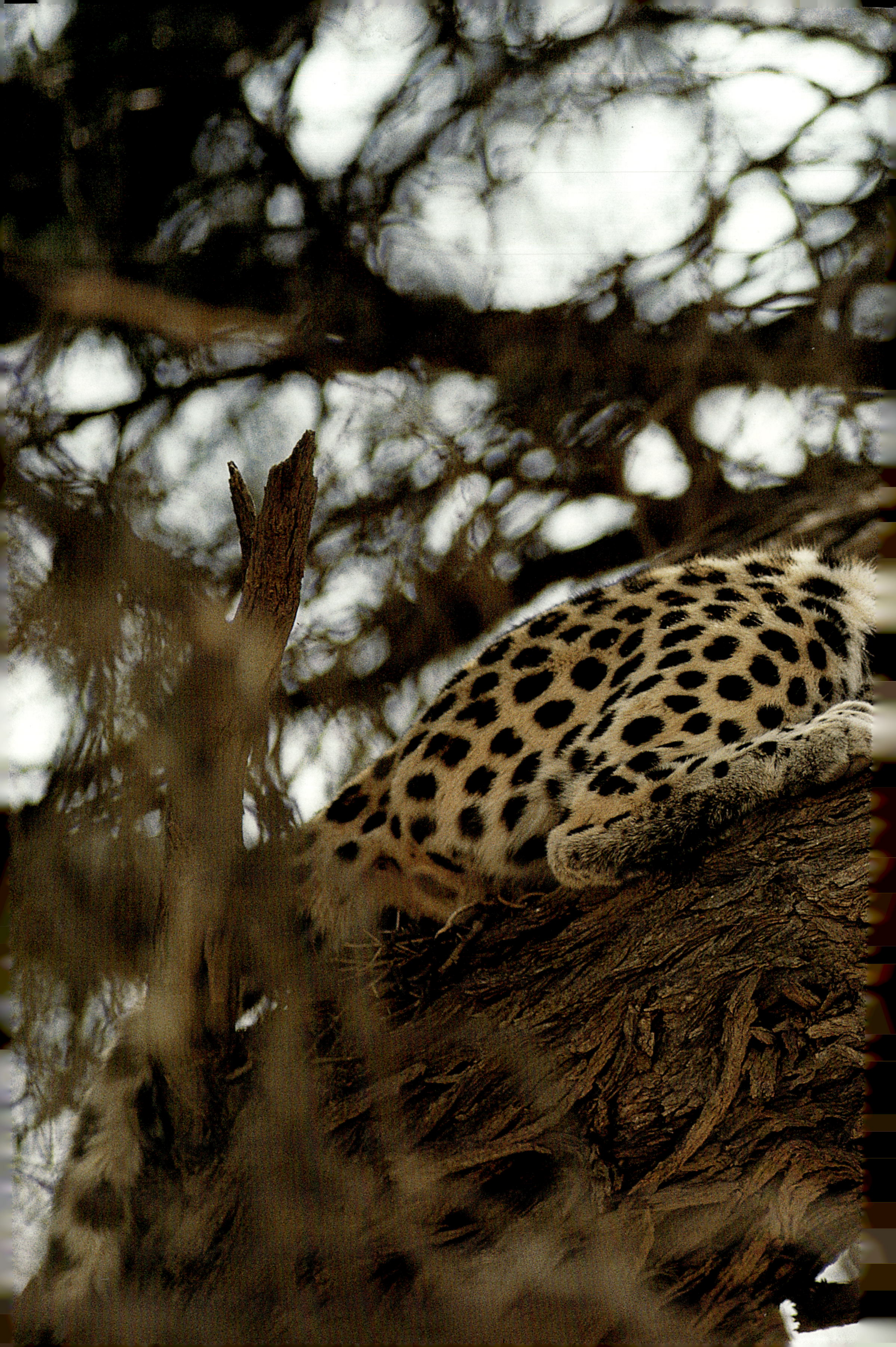

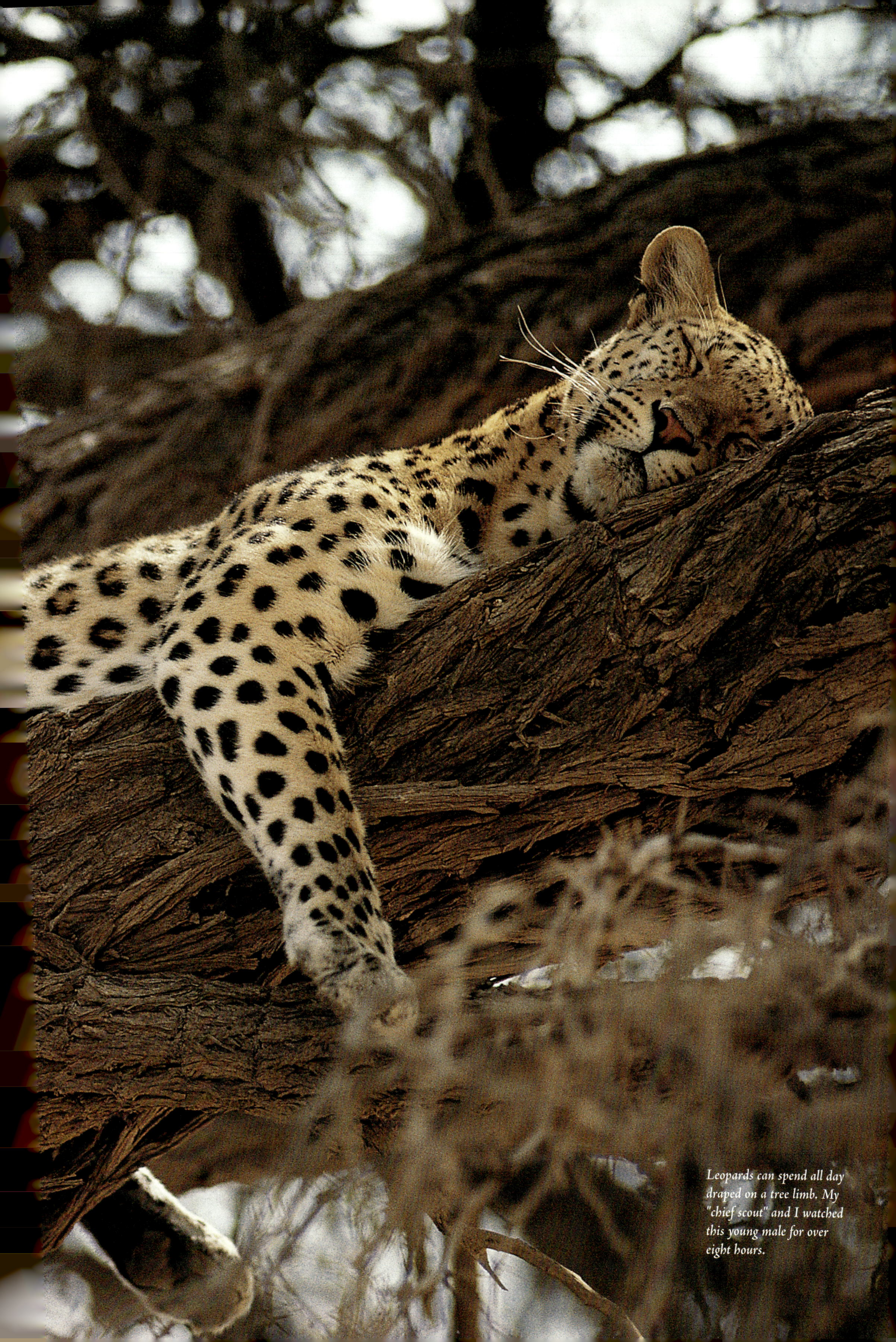

*Leopards can spend all day draped on a tree limb. My "chief scout" and I watched this young male for over eight hours.*

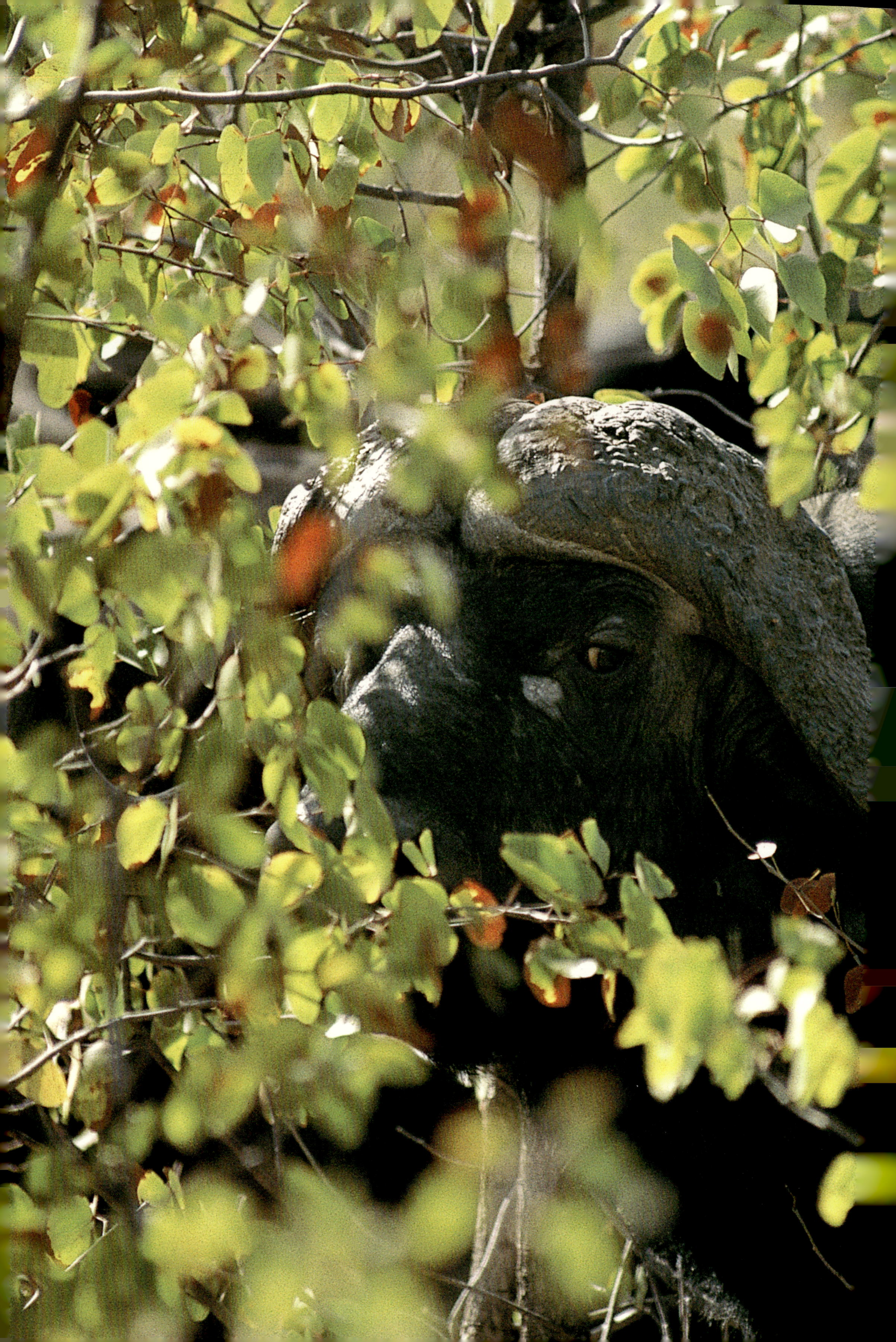

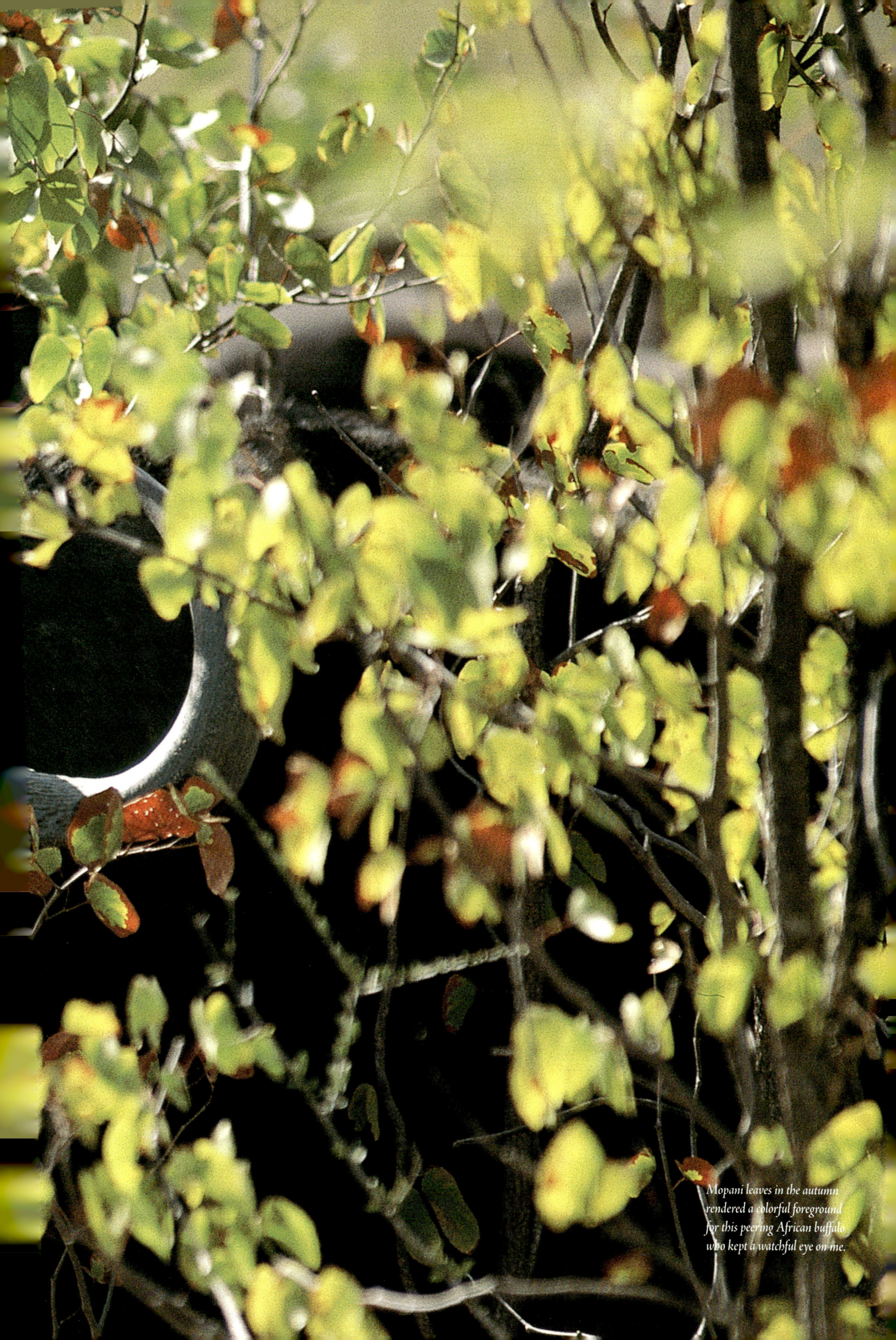

Mopani leaves in the autumn rendered a colorful foreground for this peering African buffalo who kept a watchful eye on me.

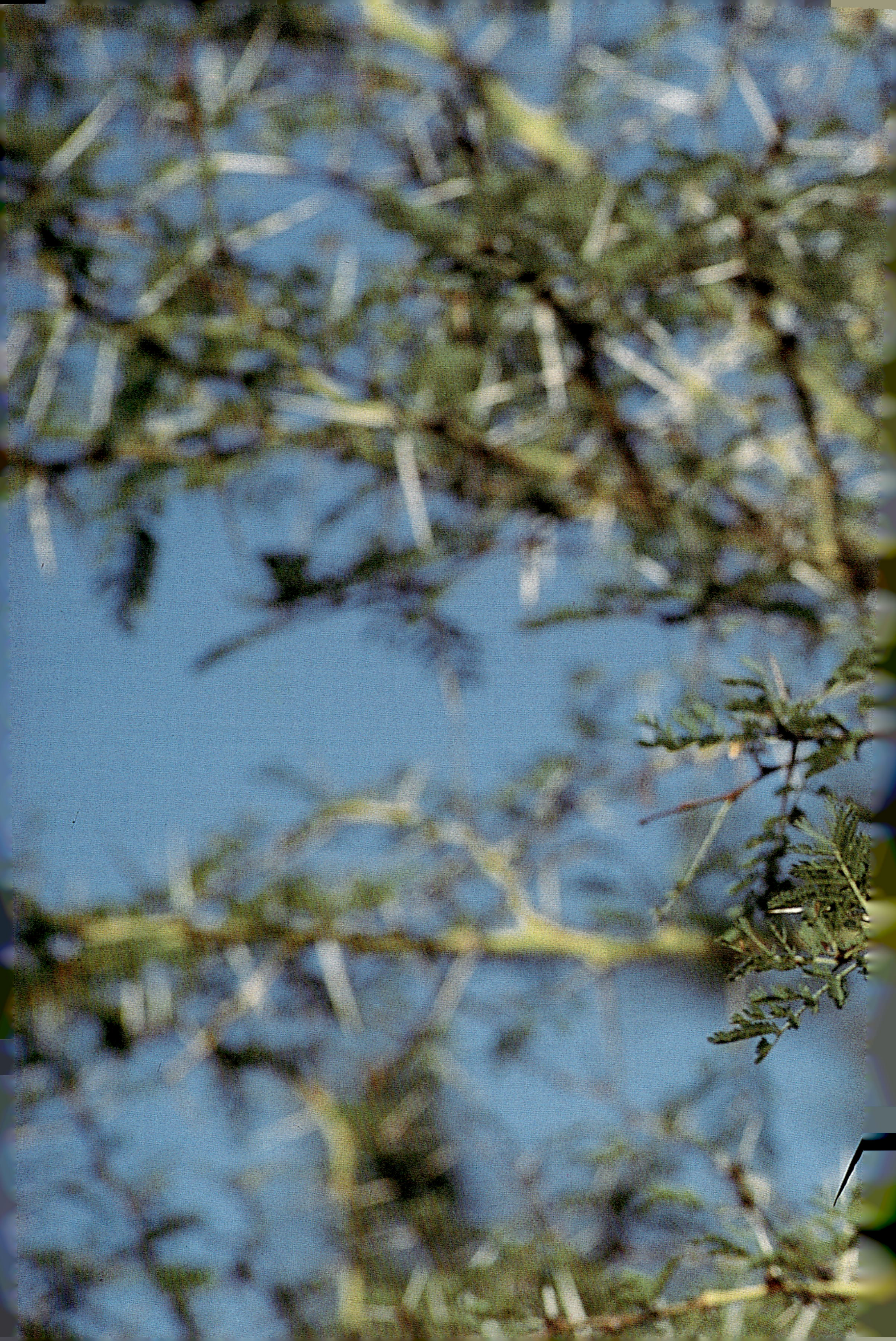

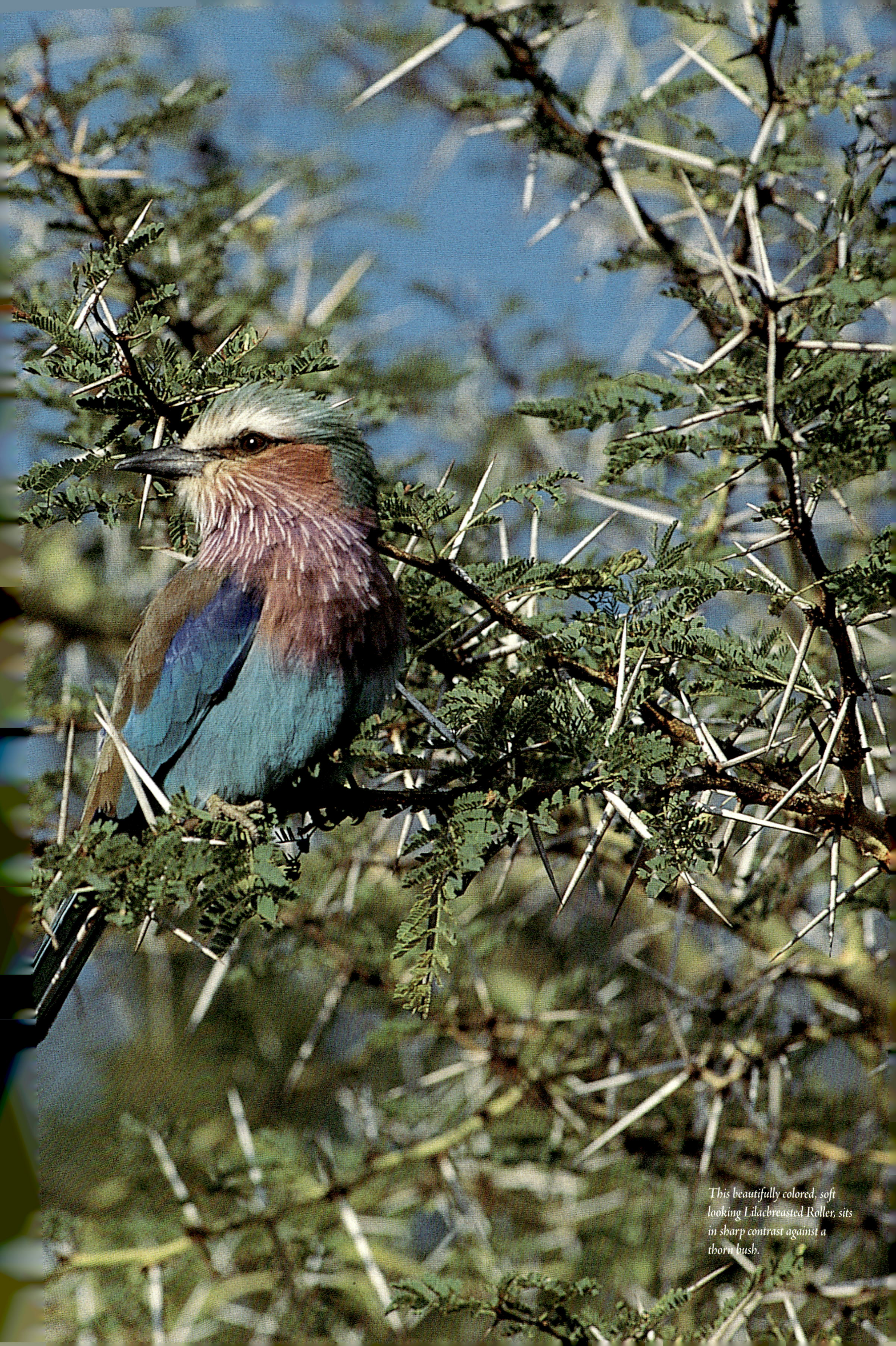

*This beautifully colored, soft looking Lilacbreasted Roller, sits in sharp contrast against a thorn bush.*

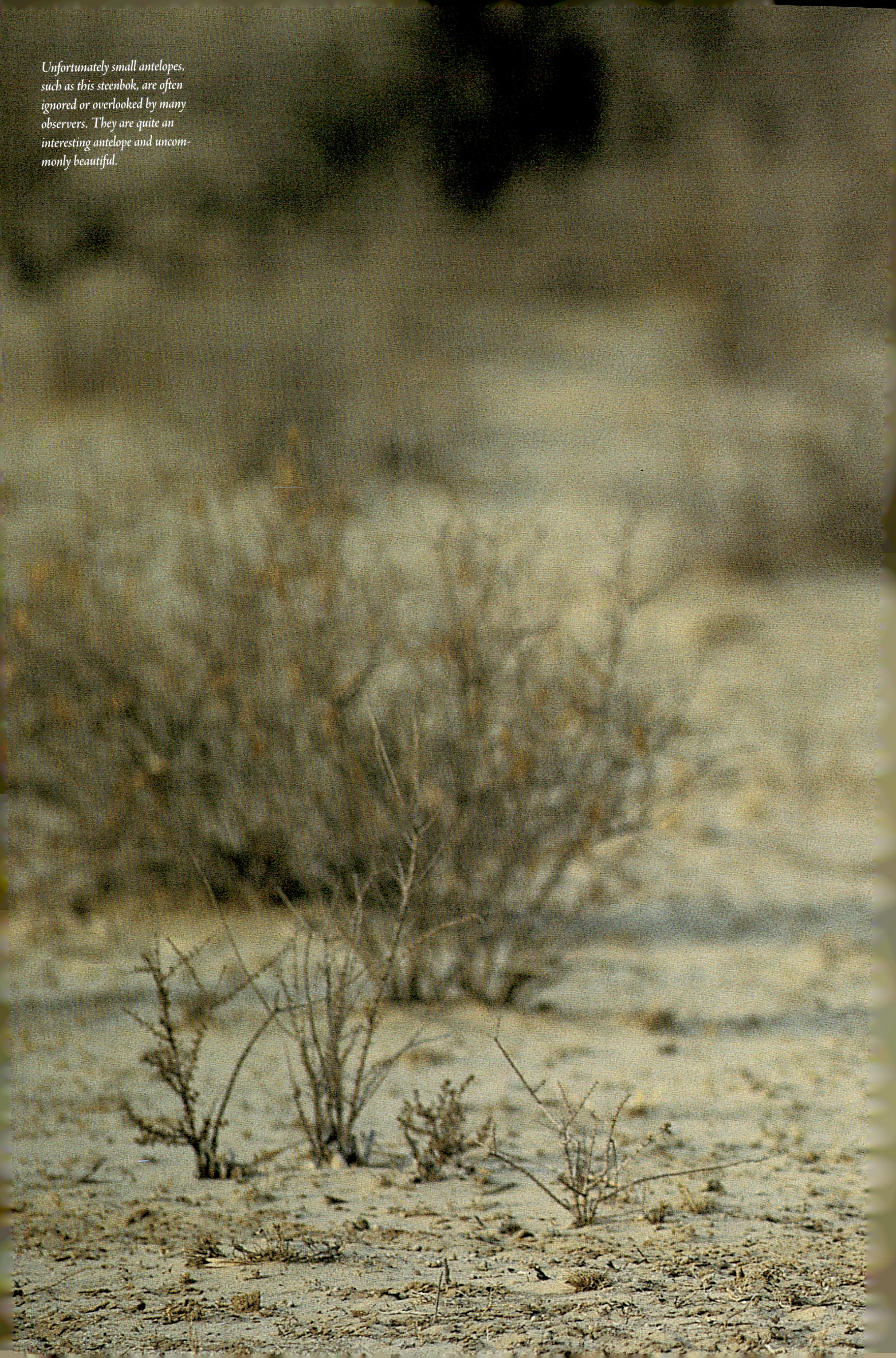

*Unfortunately small antelopes, such as this steenbok, are often ignored or overlooked by many observers. They are quite an interesting antelope and uncommonly beautiful.*

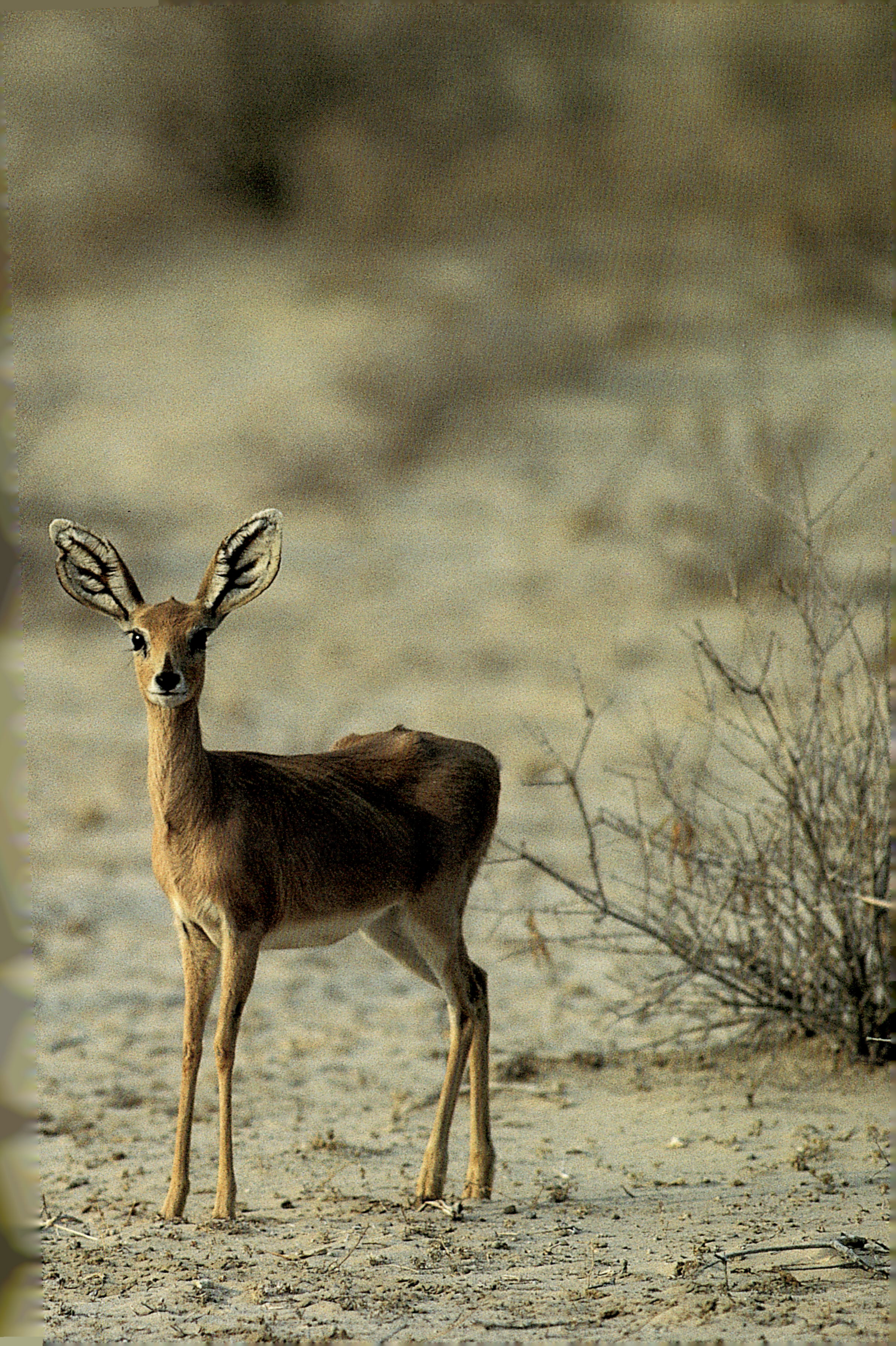

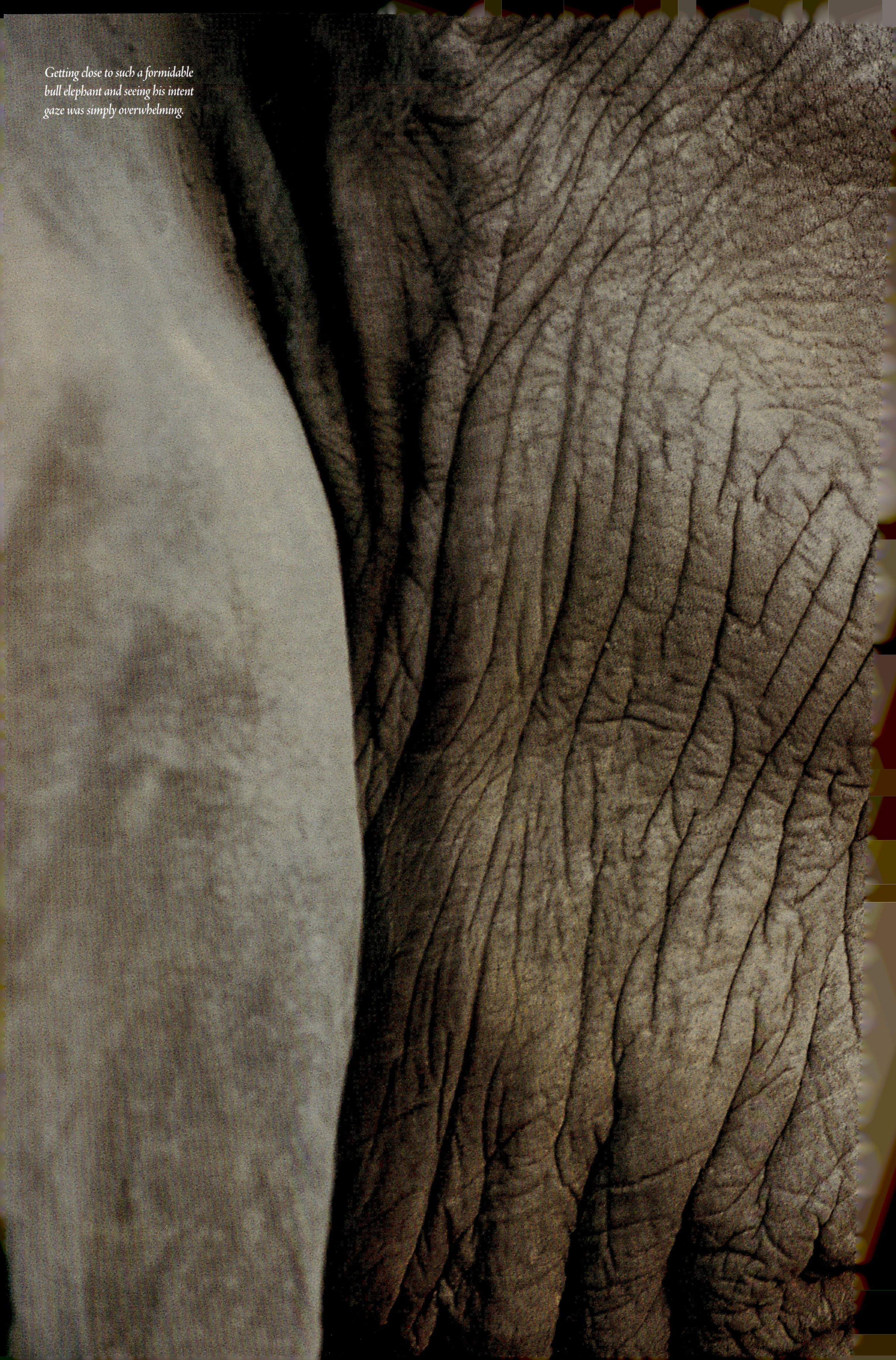

*Getting close to such a formidable bull elephant and seeing his intent gaze was simply overwhelming.*

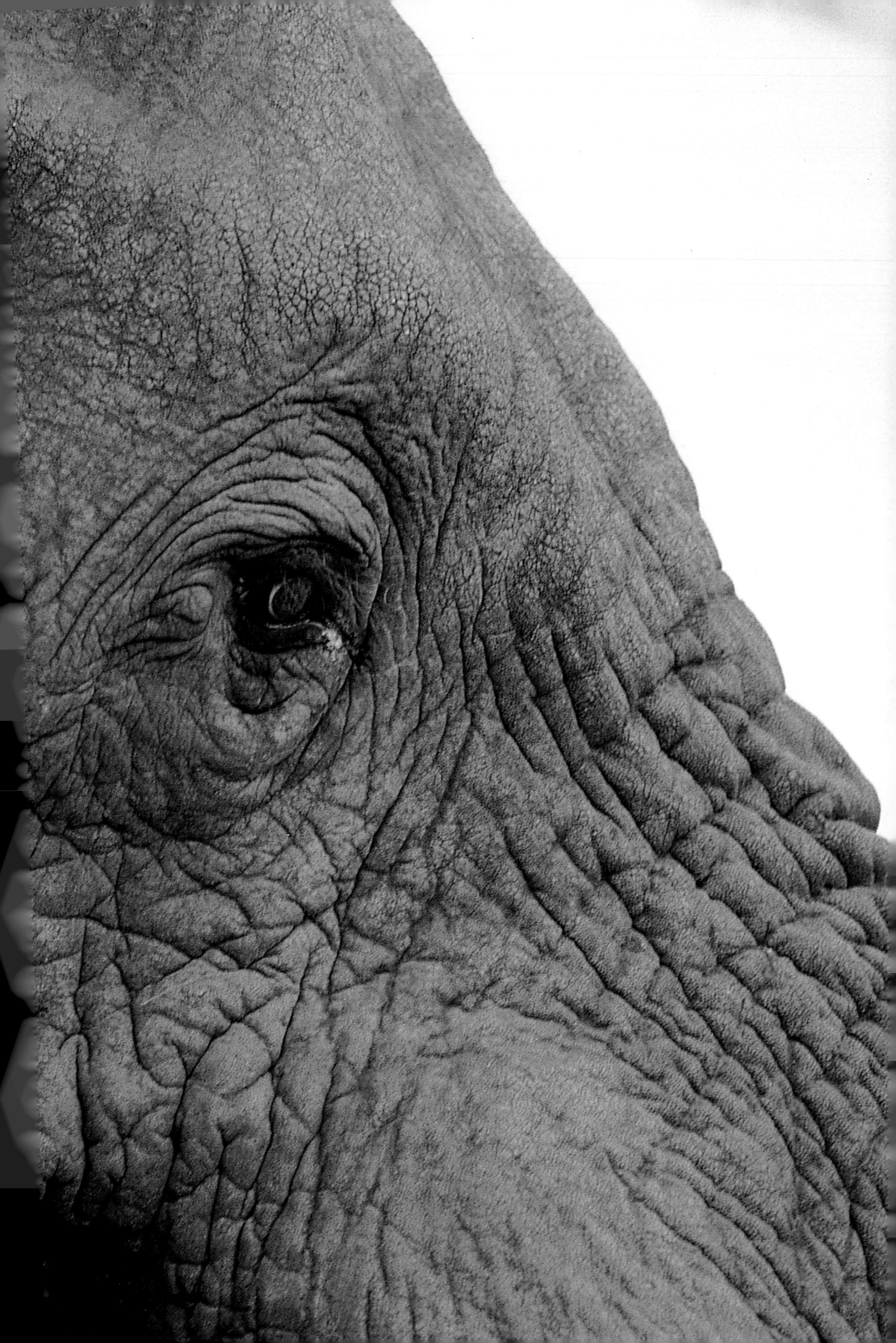

Book Design Jose Barcita



Published in the U.S.A. by PBM Graphics, Inc.
4900 Prospectus Drive
Durham, NC. 27713

Library of Congress Cataloging-in-publication data
Maratea, Ronn.
*African Wildlife: A Photographic Journal*; photographs by Ronn Maratea; text by Pamela Barcita

ISBN 0-9702253-0-X

Printed by PBM Graphics, Inc.

# African Wildlife

## A Photographic Journal

Forward:
Johann van der Merwe
*Director of Parks, Republic of South Africa*

Photographs:
Ronn Maratea

Text:
Pamela Barcita

Design:
José Barcita

Printing:
PBM Graphics, Inc.

# MAP OF SOUTHERN AFRICA

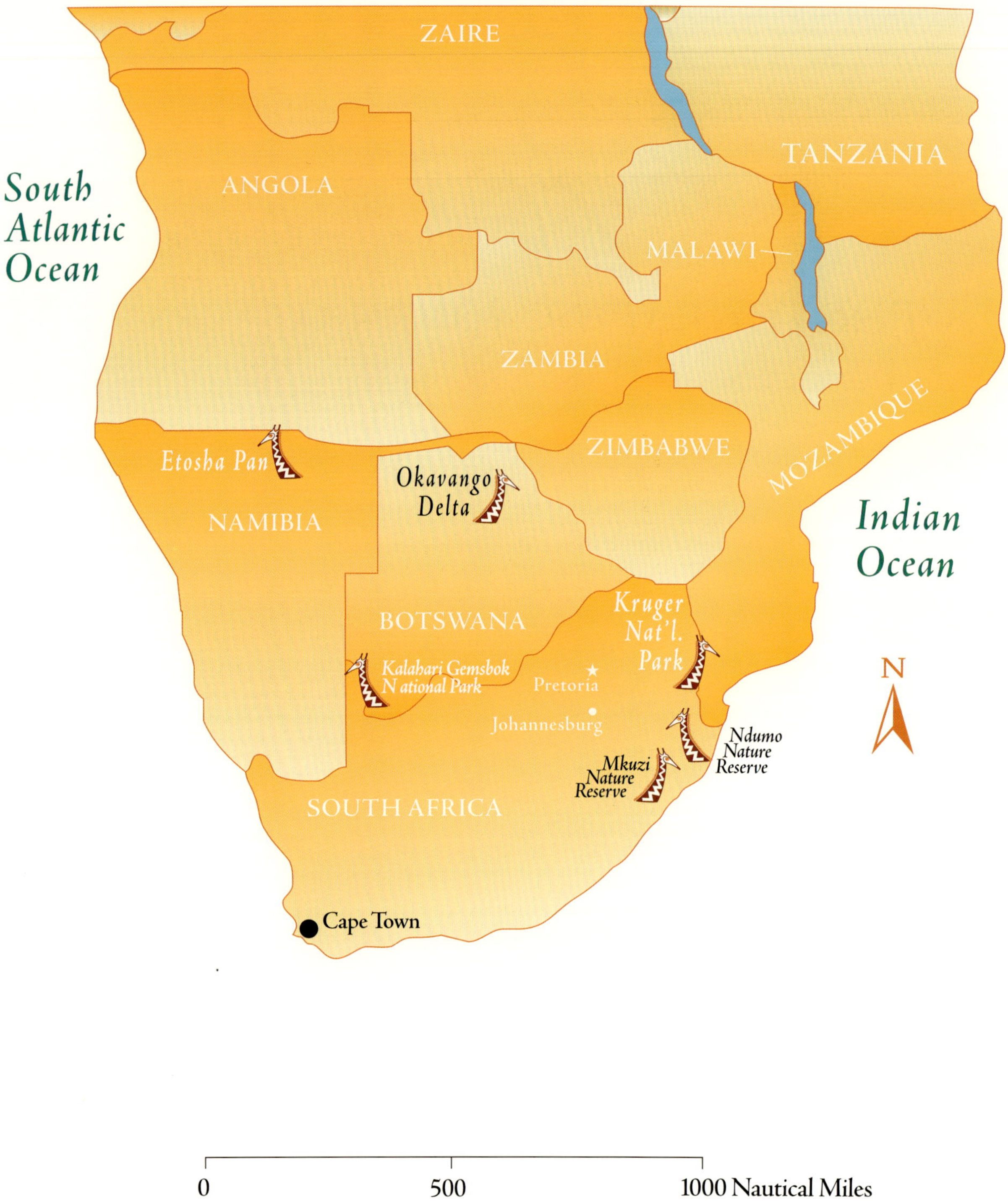


*Each of the parks indicated with [giraffe symbol] are the sites in which all photographs were taken.*

# CONTENTS

# FORWARD

I was born in Africa, and I have lived a life that most people would only dream of. Most of my adulthood is memories of a happy family in one of the wildest parts of the African bush, a place called the Kruger National Park, an immense wilderness of some 2, 000, 000 hectares in South Africa. These memories are made up of series of images, amongst others, herds of buffalo, elephant and zebra moving through the lowveld savanna. Of the synchronized birth periods of impala and wildebeest that result in thousands of calves and lambs born just after the first summer rains. Of innumerable White Storks that arrive from Europe in spring to feed on the huge outbreaks of locusts and horned crickets. Of millions of Redbilled Quelea, "the featured locust of Africa" that mysteriously and sporadically appear after good rains to feed on the bounty provided by the seeding grasslands.

Then there are also the grim images of death that accompany the regular patterns of drought. Of hippo that are stuck in mud pools, too weak to escape from the mud and too weak to escape from the hyena that have already started to feed on the stricken. Of emaciated buffalo that wander aimlessly in search of grass. Of lion that patiently await their inevitable opportunity to feast. Where Marabou Storks, Goliath Heron, cormorant and African Fish Eagles predate on fish that are trapped in rivers that periodically dry up. Where crocodile lie in wait in that last remaining pool to ambush the game that have to quench their thirst or face death. A grim choice!

Then there are images of people with us in the bush. How can I forget the expressions on the faces of my daughters when they discovered that three lionesses had slept in their wendy house. Or when they realized that two buffalo bulls were sleeping against our gate when they had to board the bus for school early one Monday, or the day they had to scramble from an elephant bull whilst they were cycling.

Furthermore, having conducted more than 200 foot-safaris in the bush, it is my opinion that the greatest wish of those who accompanied me into the wilderness, was to see one of the "Big Five" on foot. It is also my opinion that it is possibly one of the fleetest moments of joy when you do see a lion on foot. Sadly, only to see the lion run away but in some cases to see the lion turn nasty and charge at you - then the joy more often than not turns into absolute fear. Then the joy of seeing a buffalo herd semi-encircle you out of curiosity and then the fear of realizing your hopeless situation should they decide to stampede. Then there was the terror of the rifle jamming following a warning shot I fired at a charging elephant bull and the hysteria that followed when a stone to the head of the bull caused it to veer past harms way. So I can continue with this jumble of memories till boredom sets in but I want to set the scene for one remarkable relationship that emerged from this life in the bush.

Of the hundreds of people I had the pleasure to share this piece of paradise, one emerges as the most precious. It is a friendship that started in 1989 when Head Quarters in Skukuza radioed me in Letaba, a camp 200 kilometers away (we had no telephones then), requesting me to accompany an American photographer on a photographic project through the northern part of the Kruger National Park. This photographer turned out to be Ronn Maratea.

To put such a request into perspective, no one should be surprised when I say that history has taught us that a wildlife photographer on a short trip to Africa has to have a "shopping list" of "want-to-see's" that would cause any guide to squirm. This wish list would start with questions ranging from: Where are the lion? Can we expect to see a leopard? How many elephant are there in the area? Can you show us a kill? How about looking for rhino today? The story line is almost predictable. We used to call these quiz sessions "screamers" for obvious reasons. Then there was the old favorite statement, the epitome of honesty, where many would boldly claim their allegiance by stating " I always wanted to be a ranger but..." Or "if I could have my life to live over I would definitely become a ranger because I loved animals since I was a child" So it was with trepidation that I awaited the arrival of the American photographer.

Ironically, as a child I was very interested in nature, especially animals, but like so many others I ended up in a job that was killing me. I initially took up a career in marketing with a relative degree of success. But I was unhappy, I had to get out but the price of getting out was too high. With a wife and two kids, it's very difficult to contemplate a radical change, especially change that involved three years of studies plus the responsibility of a family. However, the tension of where my heart wanted to be and where I found myself was just too great. I had to do it, and three years later I was based in Letaba in a capacity that would turn out to be a dream career. So, I do have sympathy with those who proclaim their allegiance with us, but I also know what it takes to make the break.

However, back to the American photographer. The afternoon of our planned rendezvous I approached the restaurant in the camp with some uncertainty, where I was to meet the client, always excited to go to the bush but not knowing what to expect from the company. As I walked past the last vehicle in the parking bay I could not help but notice a battery of camera lenses on the back seat. I noticed the vehicle was a hired vehicle by the sticker on the fender. It struck me that this must be the equipment of the American photographer. Never have I seen such an array of photographic equipment in one place. I just knew that this was not going to be your ordinary photographic safari. This was not going to be your pseudo-professional armed with a 70-200 zoom, happily clicking away at the mundane, but ever-so-often reminding you that we must still find the "Big 5". This was going to be special, and so it turned out to be.

Having met Ronn and his wife, Jan, on the terrace overlooking the Letaba River, I got that strange feeling, that all of us at some stage or other experience, a feeling as if you have known each other from somewhere in the past. Things just clicked. After the initial familiarization I suggested we do what I know best, namely go and look for photographic opportunities. Near the camp there was a hippo pool which provided almost foolproof opportunities for high quality photography. There was a large hippo population close to the lookout with some unbelievable scenery. Often as the sun sets, elephant bulls would come down to the reedbeds to feed and to the mudflats to take baths. That day was no different, and in no uncertain terms nature started to forge a bond between Ronn and me that would last to this day. As I started to do what was expected of me, Ronn started to do what he does well, take photographs and a confidence and trust started to develop between us.. At the time, none of us could dream of the situations this confidence/trust relationship would lead us to.

Ronn's safety was always my primary concern but I used my knowledge of the bush and my intimate understanding of animal behaviour to get Ronn into the most impossible positions and situations, which he then captured in some of the most unbelievable compositions.

During his many visits to Africa we had many memorable moments that I treasure. To share some of the stories behind the pictures I wish to take you back to a few incidents that characterize our travels in the bush.

I recall the time during the 1989 drought when Ronn and I discovered the last remaining water in a dam called Kanniedood (ironically, it means, "cannot die"). But water it was not, the consistency was rather that of slow bubbling hot chocolate. This was caused by hundreds of barbel, a catfish species that was writhing, wriggling and churning up this muddy cocktail in an attempt to remain moist and thus not succumb. The biggest surprise was a crocodile that had quite happily taken up residence in the middle of the squirming mass of fish and mud. To the crocodile of course it was food and shelter. We decided to approach this spectacle on foot for a better shot. We were going to use the dry mud slabs around the mud hole as stepping-stones to really get close. In any case, that was our plan. It turned out that underneath the mud slabs, which were arranged in a puzzle-like pattern, were vast quantities of unsettled mud. Unknowing to us, sufficient weight on top of these mud puzzles would cause them to tilt and sink into a trap with such sucking properties from which escape would be difficult.

Often one would see the images of big game such as hippo or buffalo stuck in mud, unable to overcome the suction forces. That day it was not going to be game that was going to be stuck, but Ronn and I. I can't remember who sank first. I think it was I. Well, there we were, I was hanging onto my big .457 Magnum rifle and Ronn was hanging onto his expensive 400mm lens, but both of us were definitely sinking. The only sane one was Jan, who was safely in our vehicle.

It struck me that we weren't spectators anymore and were no different from the situation the fish were in and definitely very aware of the crocodile that was not far away. I proceeded to claw my way out of my mud trap, in the process losing my hiking boots as they were sucked off my feet. I then politely offered to take Ronn's camera in order for him to get out of the mud before sinking too deeply. This offer he refused and somehow managed on his own. My first reaction was that the trust that had developed between us had its limits. In hindsight, I accept that his cameras and lenses were the tools of his trade, just as my rifle and knife were my tools of the trade which often were the only thing between me and disaster, and which I wouldn't allow anyone to handle. I started to understand and associate myself with some of Ronn's views. Some people would label such behaviour as fickle but not those of us who often survive through such behavior, whether as a matter of life and death or as a matter of earning a living from unique opportunities that you want to photograph, it all made sense.

Every night, just as I would clean my equipment with great care, Ronn would sit down and clean his cameras and lenses. It would become part of a ritual. In my case cleaning kit, two-by-fours and rifle oil. In Ronn's case, lens brushes, tissues, aerosols, battery checks and film checks.

Nothing was left to chance. I noticed that discipline was part of his make-up. It was this discipline that served him well in amassing a vast library of images that truly captured the spirit of the African bush. It was also the knowledge of working with a disciplined person that gave me the confidence to start doing things in the bush that you would not usually do with other people. When there is discipline there is a degree of predictability in behaviour and I was prepared to gamble on that and as events panned out, it paid off handsomely for us.

Once we were driving along the Olifants River when we noticed fresh lion spoor. I suggested to Ronn that we follow the spoor down to the river to see if we would be in any luck. As it turned out nothing came of the lion but we witnessed the joy of life as I have seldom seen. On the bank of the river a hippo cow was sunning herself. Nothing special about the scene except for a very young calf that refused to simply just lie down and be a hippo by doing as little as possible and enjoying the winter sun. This little guy was in a mood to jump, arch and bounce. It would run around its mother, nudging her with his head and then just before the mother would react (in a manner as one would expect a mother to react) the little guy would rush off into the relative safety of the water, only to start the frolicking routine again once the coast was clear. The exuberance of youth, the joy of life was tangible that day along the banks of the river. As we crawled closer and closer I was constantly aware of Ronn being at work, his camera clicking away and methodically documenting these rare images of a hippo at play.

Generally my work was such a pleasure that I often said to people that I could not get myself to admit that I held a real job. It rather felt as if I were practicing my hobby and a good days work was like going out to play. Without fail, Ronn would ask my wife for "permission to take Johann out to play". I do not think he realized at the time how appropriate his request was. But play we did!

One year Ronn arrived, not with Jan as was usually the case, but with a friend called Spook. He was a fellow photographer with as an enthusiastic approach to wildlife as Ronn. The only difference between the two was the fact that they were complete opposites. Except for photography and wildlife there were no obvious similarities. With Ronn, always soft-spoken and well composed, Spook always bold and expansive and myself adventurous and exploratory, the "three musketeers set" out to conquer the African kingdom. What a sight it must have been.

One night we slept in a wilderness camp with no modern amenities. The food had to be cooked over an open fire, the water was heated in an ingenious contraption called a "donkey," which was a 44 gallon drum built into a framework with a wood fire at the base. The only light was a paraffin lantern and the accommodation was a wooden bed in an A-frame thatched hut. The sense of place, although primitive, was awesome. Spook loved it, Ronn reserved his judgement and I observed. The reason for being in this remote location was the presence of the largest concentration of crocodile that I have ever seen. More than two hundred animals would crowd onto a series of small sandbanks, often on top of each other, sunning themselves whilst they awaited the arrival of the fish migrations in spring. The plan was to leave camp early, just after dawn when the crocodile were still in the water. We would creep up to the big sandbank and position ourselves for the best possible view of the prehistoric beasts as they emerged from the water in mid-morning to charge their cold-blooded systems with sun energy. With adrenaline pumping and expectations high we took in our positions, having "leopard crawled" (on elbows and stomach) and having rearranged the sand bumps underneath us, we prepared ourselves for the long wait. Ronn improved his camouflage by propping up a few grass tufts and some reeds in front of him. Then he proceeded to check his equipment, the setting of the camera, the shutter speed and out came a light meter. I checked the wind, my rifle and a getaway route. You never know when you will need it. As the waterways around us started to show signs of crocodile activity the tension on the sandbank raised considerably. It is only when you come face to face with a 16-18 ft crocodile that you appreciate your own vulnerability. In that situation you are just another item in the food chain. I certainly gripped my rifle tighter. Every time I heard Ronn's camera shutter click, I held my breath and wondered how we ever allowed ourselves to get into such a precarious position. I wondered what would happen if we were suddenly discovered by the crocodiles. I knew they could not ignore us as we lay on their sandbank because we did not look like a crocodile, nor smell like a crocodile. As one beast emerged onto the bank in front of us, I sensed a hesitation in Ronn. I wondered whether he felt the same as I did. Unfortunately or maybe fortunately, the crocodiles became aware of our presence and did not emerge from the water in great numbers. We were quite content to photograph these huge monsters as they drifted past us.

Then at some point, Ronn, obviously elated with the experience, asked excitedly how often I did this? I had to be brutally honest when I replied "never before Ronn". Stunned yet innocently he asked: "Why? Again, brutally honest I replied " I have never been able to find somebody stupid enough to follow me here". I would have understood if Jan was mad at me but she was not here so I guessed it was okay.

Today, it is only the photographs that remind us of those precious moments when all that we worked for came together in moments of brilliance. I used my knowledge of the bush and animal behaviour to get us into amazing positions and then Ronn used the knowledge of his subject and artistic flair to compose photographs that captured the spirit of the African bush. Unlike the memories that I have about my bush experiences, memories that are only in minds eye and will fade from my memory as time goes by, Ronn has captured his memories on film. Today he can share these visuals with those around him who have not had the privilege of first hand experience in the wilderness. Numerous slide shows and presentations support this willingness to share. I also have an appreciation for the way in which he parted with his knowledge of photography. As we traversed the bush in search of the extraordinary, he shared with me some of the art, skill and know-how of high quality photography.

Furthermore, when you look at his work, you will be excused if you think some of the unique shots must be the result of manipulation of the subject. I know and have witnessed the patience and determination with which he pursued his subject and never did I see any evidence or attempt to photograph an animal in a position other than a natural pose. I have witnessed the greatest degree of consideration and sensitivity from Ronn when he plies his trade. He was constantly aware of those around him, he was constantly trying to be non-intrusive and never deliberately imposed himself on the animal he photographed. He would insist upon leaving a particular scene if an animal showed signs of becoming annoyed or antagonized with our presence. For this I have the utmost respect.

As the years progressed, I became aware that Ronn and I are indeed soul mates from different generations (he's old enough to be my father and he sometimes behaves as if he is). I also became aware of the fact that, like I, Ronn made a career change late in life. He was a naval aviator and at some point took up photography. As I said earlier, I can relate to this and I know Ronn knows what it takes to make the break.

In Zulu we have a saying: "Eyamagwala ayiloleki." It means a coward's spear never comes off the grinding stone. Unlike those that often falsely swear their allegiance to our cause, being conservation, Ronn's proverbial spear has indeed come off the grinding stone. The way in which his images of wildlife create awareness for the environment by all that witness them must not be underestimated. For this I salute my friend, Ronn and his ever-supporting wife, Jan. I hope that his passion for excellence and his love for the wilderness in Africa and all the animals that pass through it touch all those who read this book.

Mamba kale mackinaw
(Go well old wise one).

Johann van der Merwe

Johann van der Merwe
*Director of Parks, Republic of South Africa*

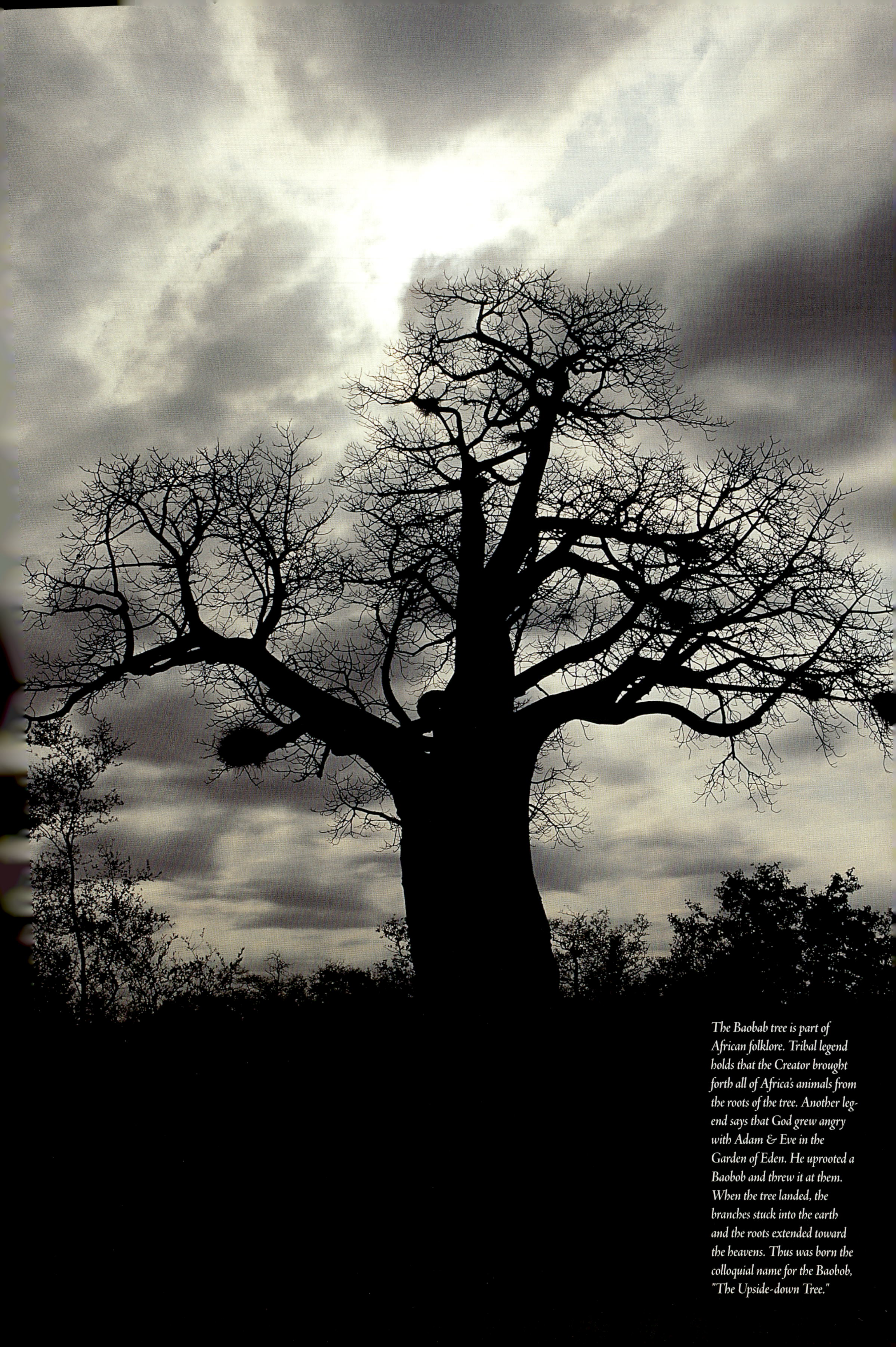

*The Baobab tree is part of African folklore. Tribal legend holds that the Creator brought forth all of Africa's animals from the roots of the tree. Another legend says that God grew angry with Adam & Eve in the Garden of Eden. He uprooted a Baobob and threw it at them. When the tree landed, the branches stuck into the earth and the roots extended toward the heavens. Thus was born the colloquial name for the Baobob, "The Upside-down Tree."*

# PREFACE

My first trip to the African continent was a photographic safari to South Africa and, in particular, the Kruger National Park. After that first trip, my wife, Jan (whom I often refer to as my "chief scout") and I have visited wildlife reserves in Namibia, Botswana and Zimbabwe.

The two of us had always dreamed about going to Africa to see wildlife, and once we got there, the excitement of realizing our dream was overwhelming. Each time that we saw animals, the excitement grew. Since that first wonderful trip, I have made 17 additional trips. My "chief scout" remained behind on a few occasions doing Grandmotherly things; but she has been with me on over a dozen trips to the bushveld and deserts. Now it has reached a point where the possibility of not going back to Africa would seem a great tragedy in my life.

People ask me what this longing for Africa is about. It's about the people, of course, and the breathtaking beauty of the natural settings; but most of all it's about the animals. Ironically, I was in Africa when I found a newspaper supplement on the country's ecology that contained a quotation from the North American Indian chief, Seathl:

*"What is man without the beasts?*
*For whatever happens to the beasts also happens to the men.*
*All things are connected.*
*Whatever befalls the earth befalls the sons of the earth."*

Those stirring thoughts were contained in a letter from the Indian chief to the seventeenth President of the United Sates, Franklin Pierce. The words, composed in 1855, are as true today as the day they were written. If any words can capture that nameless longing and melancholy I feel when I am away from Africa, that great native American's words can.

While in Africa, I've seen huge elephant cows show the most delicate and touching maternal pride and concern for their calves.

I've seen young hippos take parental scoldings and actually look shame-faced.

I've seen tiny baboons exhibit the same fierce, loving need for a mother's arms that I recall in my own children and grandchildren. I suspect there is not a single human trait, from great courage to base cunning, that I have not witnessed among the countless species I have seen and photographed in Africa. The more I see of them, the more I come to know of myself.

This is perhaps the greatest gift the animals of Africa have given me.

I hope these photographs and captions reflect, in part, the loving gratitude I feel for them.

RONN MARATEA

# ACKNOWLEDGMENTS

A special note of thanks to Janet, my wife, *"chief scout"* and partner. That's a tremendous burden for one woman to carry. This book could not have been created without her. Jan's contributions in the field were many, including spotting wildlife, driving under difficult conditions and positioning our vehicle to give me an advantageous camera angle.

Jose Barcita's creativity and perception are indeed priceless gifts used in the formation, design and production of this book. Our long friendship was instrumental in creating a wonderful working atmosphere. We look forward to working together on many more wildlife projects.

Pamela Barcita's appreciation and deep interest in all living creatures are exquisitely displayed in the text she has written. She gave her heart to every animal researched and her flowing prose is truly a treasure to this book.

Terry Pegram, Chairman and Chief Executive Officer of PBM Graphics, has indeed created a standard of excellence in graphic arts production, for himself and his entire organization. His magnanimous support for this project is immeasurably appreciated, as is the help, enthusiasm and professionalism of his entire staff.

Johann van der Merwe is my mentor, but more importantly, my friend. We have shared unforgettable experiences and he has selflessly taught me much about wildlife and the bushveld. He is a big part of this book. To his wife, Sonea, I say, "Thank you again for allowing Johann to come out and play."

Irene Grobler is the chief librarian of the Stevenson-Hamilton Library in the Kruger National Park. She is always ready to lend a helping hand with information and research projects.

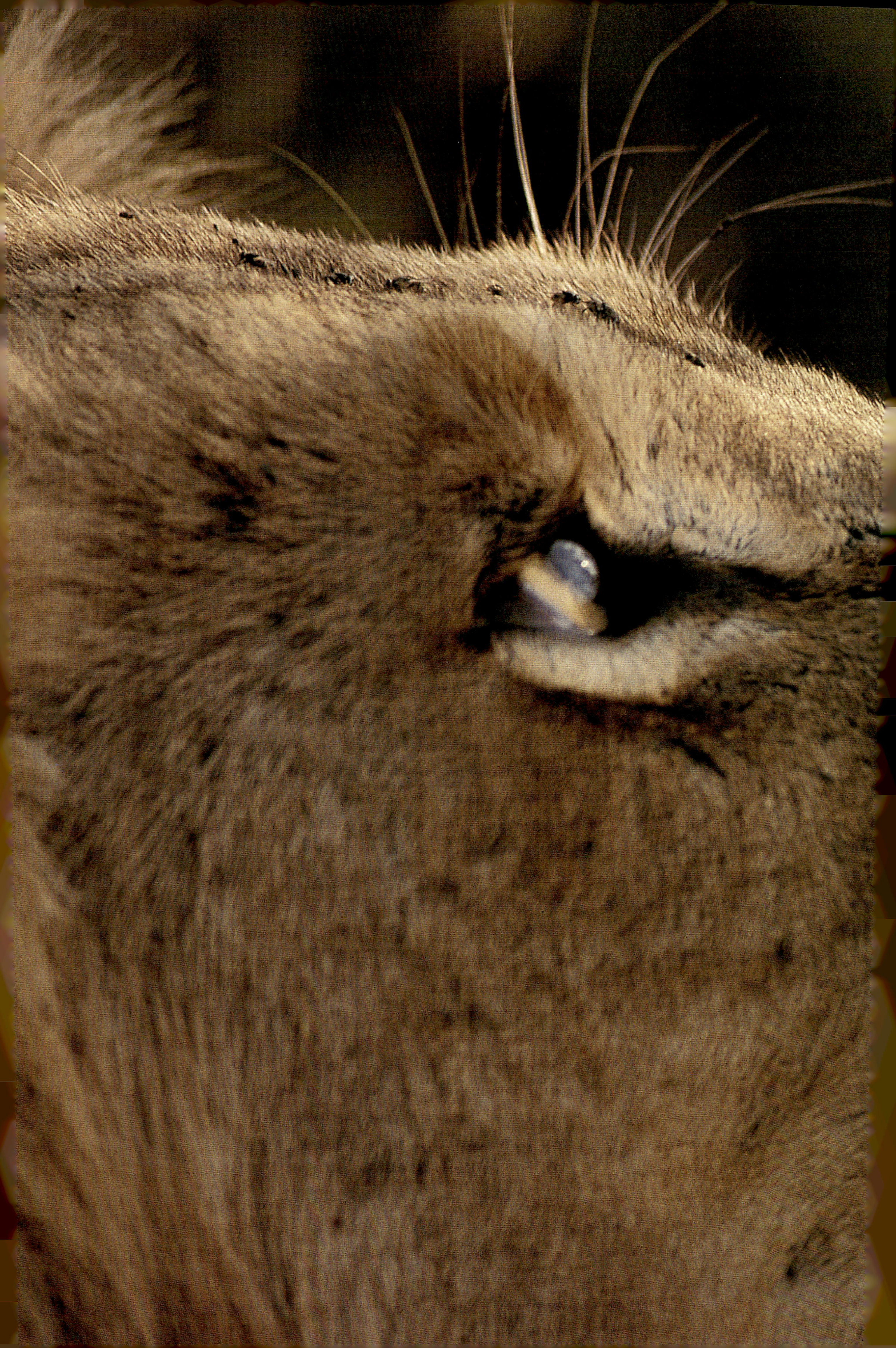

# CHAPTER ONE

# The Royal Family of the Bushveld

## *Lions*

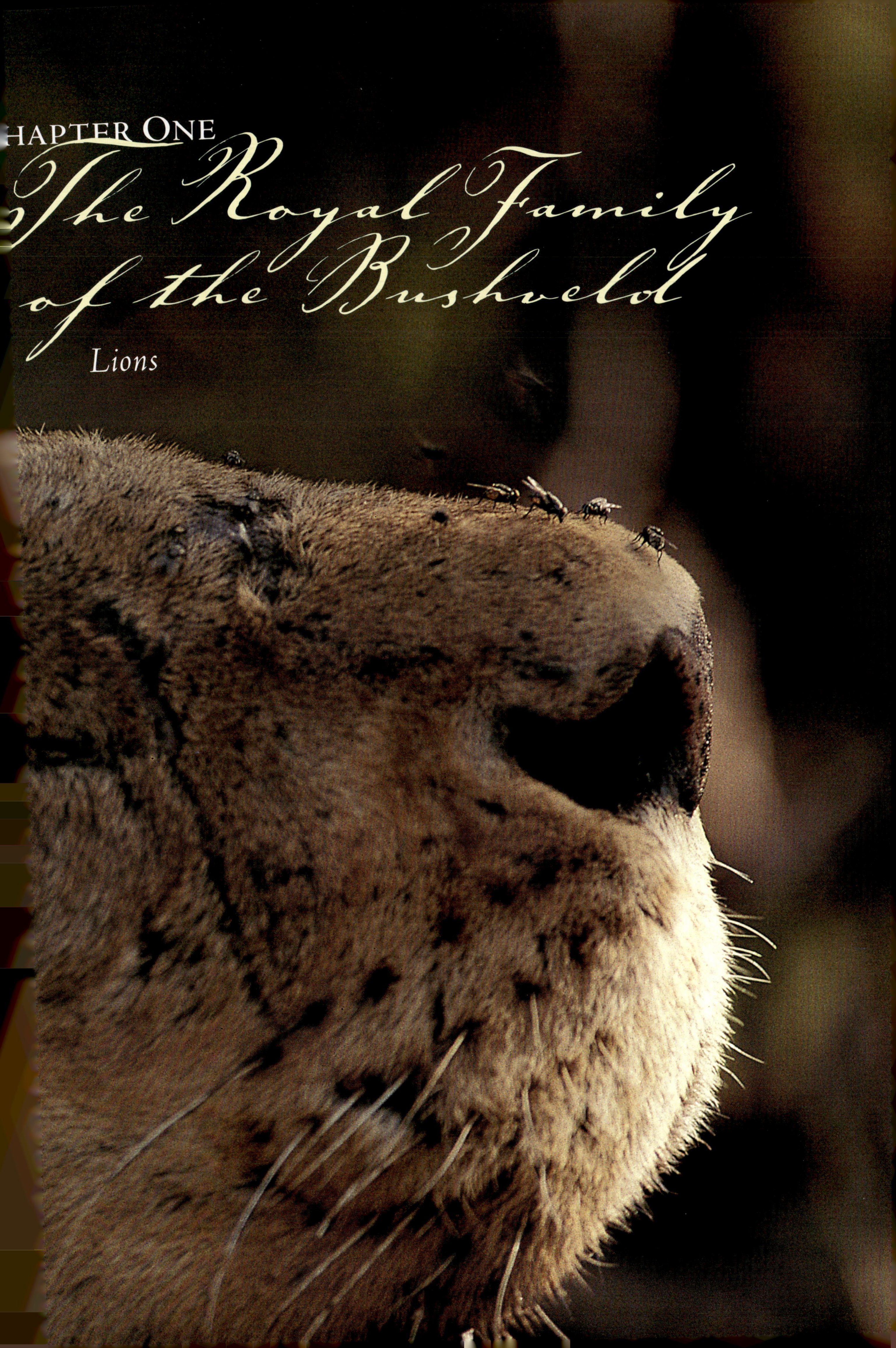

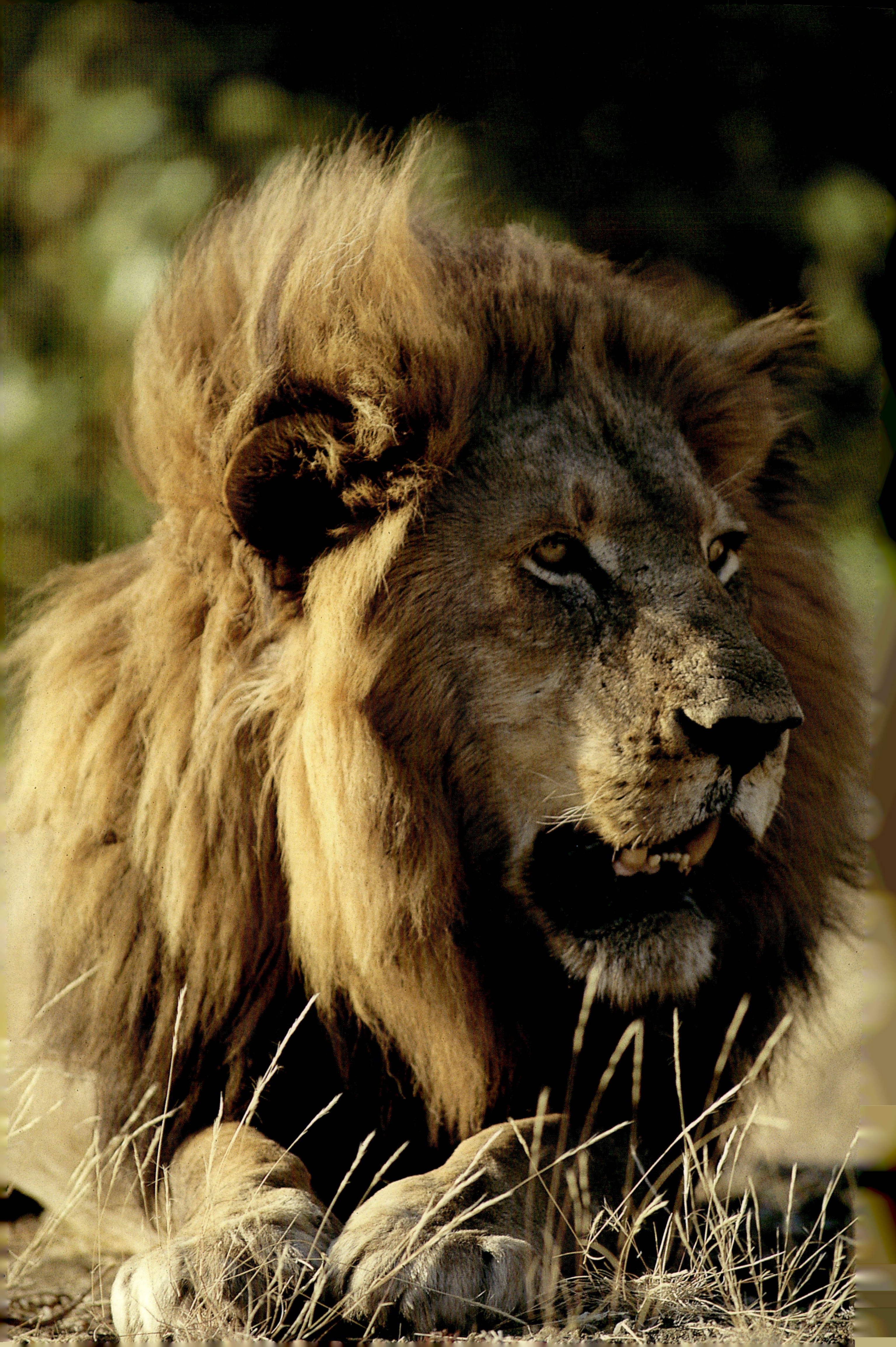

Any journey into the world of African wildlife should rightfully begin with the royalty. Lion. The word stirs many emotions; fear, respect, reverence, awe. And pictures come to mind of a proud, powerful, roaring, king of beasts. The lion has been a fascination of man through the ages. Man has hunted lion, been hunted by lion and has symbolically aligned himself with its power. Around smoldering campfires, myths and legends and fables have been told to generations of rapt listeners who share the fascination with the proud ruler of the grasslands and savannas.

In the wild, lion is just another animal in the circle of life seeking to survive and pass its genes on to the next generation. It's not an easy life for a creature who must live up to the reputation that he owns all that he surveys. Even though it is the largest African carnivore, lion must fight for its food, its family, its own life... sometimes on a daily basis. And so it has developed many ways to survive... ways of hunting, ways of structuring family, ways of defending itself.

Of the many feline species lions are the only ones that are social. They live in groups called prides. A fitting name. A pride can number in members from a solitary female and her cubs to a community of forty individuals. The average pride consists of about fifteen lions; four to five of which are related adult females, a few are males which may or may not be related and most often there are cubs of various ages. Sometimes the pride splits into several groups. The lionesses establish the pride and maintain its stability. They are the hunters, the nurturers and the defenders of the prides. Male lions defend as well but repel only male intruders while the females will attack both male and female interlopers. Lionesses determine the family's hunting range. Between dusk and dawn they stalk their quarry. Three quarters of the prey are felled by females hunting cooperatively even though the males demand to eat first. But sometimes, after they have eaten their fill, the males will share with cubs.

PRECEDING PAGES

*Despite being a member of the "Royal Family of the Bushveld" this lion still has to put up with flies.*

FACING PAGE

*This portrait says, "king of the beasts." His large mane, heavy muzzle and massive size make him quite regal.*

Scouting a range which can be from two to one-hundred square miles, lions stalk prey of many types. Strictly carnivores, their palates prefer creatures like wildebeest, zebra and impala. Felling large prey nets the best reward for the least energy; one zebra can feed a pride for three days. But they are not adverse to snacks such as birds, fish and reptiles. Sometimes, when driven by hunger, lions will attack larger prey such as crocodiles, hippos, buffalo and even elephants. However, such brazen efforts do not often pan out. And for all hunting attempts, success is only one in ten even when they mainly target the solitary young, old, lame, sick or wounded and the unobservant. When a kill is made, it is accomplished by attacking and pulling down the quarry with sheer body weight; then seizing the throat and strangling it. The kill is swift and efficient.

All hunts are in ernest, for an adult lion requires four-thousand to five-thousand pounds of meat per year to survive. That amounts to from ten to thirty-five medium sized prey. At one sitting a lion will consume 25 percent of his own body weight. A male lion can grow to over five-hundred pounds and nine feet long. A female often reaches over three-hundred pounds. That weight is chiefly muscle, compacted into wiry bodies capable of jackrabbit starts from a standstill. They aren't particularly fast though, and are

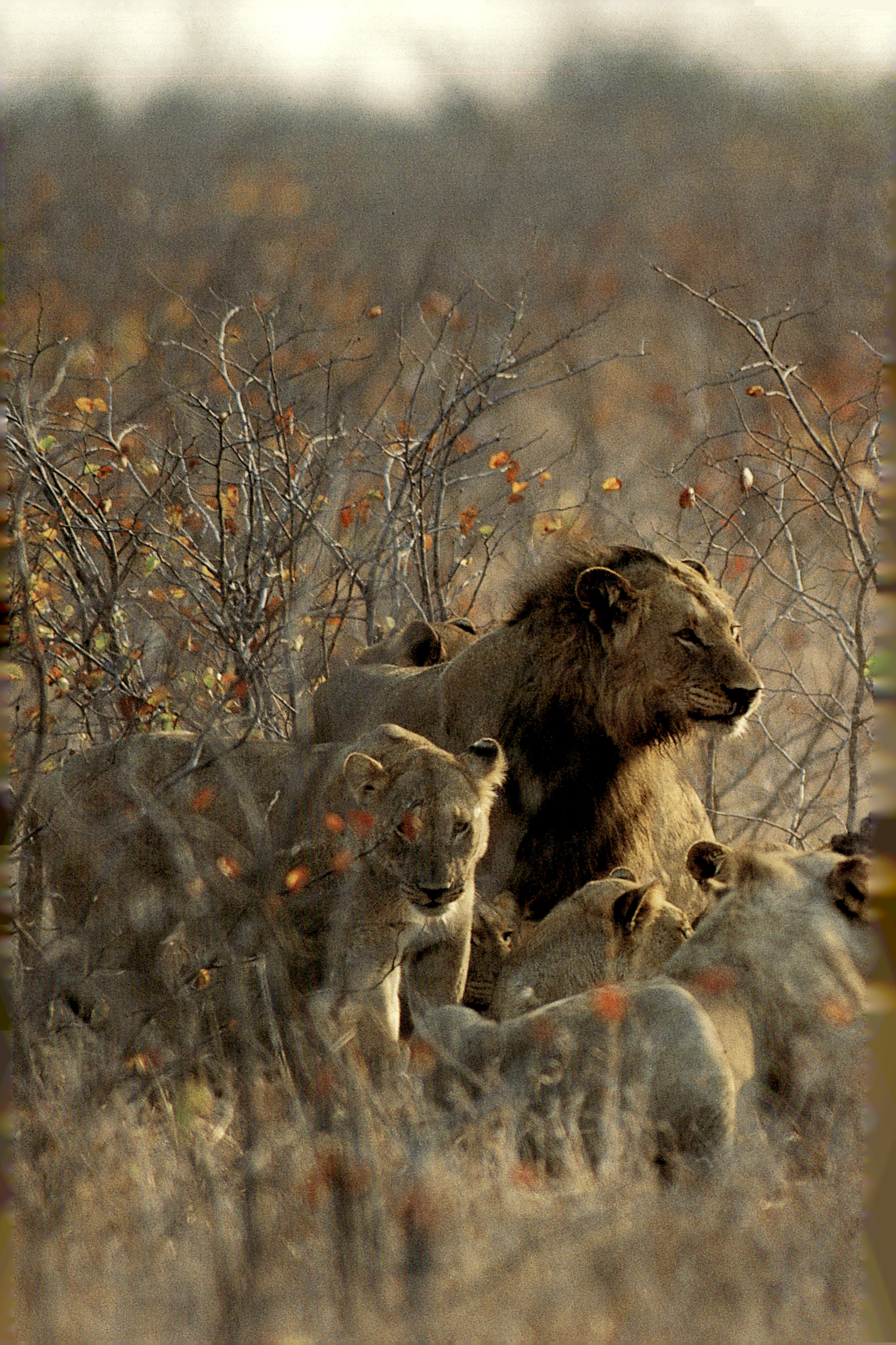

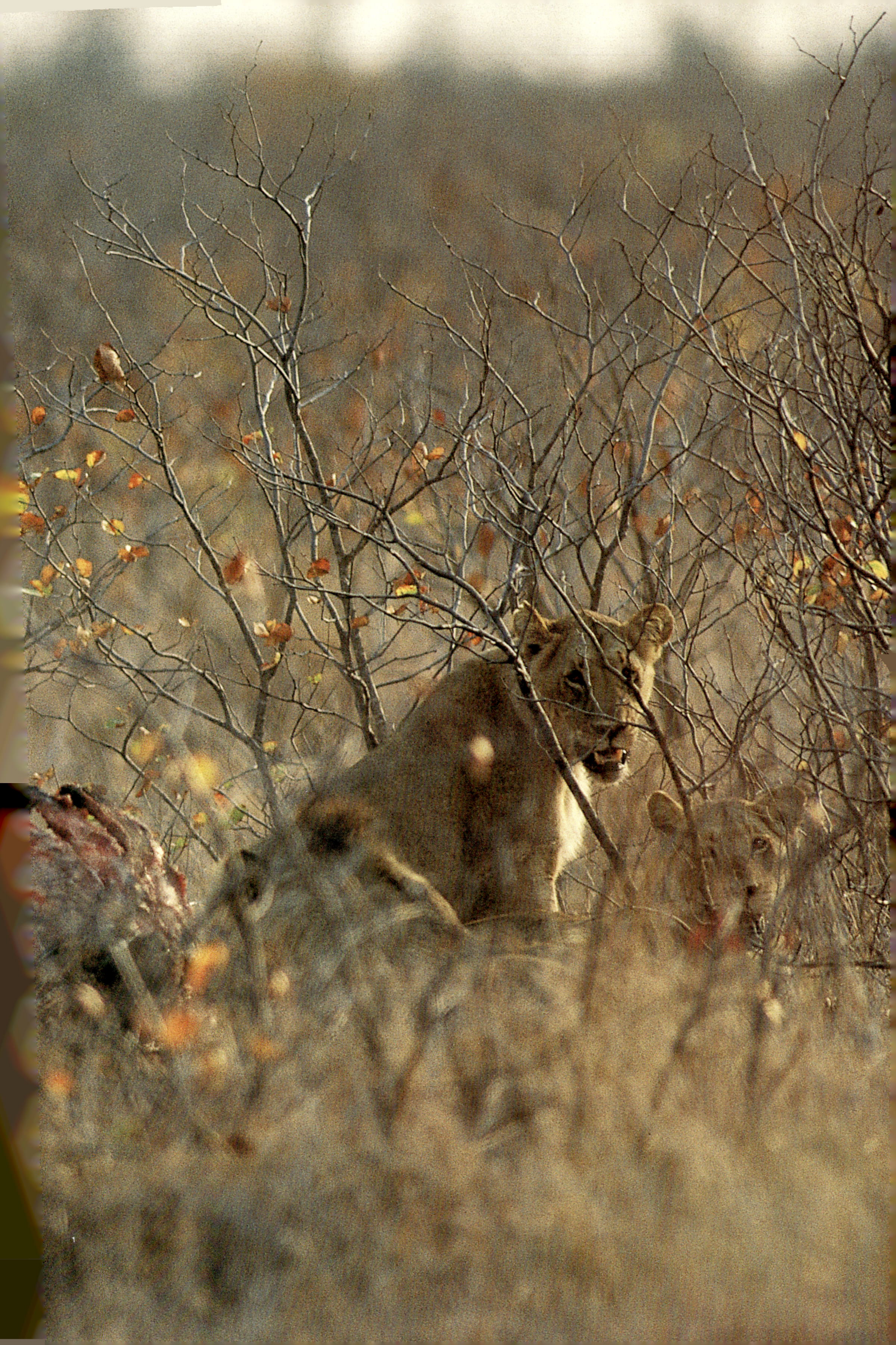

sprinters rather than long distance runners. Therefore, they rely on skill, experience, and patient stalking as a team, knowing they must get very close before they can attack their fleet-footed prey. Camouflage too, comes in handy for the element of surprise. The lion's tawny coat is well suited to its grassland habitat where it blends in with the tall dry grasses, enabling quick ambush.

But lion's sustenance does not come only from hunting. Scavenging accounts for over 50 percent of their diet. Lions are attuned to the movements of vultures and the excited calls of hyena. They will move into a kill, assert their royal privilege with an intimidating roar and take over. The less energy spent, the better.

Pride members often 'copycat' behavior after eating. If one yawns, the others will yawn.

PRECEDING PAGES

*The entire pride of 9 lions is here and today is a feast day. There is plenty of food for all.*

BELOW, LEFT TO RIGHT

*During the time of mating a male will stay close to watch the female.*

*Females will stride in front of a male during breeding and "flirt." This behavior by the lioness is to insure that the male does not lose interest.*

*A male lion will lick the female's neck in order to induce mating.*

*The male seems to be biting the neck of the female during mating. This is a type of ritual behavior.*

*It appears that these lions are fighting. However, the snarling is also ritualized behavior which occurs after mating.*

*A lioness is taking a "cat nap" while the male is guarding her. Often he too, will nod off.*

*The male and female often roar. I like to think of this as an impending "birth announcement."*

## *The Lion's Share* (An Aesop's Fable)

*Long long ago Lion went out hunting with Wolf, Fox and Jackal. Together they killed a gazelle. All were hungry and immediately began arguing about how to divide the kill. Lion asserted his authority and instructed them to divide it into fourths. He then took the first section because he was King of Beasts. The second part was his, he claimed, because he judged the dispute, and the third portion was his due for being a partner in the hunting party. Only one piece remained. Shocked and intimidated, the others were silent. Well then, spoke Lion, if no one claims the last then I shall take it ... unless one of you wishes to challenge me!*

*Hence was born the term, "the lion's share."*

If another grooms the rest will groom. Then they will roar, one after another. One gorging meal means a lion won't need to eat for several days, giving it plenty of time to pursue its favorite pastime ....snoozing .... which can take up to nineteen out of twenty-four hours. The hot daytime hours are spent lying in shady spots and dreaming, no doubt, of the evening hunt to come.

While the picture of one big happy family may come to mind... the reality is that the pride relationship is full of the stuff of a notorious crime family. Power plays, changing loyalties, genocide. Family comes first, but at a price. Nomadic groups of males will challenge the reigning males of

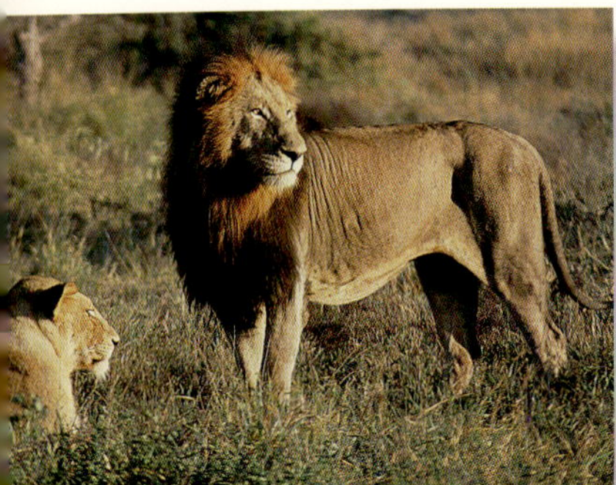

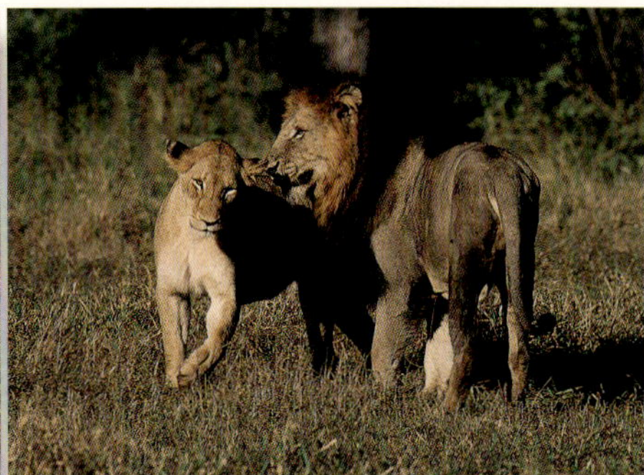

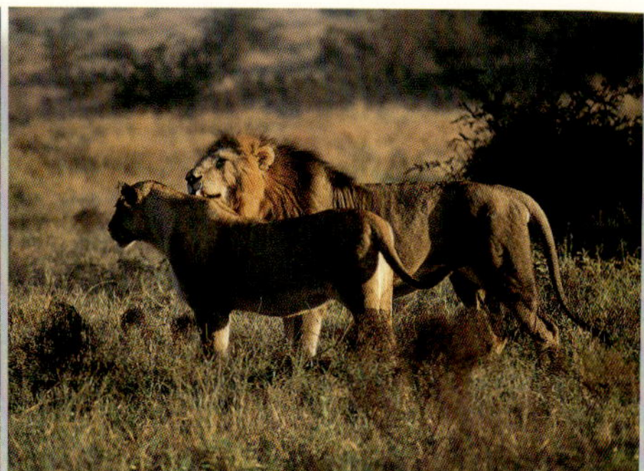

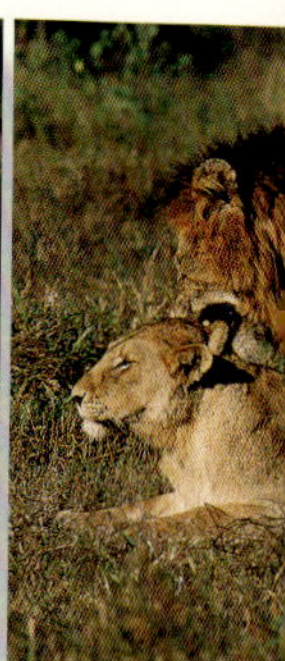

any given pride. The challenger emits a coughing sound to signal an attack. The fighting is furious and sometimes, to the death. Then, if the intruders win they soon begin the grim task of killing all the young cubs fathered by the vanquished males. They even kill cubs born in the months after taking over the pride. If the lionesses try to protect the cubs they can be injured or even killed.

In the beginning of the new order, adjustments need to be made to the relationships within the pride. After the loss of their cubs the lionesses will initiate mating behavior with the new males. They seek to establish bonds to keep the males from deserting the pride. Then, when all the lions are familiar with each other, the females are ready to begin mating in ernest.

The mating ritual has many stages and is repeated often during the female's receptive time. There is no real fighting among males for the right to mate. It is given to whichever lion is there first. The male will follow and stay close to the female for the four days she is fertile. At first the female will play hard to get, growling and slapping him away. But, she will keep him interested by 'flirting' and when she is ready she will crouch and present herself to him. The male may lick and gently bite the female's neck during the mating. She will snarl during the process, probably annoyed by being held down. Copulation lasts only fifteen seconds. Afterward, both lions will roar, announcing their success.

The lionesses in the pride will all mate at the same time. Consequently, one-hundred ten days later, they all give birth to litters of from one to four cubs. The cubs are born tiny and helpless in well-concealed lairs. They weigh only about one pound and are blind. Their eyes open about a week after birth. They are able to see well in two weeks, just about the time they are first walking. And by the time they can run, in thirty days, they are biting with newly erupted milk teeth. Not long afterwards the mothers bring them to the kills and they are weaned at ten months when they make their first kill.

Synchronized births are an advantage for both cubs and lionesses. The cubs grow up with lots of siblings and cousins of their own age with whom they can play and learn. The lionesses share nurturing duties, allowing time for activities other than cub rearing. Cubs will stay with their mothers for two years, all the time learning how to be lions. By then there are probably new cubs to be nurtured and the older cubs are on their own. After that time the adult female cubs will stay with the pride. Around three and one half years of age the males will leave the pride voluntarily or may be ousted by new pride leaders. Littermates will often stay together. Having already formed a family bond, they are more capable of surviving when they can hunt and defend themselves as a team.

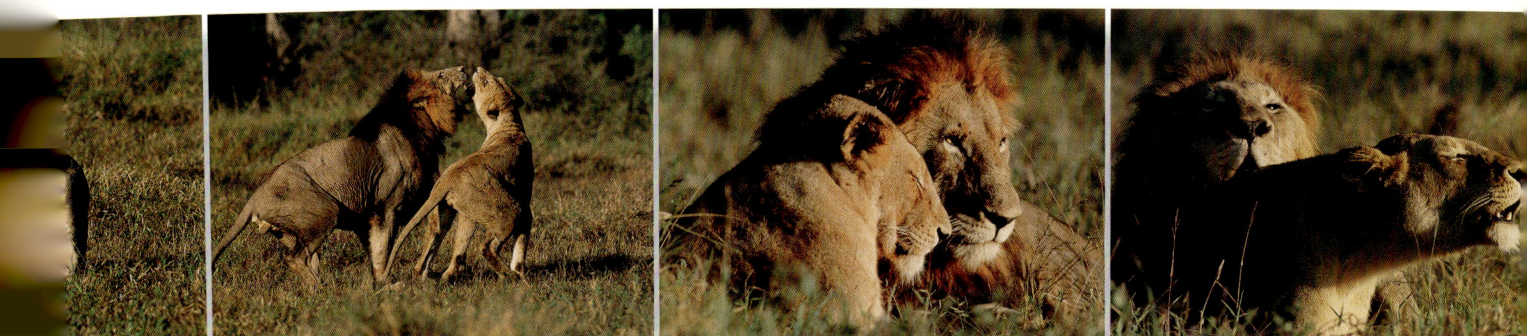

The hostile environment into which young lions must venture requires survival characteristics and behaviors. They come fully equipped with tools for hunting and defense. Their paws contain sharp, curved protractile claws, which extend when needed to catch and hold prey while the killing bite is carried out. Their teeth are designed for slicing and dicing flesh into small pieces for swallowing.

The terrible roar for which lions are famous lets ears for miles around know where they are. As a prelude to hunting time, they roar after sundown, intermittently, for an hour or more. They roar when a kill is made, after eating, and sometimes even during their feast. Staking claim to territory is occasion to roar. And they use it as a warning to any lions that dare to intrude upon that territory.

If any do cross the line, and confrontation ensues, male lions have a sort of armor that protects them from the slashing teeth and claws of their challenger – the massive mane. The lions mane begins to appear at eighteen months of age and is fully developed by five years of age. It grows up to ten inches long. Starting as the tawny color of the body hair, the mane darkens with age and they acquire a dignified look. The dark mane is a status symbol that attracts females and intimidates other males. Lions this age are in their prime. They fight for their right to a pride, to mate, and to take the lion's share.

Life is grand for lion between November and May, in the rainy season. Prey is abundant and water is readily available. The dry months from June to October are when the struggle to survive causes prides to split up for easier hunting. Hunger is a constant companion. Times of drought are hardest on the cubs. Their mortality can be up to 50 percent.

For lions that survive their cubhood, adolescence, and the many hardships of existence, their life span in the wild is moderate. Males live twelve years on the average. Females outlive males by about four years.

Overall, the range of lion habitat has diminished over time. Once widely distributed over Europe and Asia as well as Africa, they can now only be found in a small area in India and in the game reserves and National parks like Kruger. Habitat loss and hunting have taken the toll. At times hunters have paid thousands of dollars to shoot a lion. Today, as African tourism's top attraction, a lion that lives fifteen years will draw millions of tourist dollars, making it more valuable alive than dead.

Now, these majestic beasts are more often 'shot' by professional wildlife photographers and camera carrying visitors. Photographs are the trophys. Pictures of royalty. Long live the king.

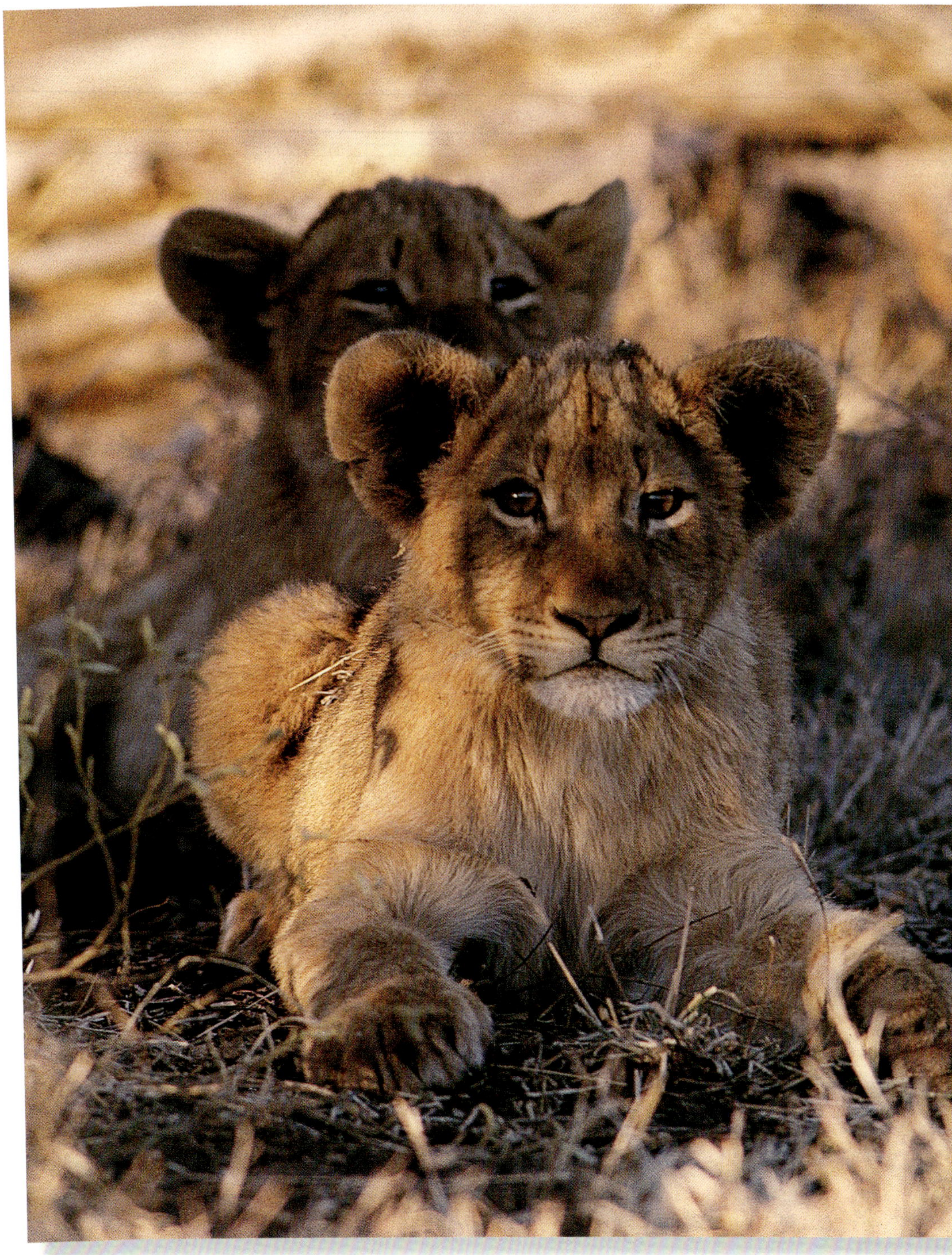

*Lion cubs always elicit smiles and a feeling of wanting to hold and pet them.*
*These cuddly images are somewhat bittersweet since about half of all cubs never reach their first birthday.*

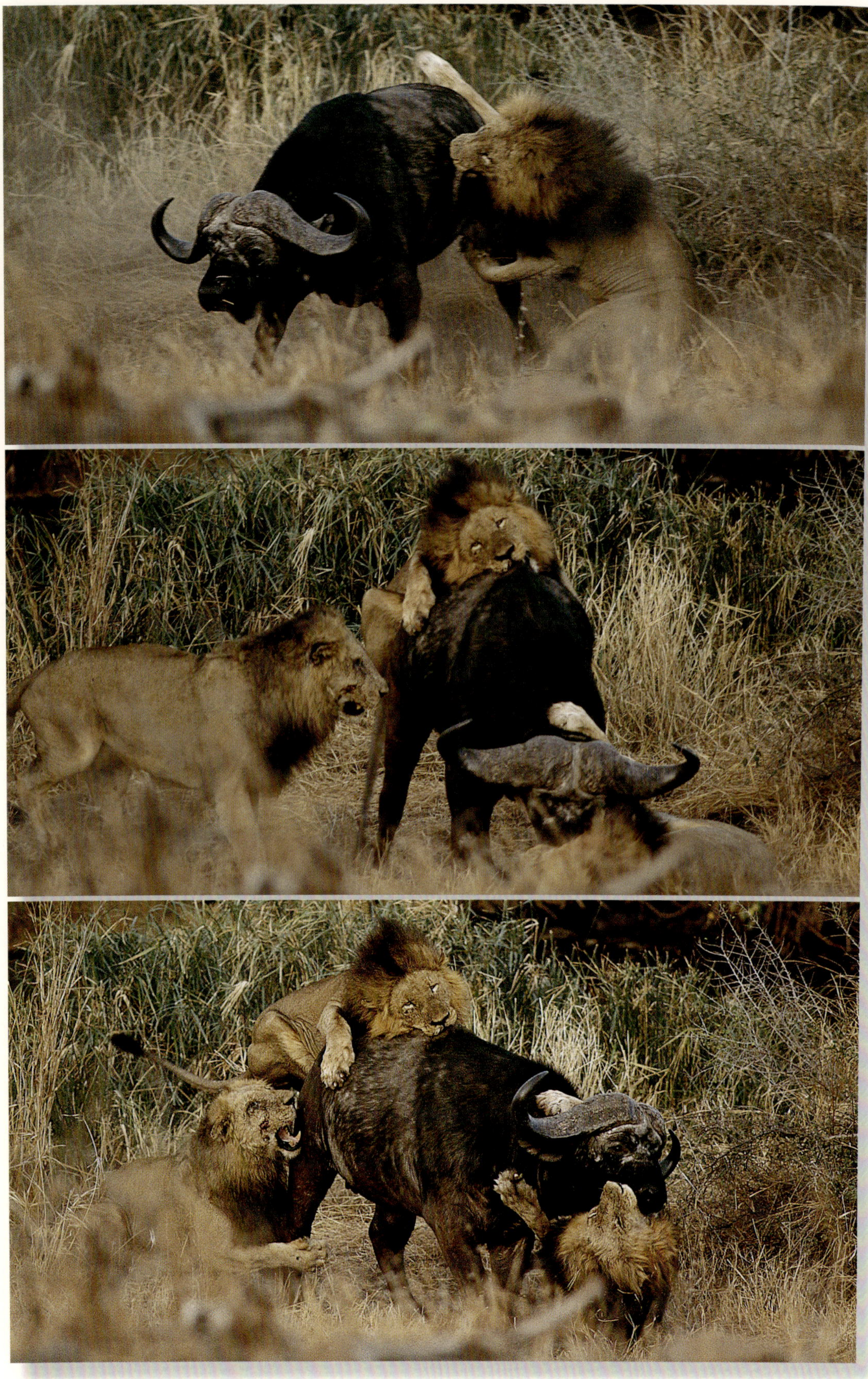

*This sequence is a testament to the brute strength of the African buffalo.*
*I did not think it possible for this bull to withstand the onslaught of three lions.*

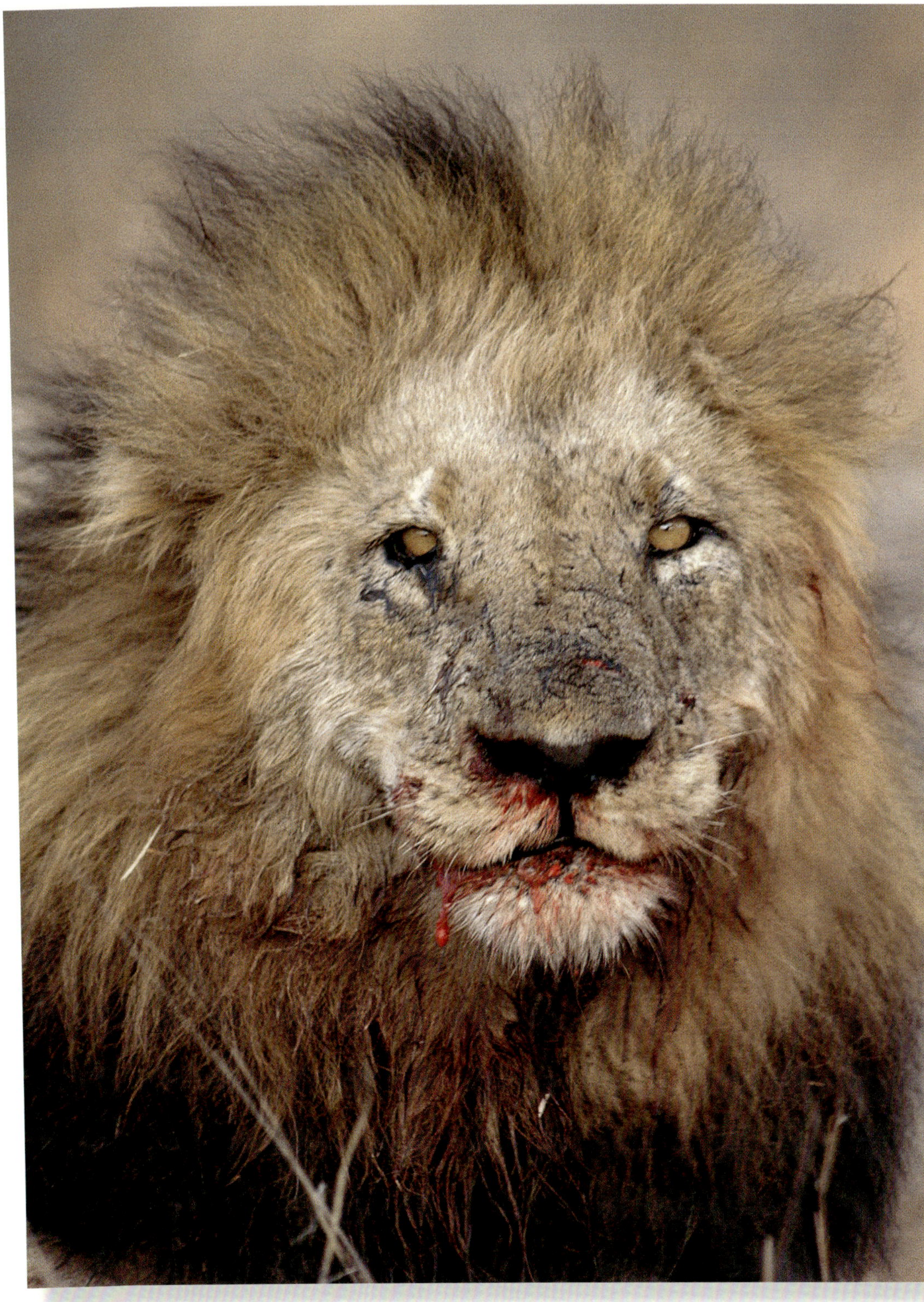

*One of the lions that could not bring down the buffalo is lying in total exhaustion. Just look at his glazed eyes.*

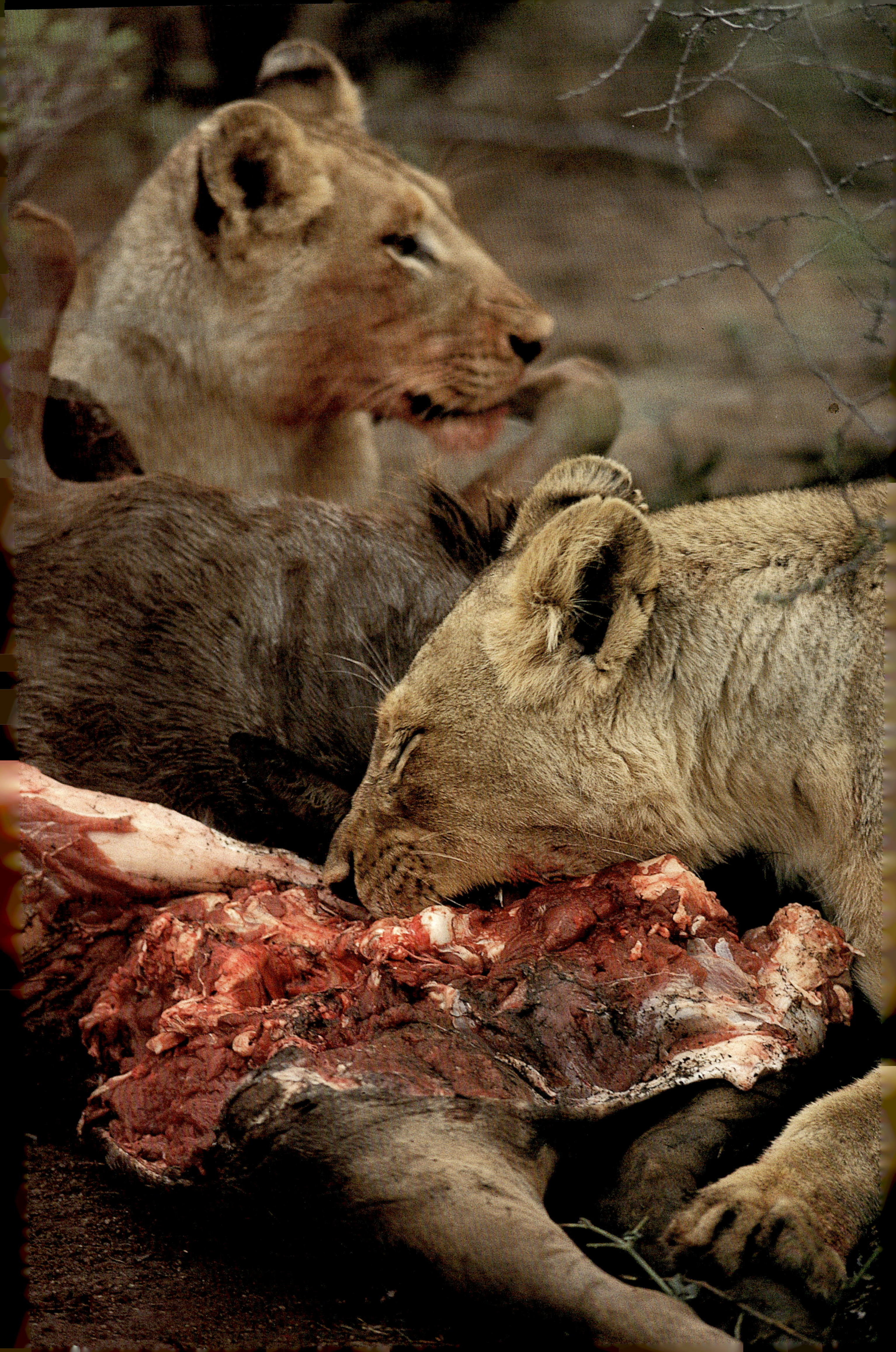

Facing page

*It is possible that this is the first kill for these two young females. If so, they are becoming survivors!*

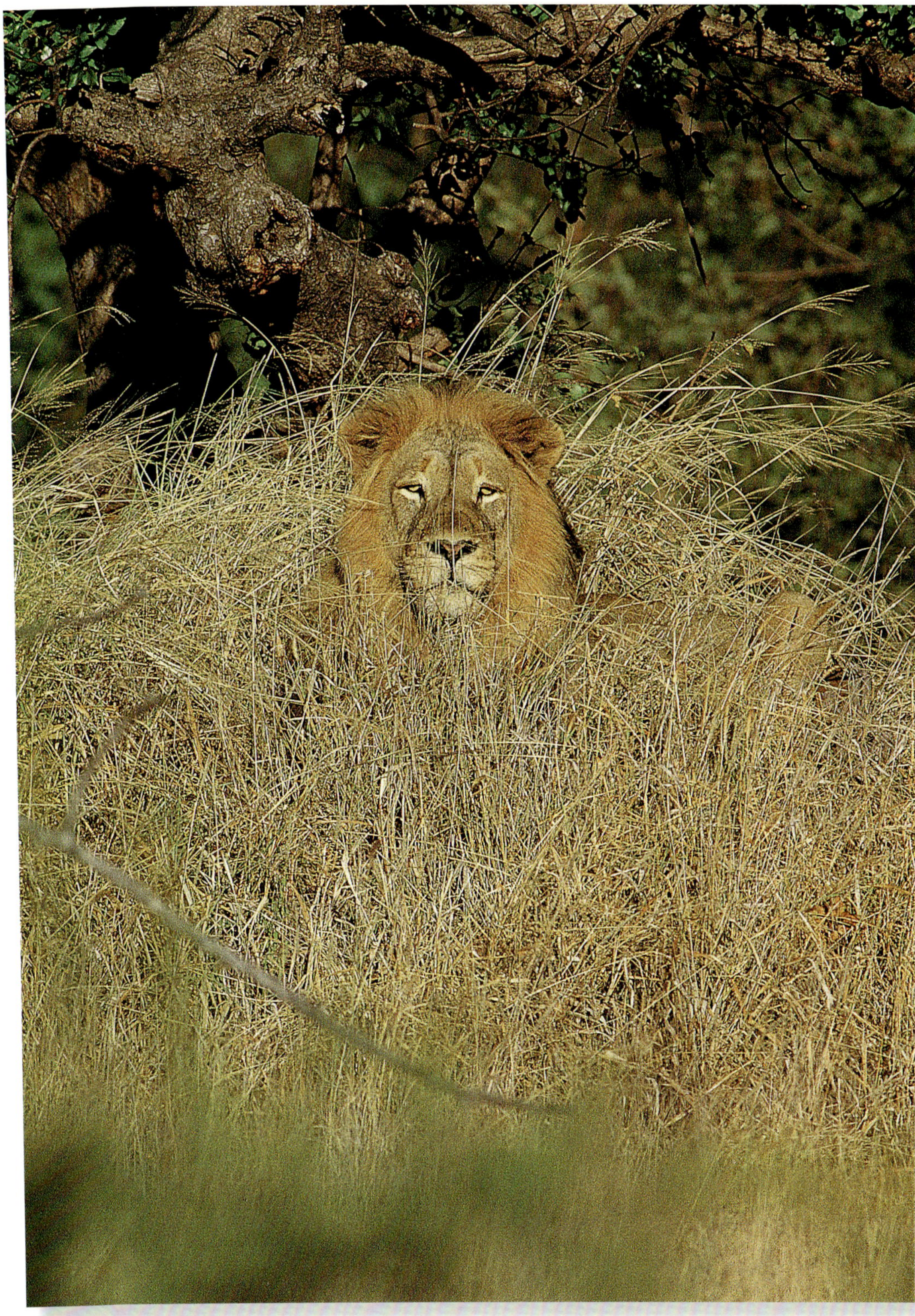

*It is easy to see why prey animals fall victim to lions. They blend so well with their surroundings, often disappearing in tall grass.*

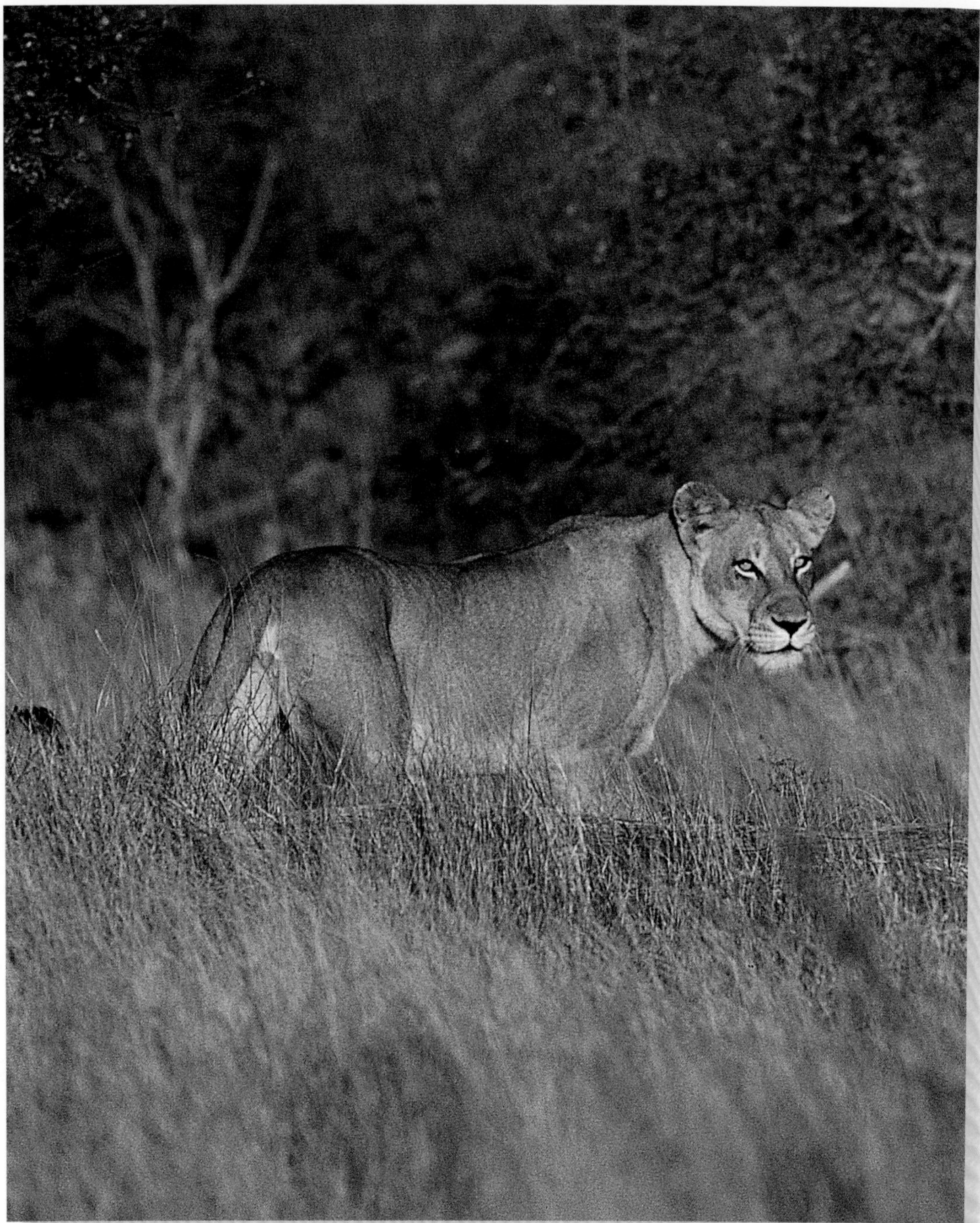

*The first big cat I ever photographed was this lioness. She was wending her way through the bushveld with resolve, probably to rejoin her pride.*

FACING PAGE

*This older beast represents the life cycle of lions. Regardless of his age his regal bearing is still intact.*

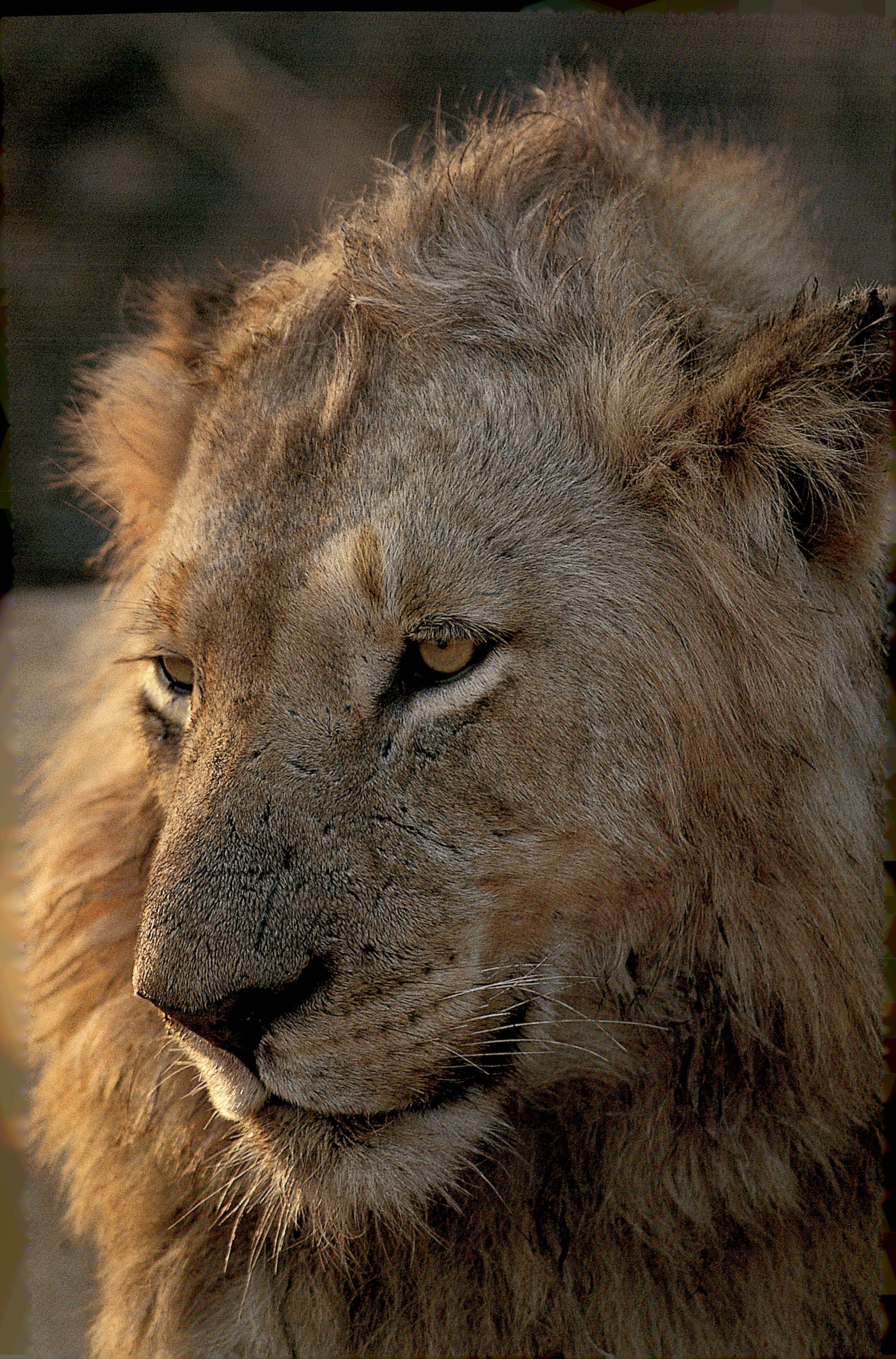

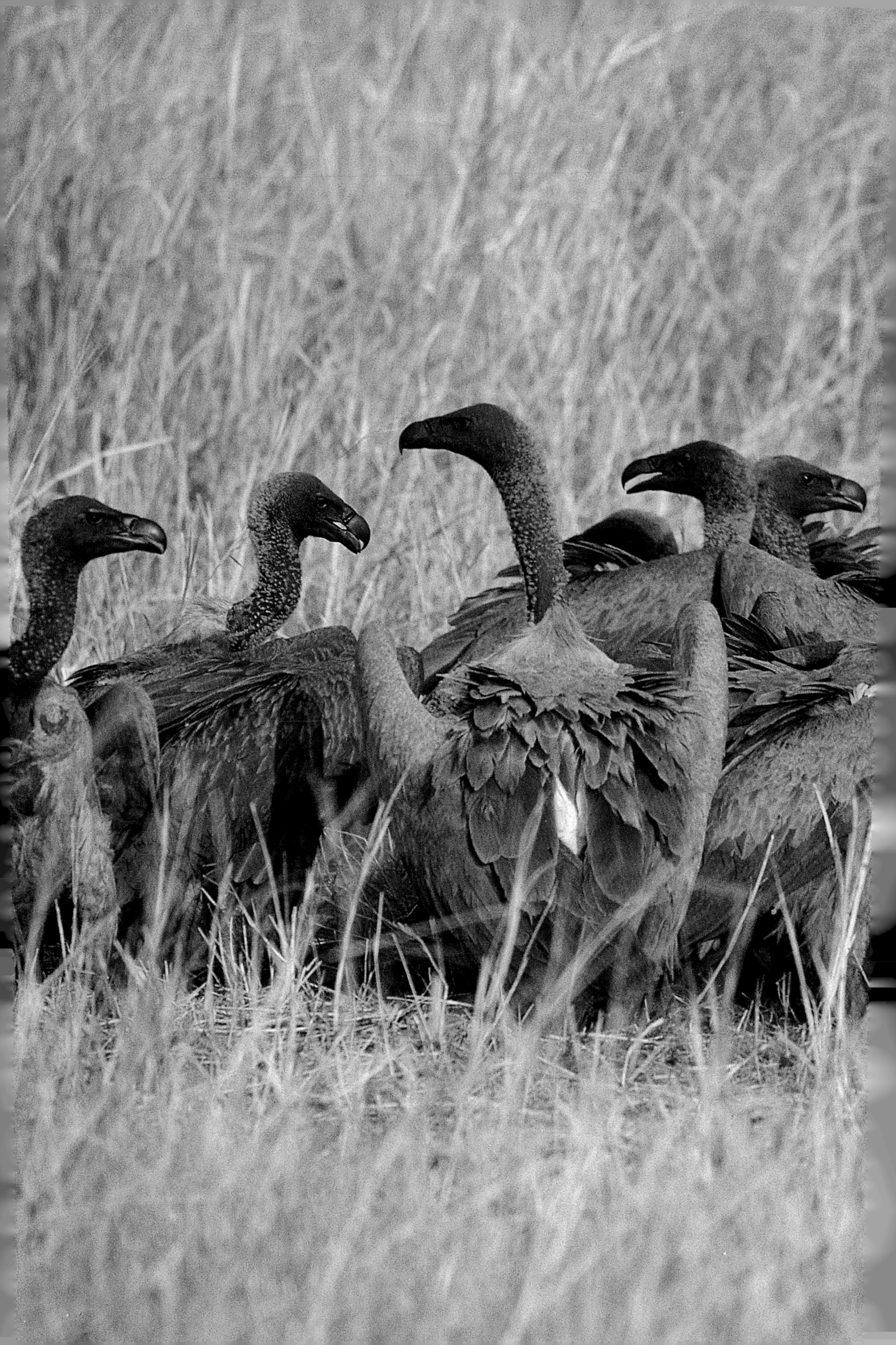

# Chapter Two

# Soaring the African Skies

## *Raptors*

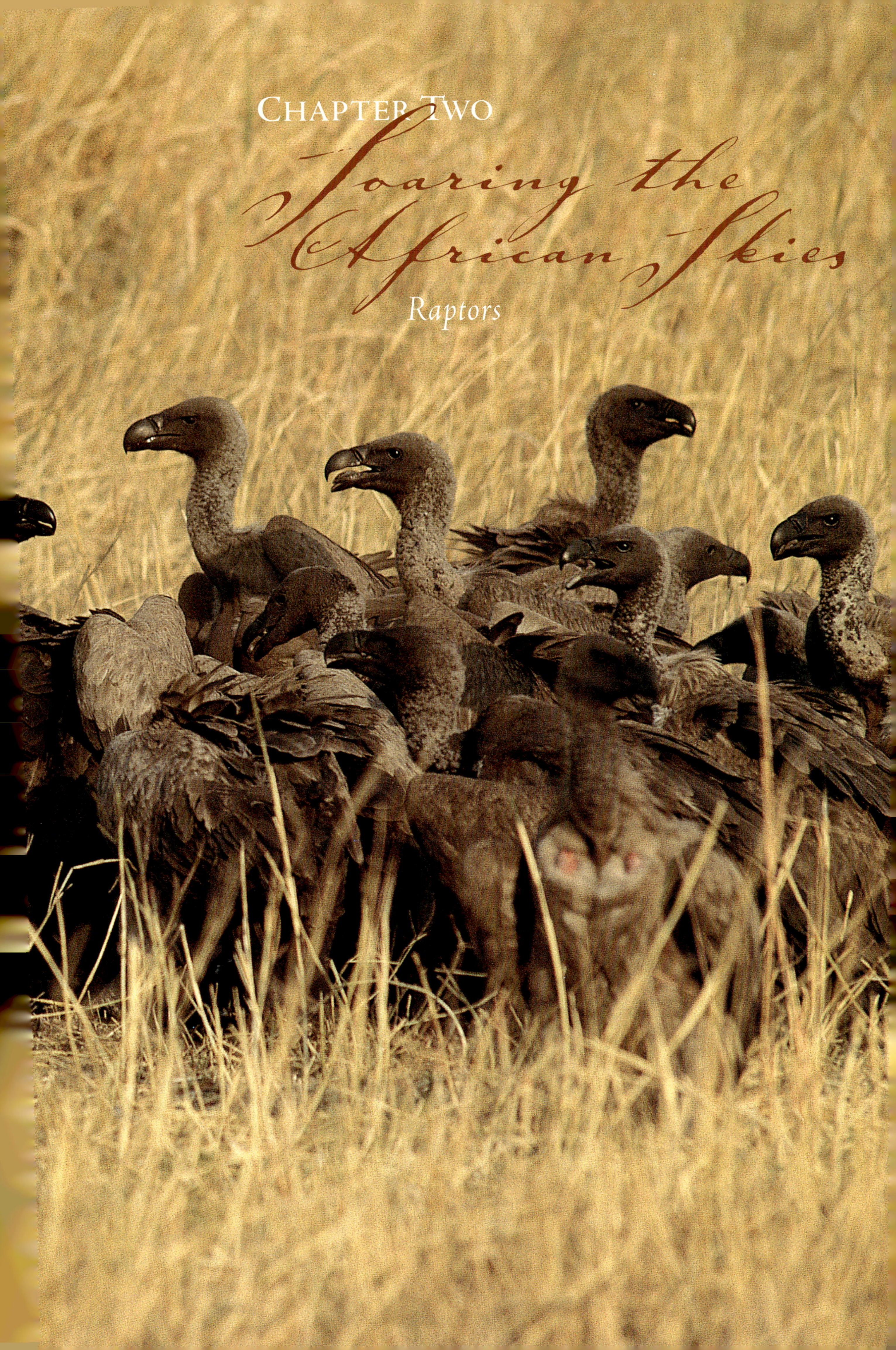

If lions rule the grasslands ... birds of prey rule the air. Wheeling and soaring through African skies, they search the savanna below, ever vigilant, for any sign that might signal when to dive for a meal.

PRECEDING PAGES
*Who could know that there is a zebra carcass beneath this horde of vultures? They were quick to clean the carcass down to the bare bones.*

FACING PAGE
*Sometimes images of a blood red sky and vultures roosting at sunset are quite foreboding.*

They are called raptors for their rapacious way of life. With keen 'eagle' eyes, raptors can accurately spot prey and swiftly swoop in with strong taloned feet outstretched to seize and kill. Equipped with sharp curved beaks, they can voraciously shred meat to eat quickly. No wonder they evoke strong emotions in people. Raptors are symbols of freedom, speed, power, and strength. They are eagles, hawks and owls. And, they are vultures.

Vultures may not be the first bird that comes to mind when we think of raptors but they share most characteristics with their fellow birds of prey. They soar the skies. They have keen eyesight. Their hooked beaks efficiently tear flesh. And although they do hunt like other raptors, most of their food is carrion. Unlike other raptors, though, their feet are flat – adapted for standing and eating rather than killing.

Never, it seems, was there a bird that enjoys a meal more than a vulture. Social birds, dozens of them will gather excitedly around a zebra carcass and gorge themselves like starving guests at a buffet. They have no table manners, though. Feeding is a furious free-for-all complete with food fights and feathers flying. And they talk with their mouths full! Usually silent birds, vultures squeal, whistle, hiss and grunt while eating.

They eat fast, filling a crop at the base of their neck with food that can be saved for a later meal that they access by regurgitation. Speed counts too when other birds, jackals and hyenas are competing for the same carcass. Vultures can finish an impala in a short ten minutes. The older birds eat first, after which the younger ones are left to pick bones. Females eat more when forming eggs.

They aren't picky about what they eat. Any type of carrion will do. The savanna is a smorgasbord of elephants, buffalo, zebra, wildebeest, impala, monkeys – and even rats. Vultures will employ many means to find food. They seek and find dead animals, follow predators, steal from other raptors and wait for sick animals to die. At times they will kill small animals and resort to eating eggs, fish or termites. Contrary to popular belief, they will pass up rotten, decomposing meat.

The best season for vultures is the dry season. Opposite of what is good for lions and other creatures of the bushveld and plains. The dry season kills more animals and so carrion is more plentiful. Grasses are not grown high, hiding food from view. To everything there is a season.

After a meal, vultures will seek out water holes where they will drink and bathe together, putting to rest another misconception that they are filthy birds. This social event is a relaxing one where they take their time resting and drying their wings before they can fly. Vultures spend a lot of time on the ground, whether eating or bathing. They partake in an activity called 'sunning.' In early morning as the first rays of sunlight appear, they stand in a spread wing posture, soaking up the rays. As the day warms, they then prepare to take to the sky.

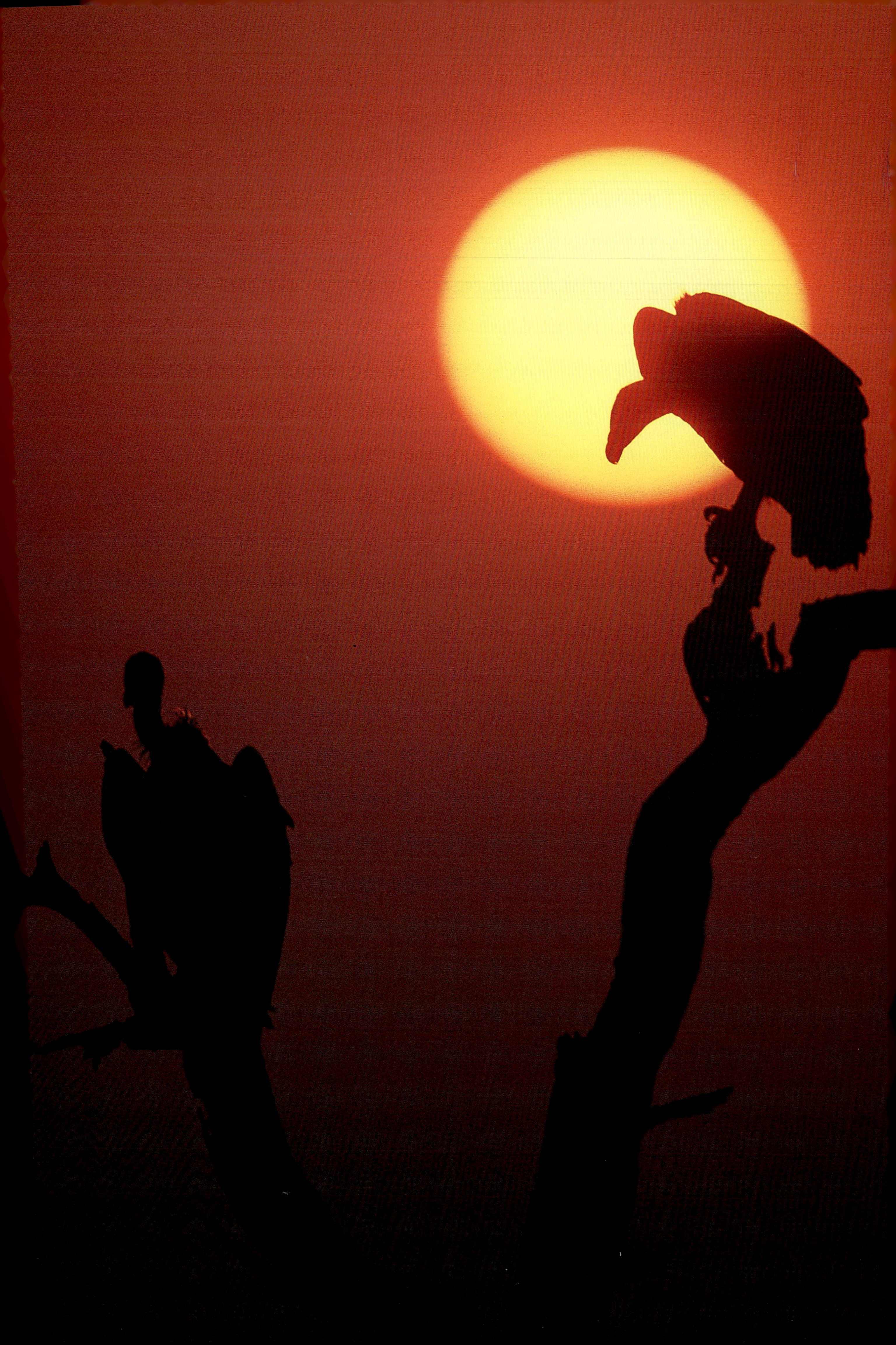

Because vultures are heavy birds, they depend on rising warm air currents called thermals to soar and stay aloft. Their flat feet are adept at running for take off into a rising thermal, where they spiral up, up, up to soaring height. They must soar, for they are unable to flap their wings for more than a couple of minutes. From their lofty vantage point they can circle for hours and use their keen vision to survey the savanna below for their next meal. Sometimes they use jet stream currents at altitudes of over twenty-five thousand feet for traveling.

BELOW
*Their size and power make these Lappetfaced Vultures the dominant vultures at a carcass.*

Vultures are very visible creatures; plentiful, easy to spot and study. Most are dark in color. Some are light colored. There is no difference between male and female except that females tend to be a bit larger. For warmth, they have a thick undercoat of down which easily sheds. That accounts for the cloud of feathers surrounding a feeding site. Tales are told about vultures drooling and slobbering. Close observation reveals that it is, in fact, a secretion from salt glands located in the nostrils. They must excrete excess salt that accumulates in their bodies from the food they consume.

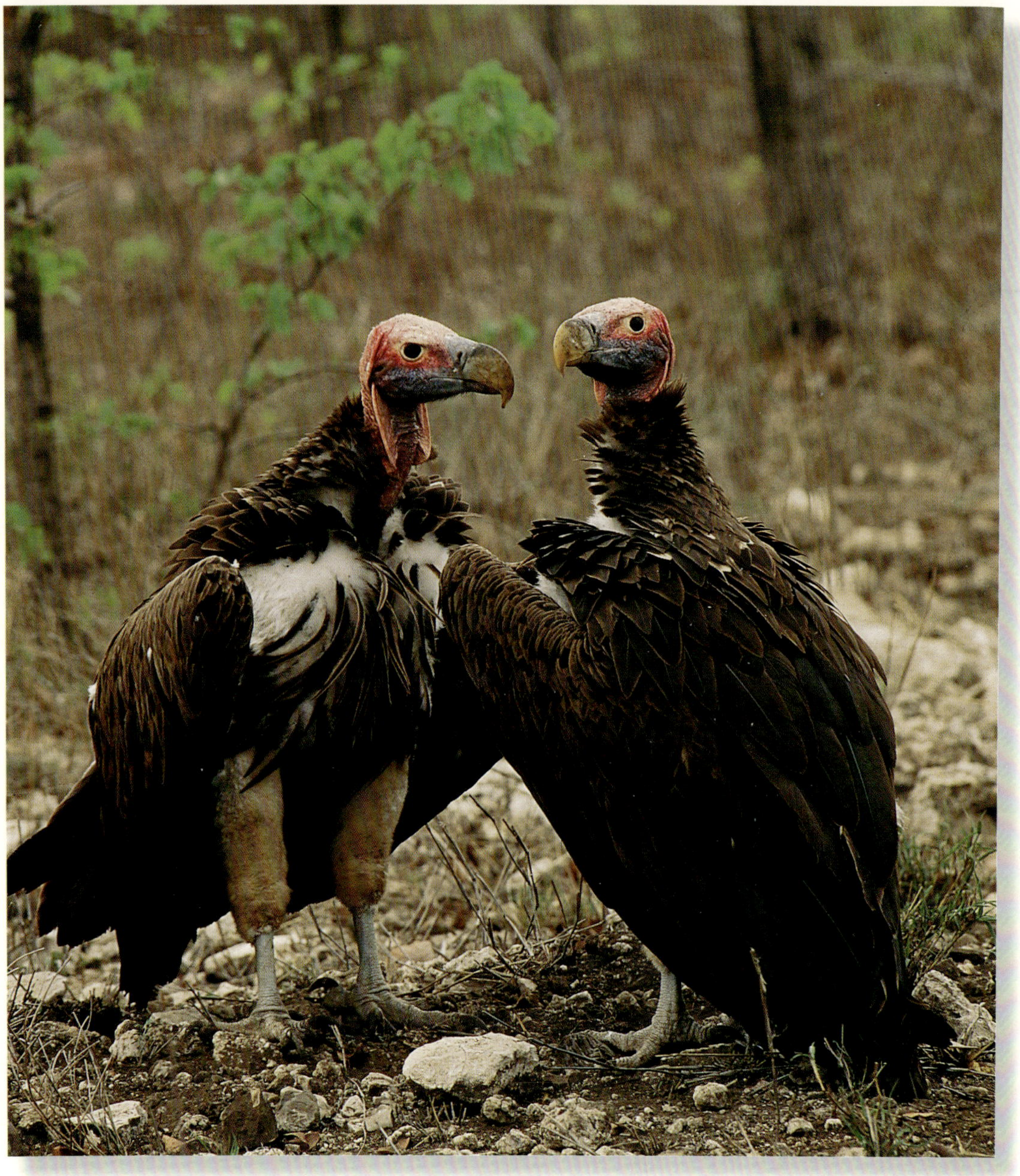

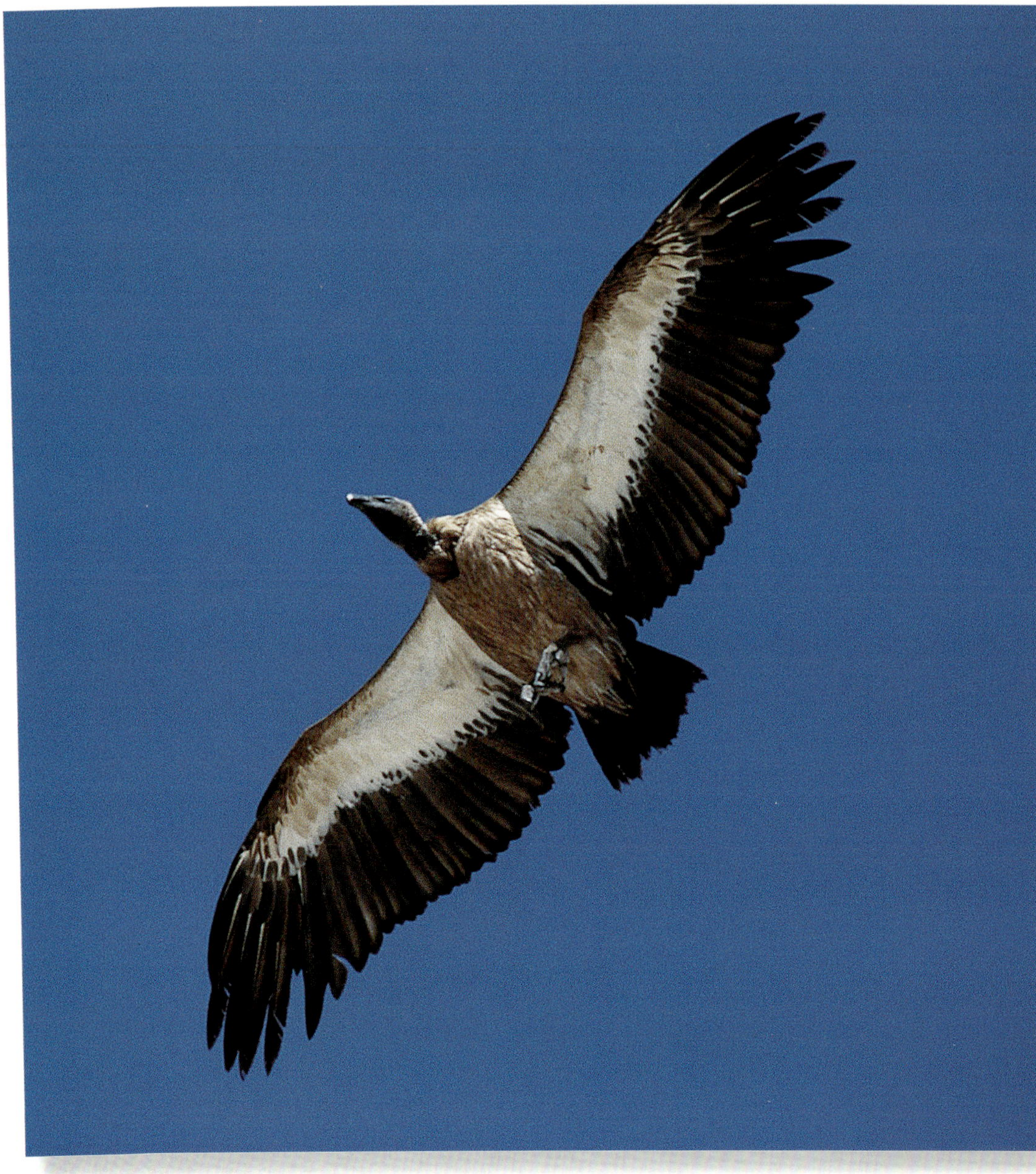

Many species have the naked heads and long necks with feathered ruffs. This odd look and their association with death has made them popular for cartoon characters. But these features have practical uses. The heads and strong necks have only short downy feathers that are more easily cleaned after diving into a carcass. The feathered ruff helps keep them warm in cold weather and is also used in courtship.

At three to four years vultures reach breeding age. Mating behaviors include terrestrial rituals and balletic aerial displays. On cliffs or in trees they build nests made of sticks and lined with grass. For some species these nests can be up to six feet in diameter. Some also line nests with found objects such as rope, rags, wool or even garbage.

Above

*Few wing flaps are required to keep this Whitebacked Vulture in flight. It takes advantage of thermal lifting action.*

A single egg is laid by most species. Some lay two. Both parents take turns incubating the egg for an average period of fifty-six days. Nestlings are fed regurgitated flesh and bones from the parents' crops. Fledglings will stay in the nest for several months - even after learning to fly.

The most common vulture in Africa is the Whitebacked Vulture, with a wingspan of over seven

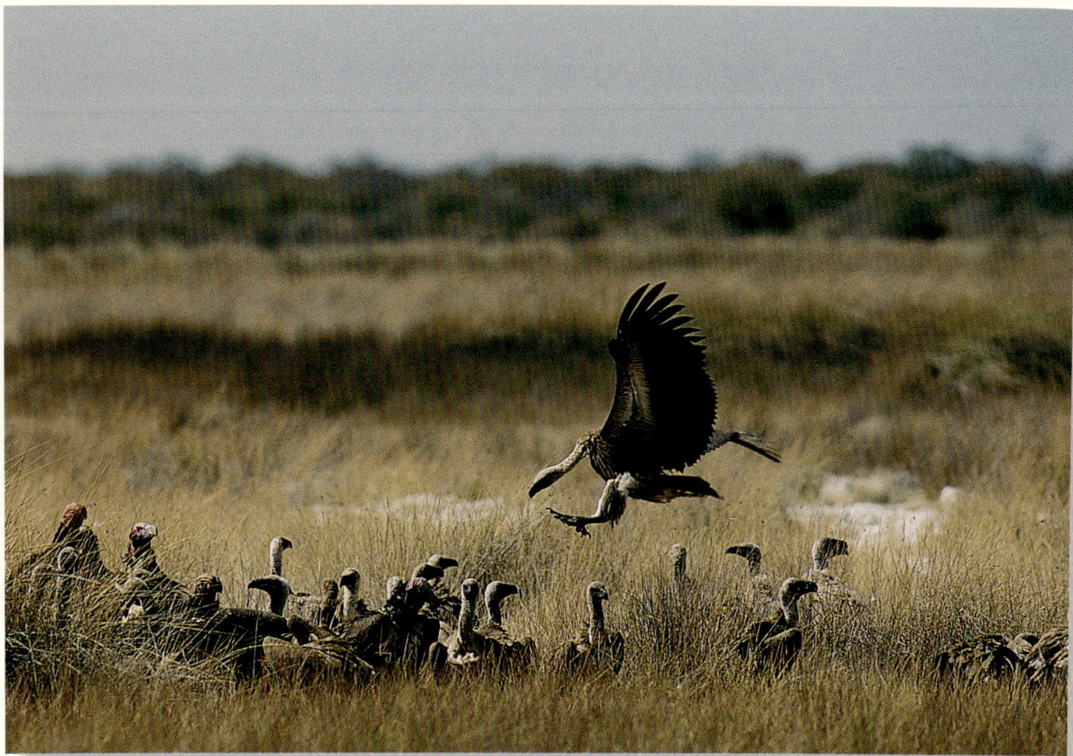

*As one vulture lands close to a carcass, one can almost hear the cackling and squabbling that goes on.*

feet. Small colonies can be seen nesting in tree tops. They are very social – comprising the largest numbers at water hole baths.

Lappetfaced Vultures are less social. Their naked bright-red heads and necks are most visible at a feed even though they are often seen only in pairs or alone. With a wingspan of nine feet, they are the largest and most powerful of vultures. They are dominant participants at feeding time – often the first to open a carcass. They can also eat the tougher parts like the skin and ligaments. These massive birds will also hunt and kill small mammals.

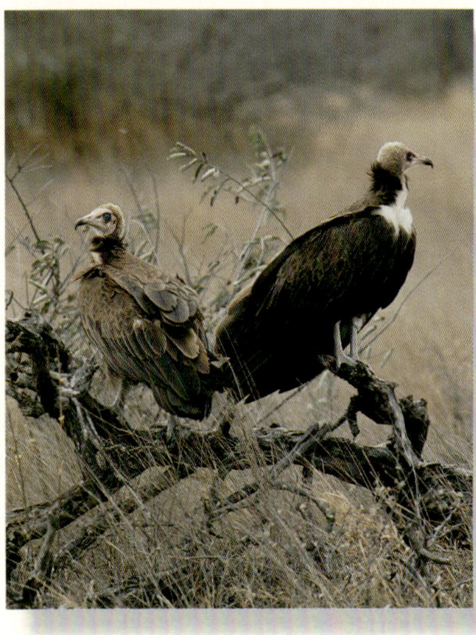

*This pair of Hooded Vultures is unable to compete with larger vultures due to size, but they always manage to get their share at a kill.*

The smaller Hooded Vulture has a five and one half foot wingspan and is identified by a grey feather cap, or hood. It is a gregarious bird, and a scrappy type at dinnertime. They fuss and fight over whatever they can take – though they are unable to compete with the larger vultures. Their smaller size is an advantage for flying since they don't have to depend on thermals like the heavier birds. This allows them to scavenge earlier in the cool part of the day.

Other daytime raptors, like eagles and hawks, are chiefly hunters. Solitary silhouettes, they circle the skies casting undulating shadows on the hunting grounds below... searching out their unsuspecting prey. These raptors hunt, but they are not above eating carrion and robbing other birds.

One of these, the Bateleur, is an eagle easily recognized for its unique characteristics. Named for its unusual flying style, Bateleur is French for acrobat. As it soars, the Bateleur rocks its wings back and forth and some have been observed performing 360 degree rolls. The shape and coloration of the adult Bateleur is striking, but it takes seven or eight years to attain maturity. The plumage, beak and feet go through many changes from hatchling to adult. The immature bird is light brown with a greenish-blue face and feet. The plumage becomes progressively more brown as the face and feet turn more orange. The mature Bateleur is identified by black and white wings and mantle and a chestnut back. By then the face and feet have turned a striking scarlet red. Another unique characteristic is its short tail, which makes it look almost tailless in flight.

This stand-out eagle is renowned. It is the frequent subject of native folklore. One of its tribal

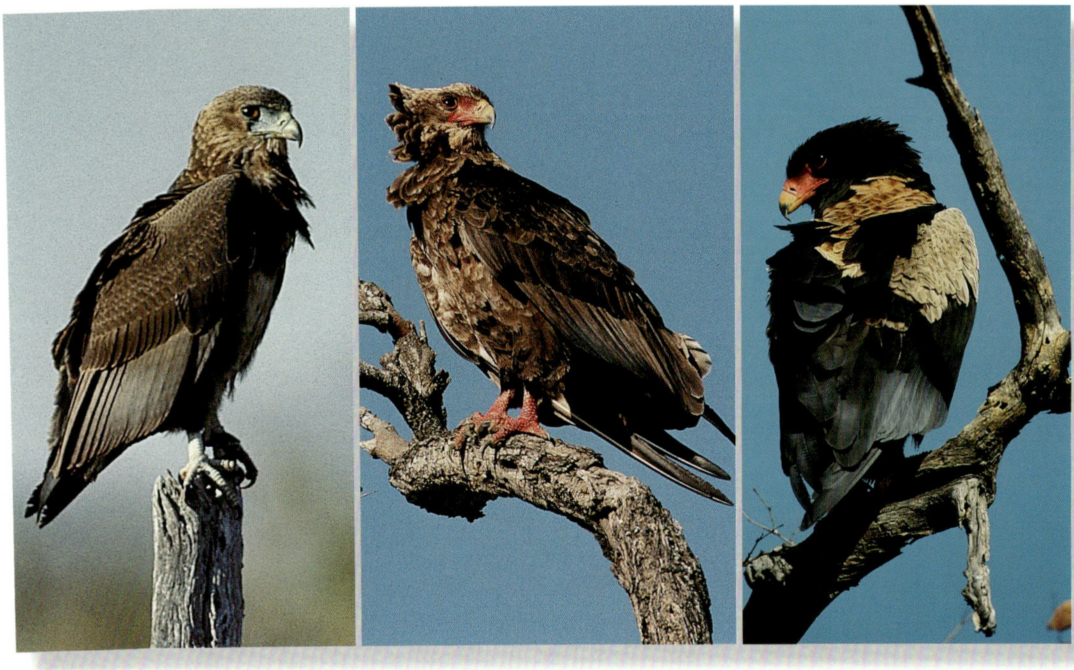

*The color changes in plumage, talons and beak during a Bateleur's maturation are quite astonishing.*

names means 'warrior bird' for its battle cry – 'kow-wah!' For some, it is the spirit messenger of the grasslands... carrying the spirits of departed kings. To the naturalist, it is but another raptor hunter. Much of its day is spent in the air, swiftly soaring at between forty and fifty-five miles per hour, searching out a varied menu of mainly birds, small mammals, and snakes.

While the Bateleur soars the skies, the African Fish Eagle perches hour upon hour on end in tall trees overlooking its favorite fishing spots. Alone or in pairs, they can be observed near large bodies of water such as lakes, rivers and dams. A cousin to the Bald Eagle, the African Fish Eagle also sports a white head. Added to that is a white breast and tail... a stunning contrast to black and chestnut wings and body. At dawn its loud call breaks the silence. Frequent vocalizations can also be heard while it soars over the waterways.

When the fish eagle is hungry, it will hunt from its perch ... swooping down to snatch a fish with its talons, barely getting wet. It can catch fish up to eight wriggling pounds.

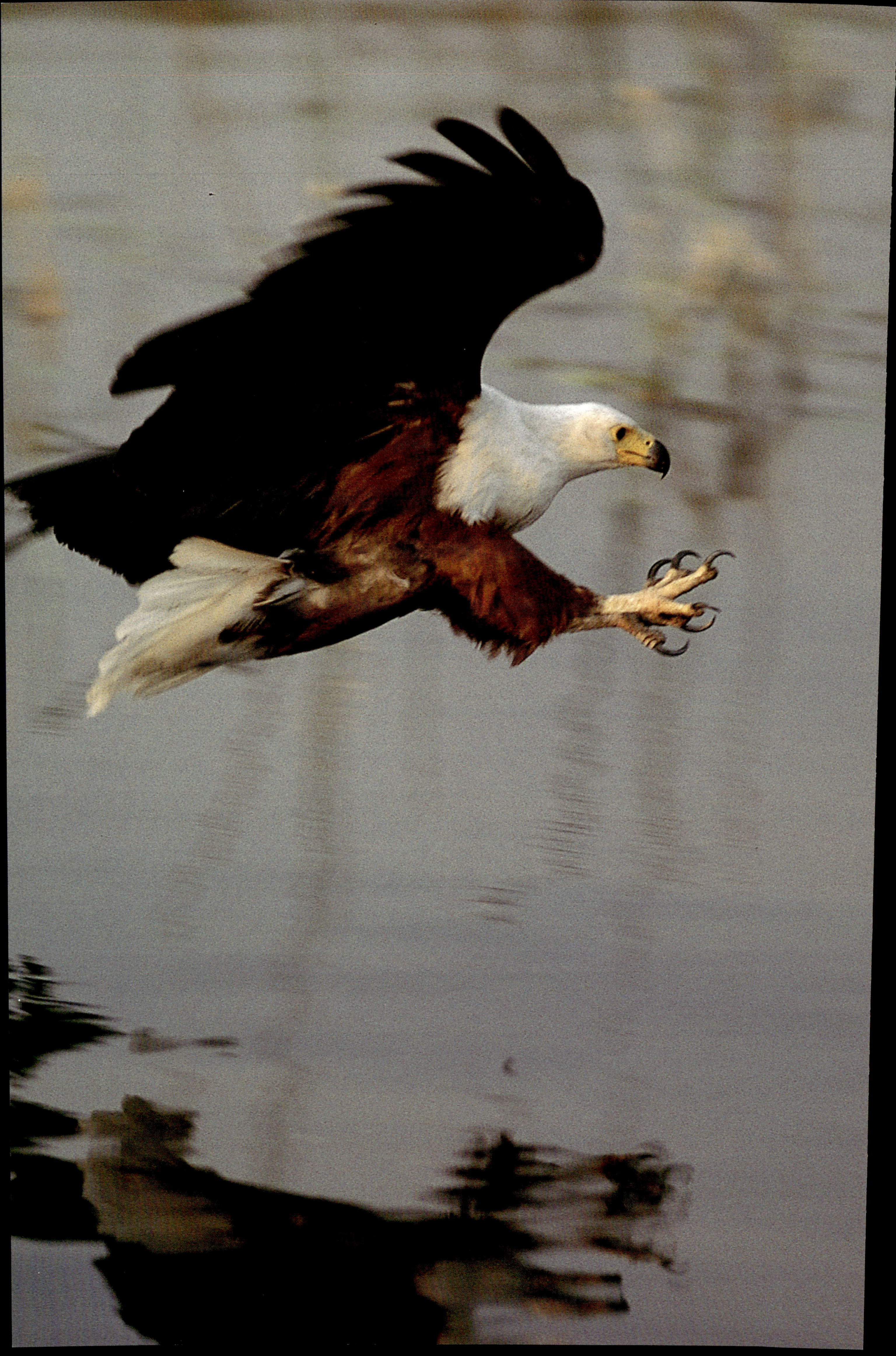

Another raptor found perching in tall trees is the Martial Eagle. This large brown bird will hunt from his lofty perch too, but also soars at great altitudes, all the while scanning the savanna and forest edges for his favorite food. This great eagle can take birds of all sizes, small mammals and antelope. It also likes reptiles up to the size of the Monitor lizard.

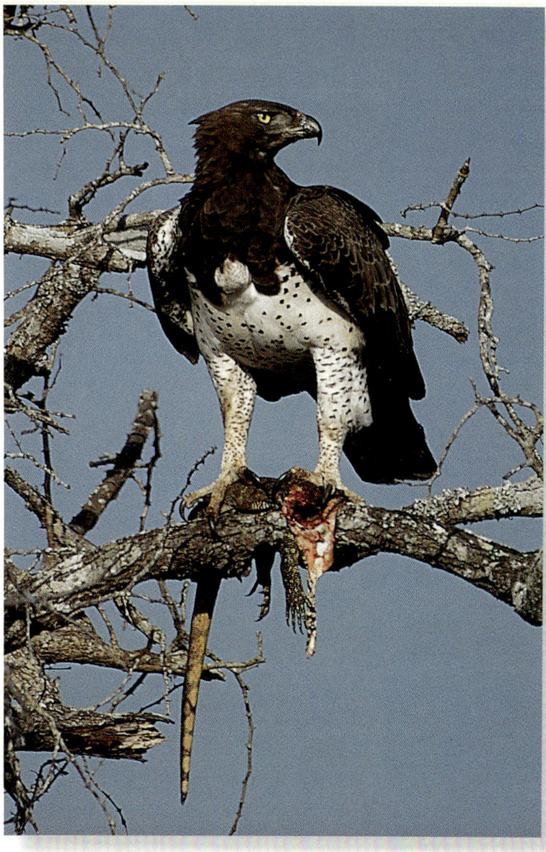

The Monitors fate is doomed because it, too, is a diurnal animal living in the same habitat as the Martial. It is mostly terrestrial, but climbs and swims as well. The Monitor's best defense is its intimidating behavior. Imagine for a moment the confrontation between giant lizard and hungry bird. The eagle plunges with outstretched legs, talons poised. The Monitor thrashes its tail while hissing loudly and inflating its leathery throat. The menacing posture doesn't always work, however, and the Martial Eagle may victoriously make off with a tasty meal.

A smaller eagle, the Tawny Eagle, can be identified by its rough appearance... scruffy brown plumage mottled with tawny patches. The Tawny is usually solitary, perching on treetops overlooking woodlands and savannas. Sometimes it is found in groups. Like the African Fish Eagle and Martial Eagle, it hunts both from its perch and from the air. Tawny eagles prefer small mammals like hares, but will sample birds, reptiles and fish. It has a reputation for robbing from other birds as well.

Another group of raptors, hawks, must share the daytime sky and game with vultures and eagles.

From within the canopy of trees the Little Sparrow Hawk may be found. This small bird is grey backed with yellow feet and is rather bold for its size.It watches intently for its favorite food – birds. It will snatch prey off the ground or as it flies past and can take birds that match its own eighty gram weight.

The Dark Chanting Goshawk, a mostly grey bird with red feet, perches in trees. It is always looking for snakes and birds up to the size of Guinea Fowl. It will also enjoy the occasional small mammal.

The Gymnogene is a rather large bird with black and white barred plumage and with yellow face, legs and feet. Its head is small and adapted for pulling prey from small crevices. Gymnogene's tarsal joints swivel in all directions so it can reach into holes with its feet. Versatile hunting habits allow for prey to be taken from the air, from its perch or from the ground. A notorious robber, the Gymnogene steals from nests of other birds.

As the sun sets, the diurnal birds of prey settle into their roosts to rest for the night.

Above

*His prey is often a Nile Monitor Lizard. This Martial Eagle has captured a lizard that is about three feet long.*

Facing Page

*With talons extended, this African Fish Eagle swooped toward its prey. This scene occurs repeatedly over the tributaries of Botswana's Okavango Delta.*

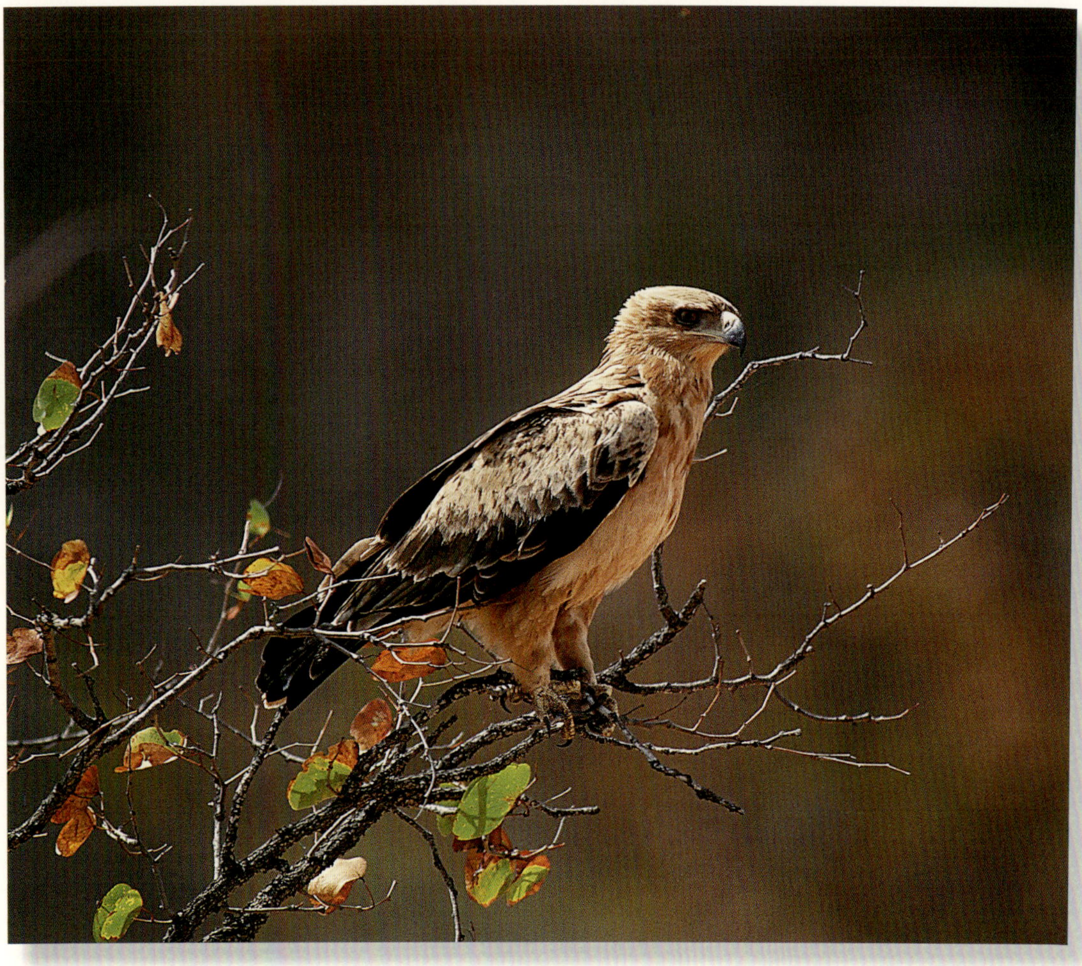

*A Tawny Eagle is hunting from its perch.*

It is then that the nocturnal raptors, the owls, begin to stir.

FACING PAGE

*As this Little Sparrow Hawk approached the water hole, a flock of Crested Guinea Fowl exploded into the air and flew off. Though several times larger than the hawk, the Guinea Fowl used a great deal of discretion around this raptor.*

One of these is the diminutive Scops Owl. With greenish skin and grey mottled plumage the Scops Owl has a unique talent for camouflage. It will close its eyes and elongate its body until it appears to be just another branch of the tree on which it is perched. From its nearly invisible posture it can easily drop below to catch its favorite food... insects and scorpions.

The deep booming night sound... *hru hru, hru-hru-hru, hru hru* ... sometimes heard in duet, is the voice of the Giant Eagle Owl. This largest of owls is a pale grey and has distinctive pink eyelids. By day it roosts in large trees by waterways or sometimes in woodland or savanna areas. By night it hunts, with an appetite for birds, reptiles and mammals.

When the night passes and dawn reappears, the owls close their full moon eyes and sleep. The morning light signals the daytime raptors to awaken. The fish eagle calls. The cycle repeats.

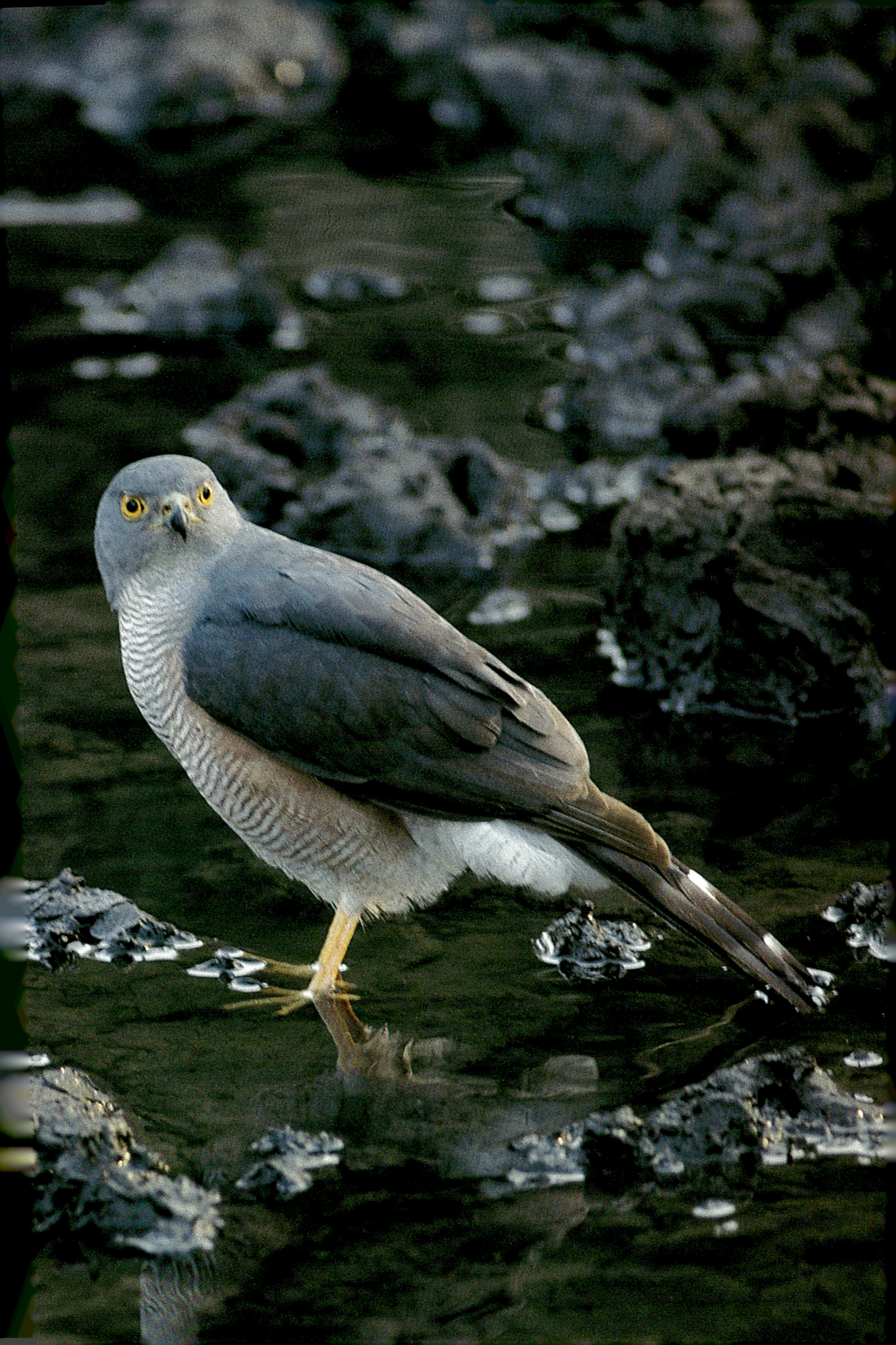

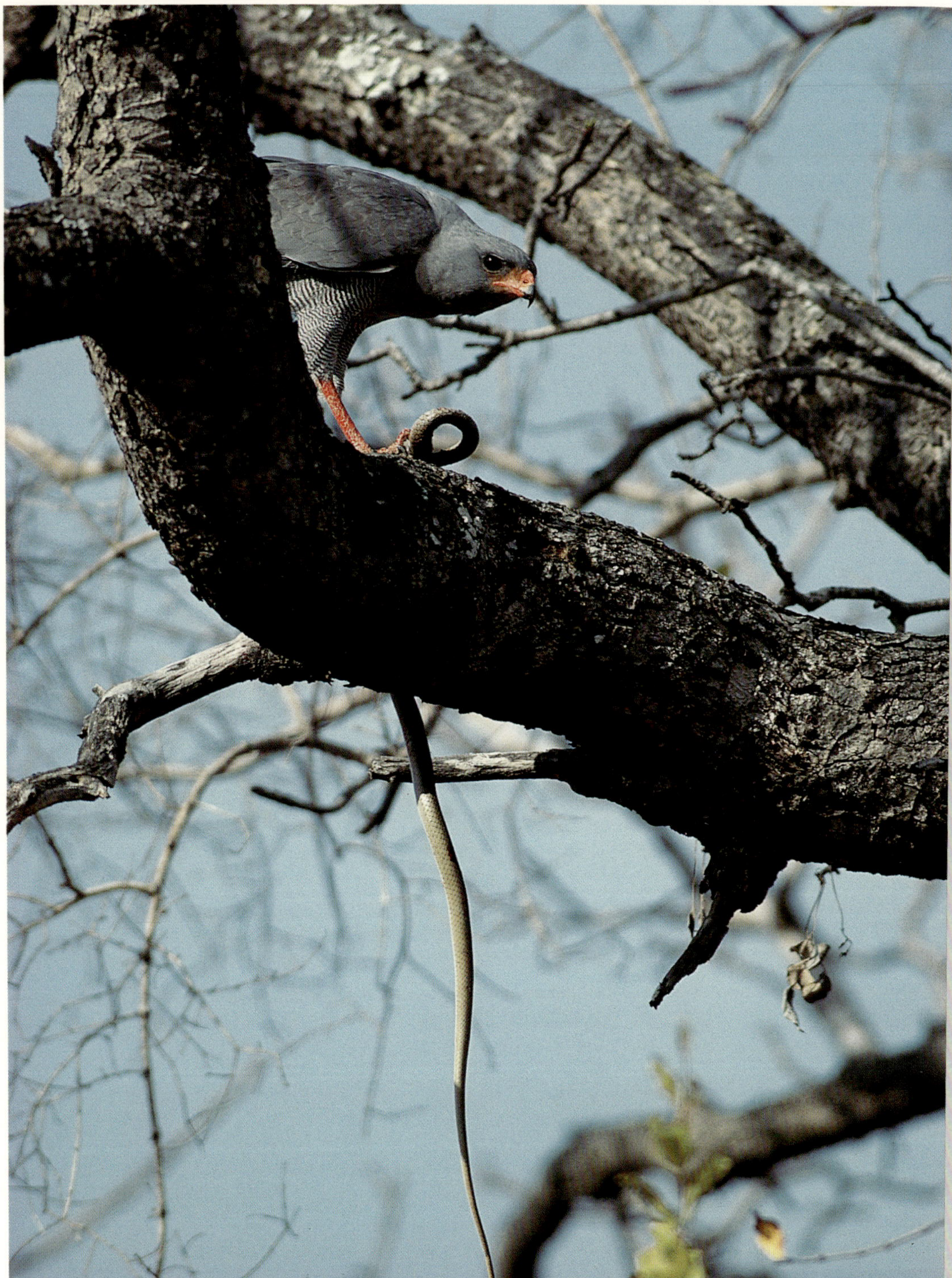

*When we came upon this scene, the Dark Chanting Goshawk was perched in a tree. He took one look at us and high-tailed it to a tree further into the bush. Maybe he thought that we might try to steal his meal.*

FACING PAGE

*I almost confused this Gymnogene with a Dark Chanting Goshawk. This was the first time I had seen one.*

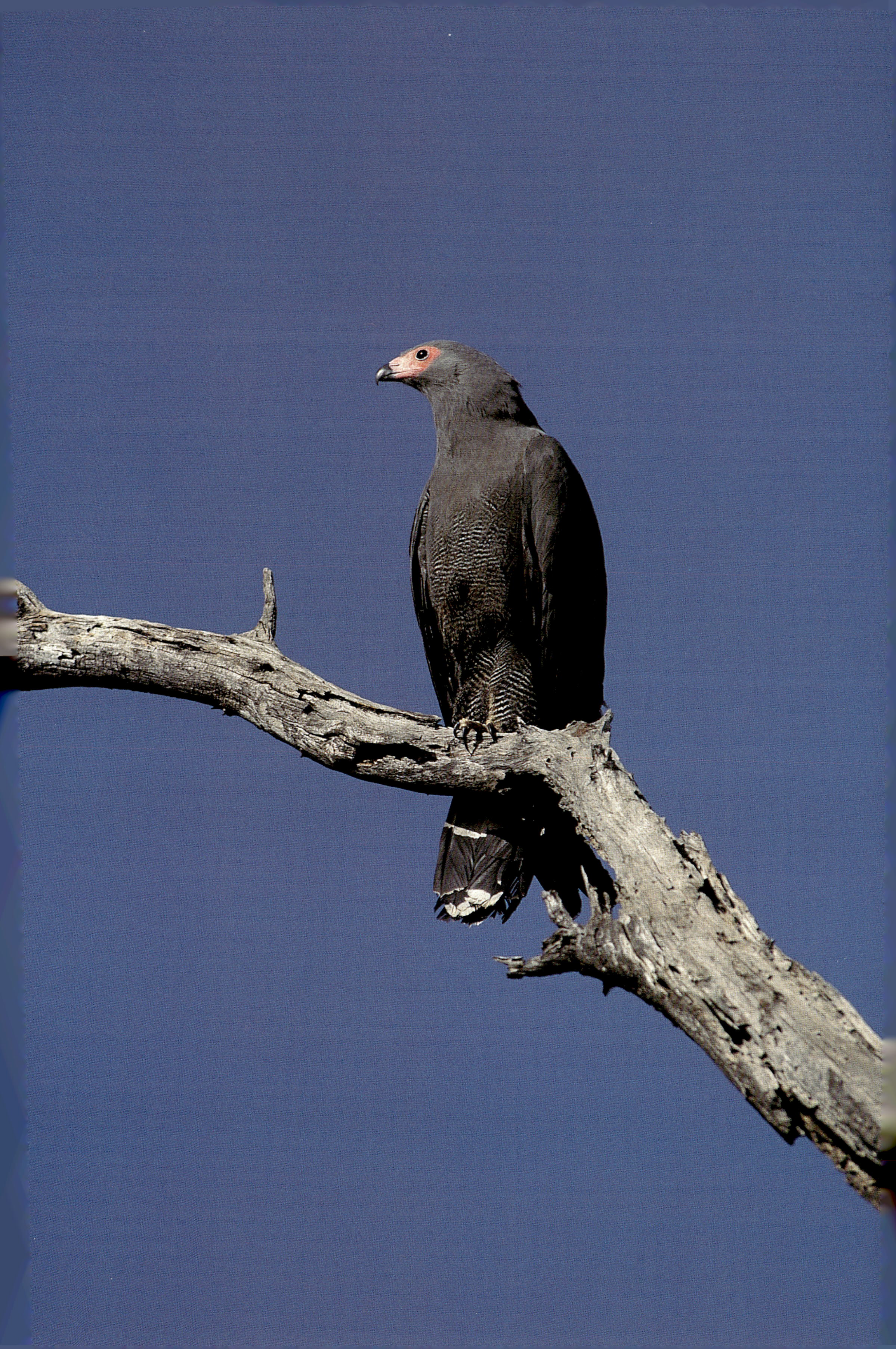

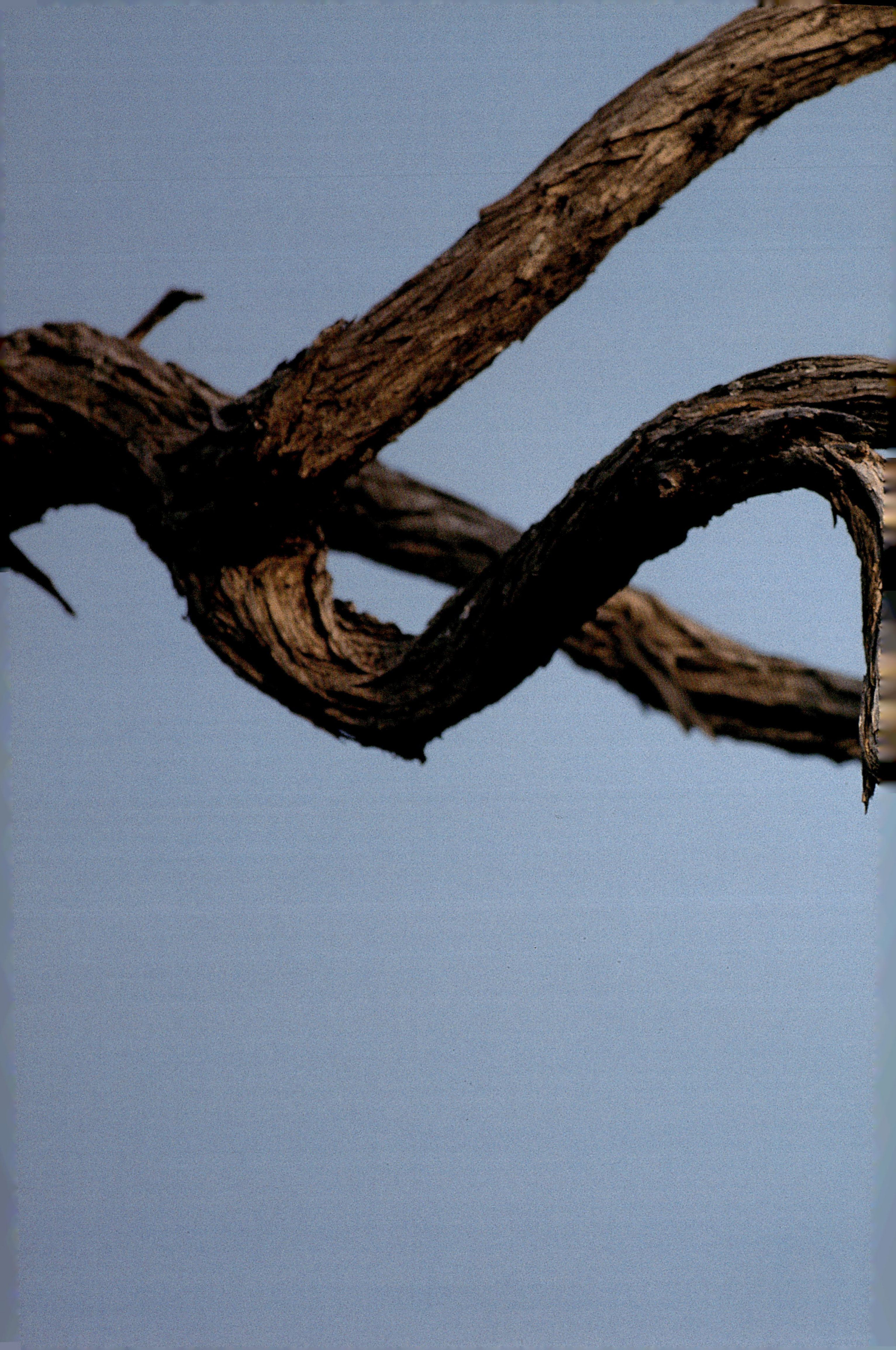

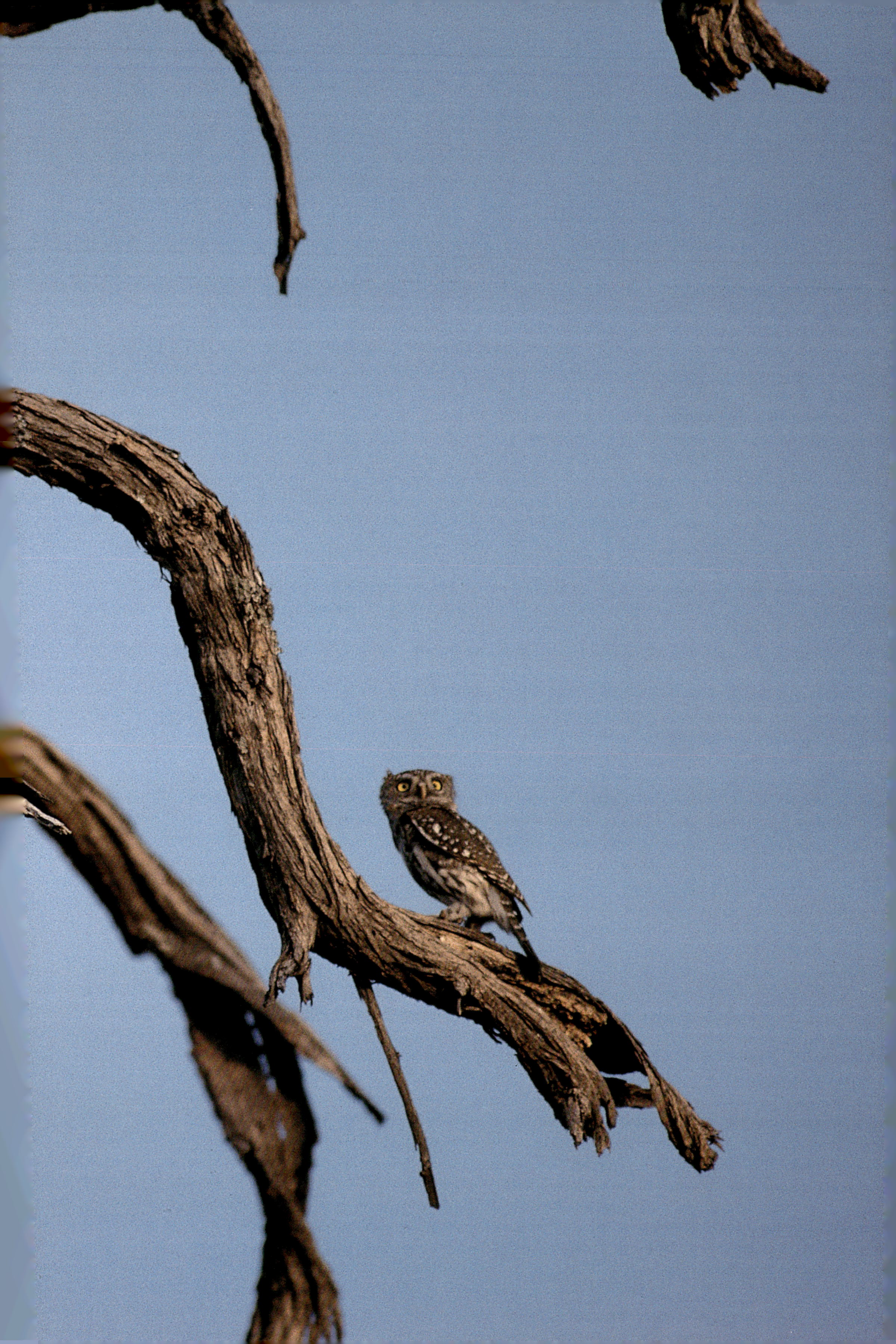

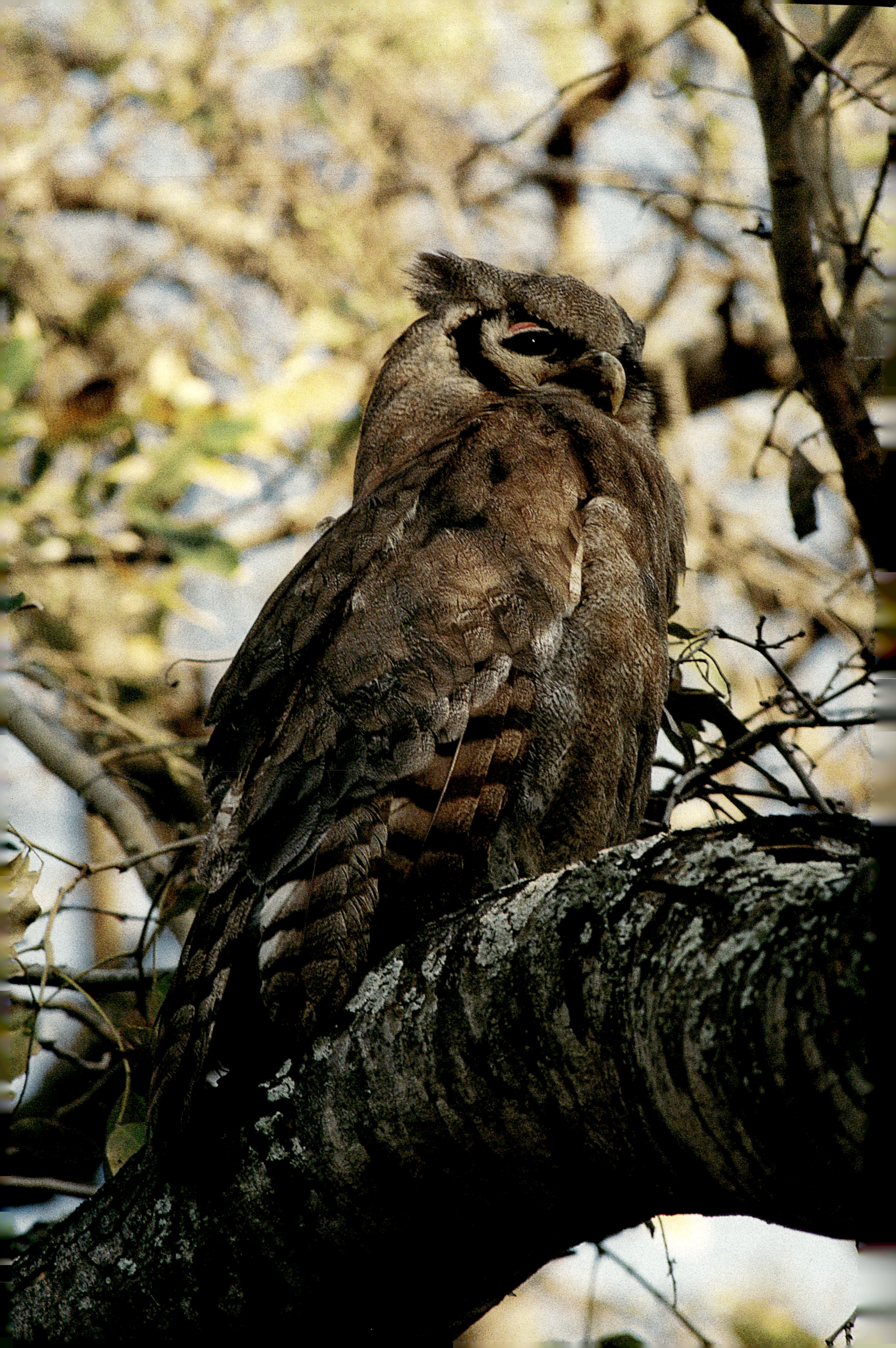

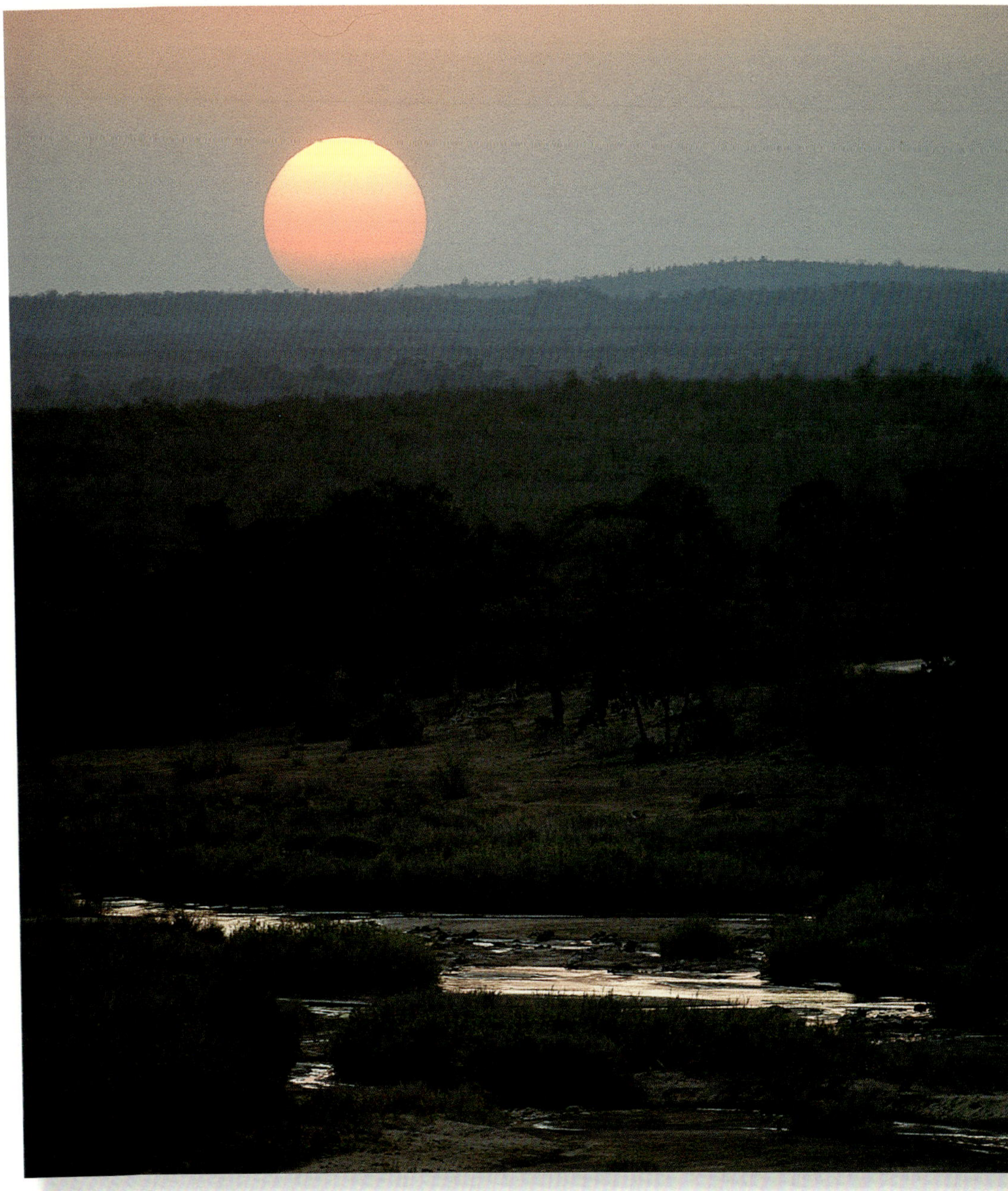

*This sunset over the Olifants River is a favorite of mine because it always brings back fond memories. As in other riverine areas, it is rich in bird life as well as mammals. When darkness covers the river, it is time for nocturnal creatures to become active.*

Preceding pages

*This Scops Owl, barely 8 inches tall, has an ideal perch on this camel thorn limb.*

Facing page

*Being a creature of the night, a Giant Eagle Owl does not present many photo opportunities. I photographed him in the dense forest along the Shingwedzi River in the Kruger Park.*

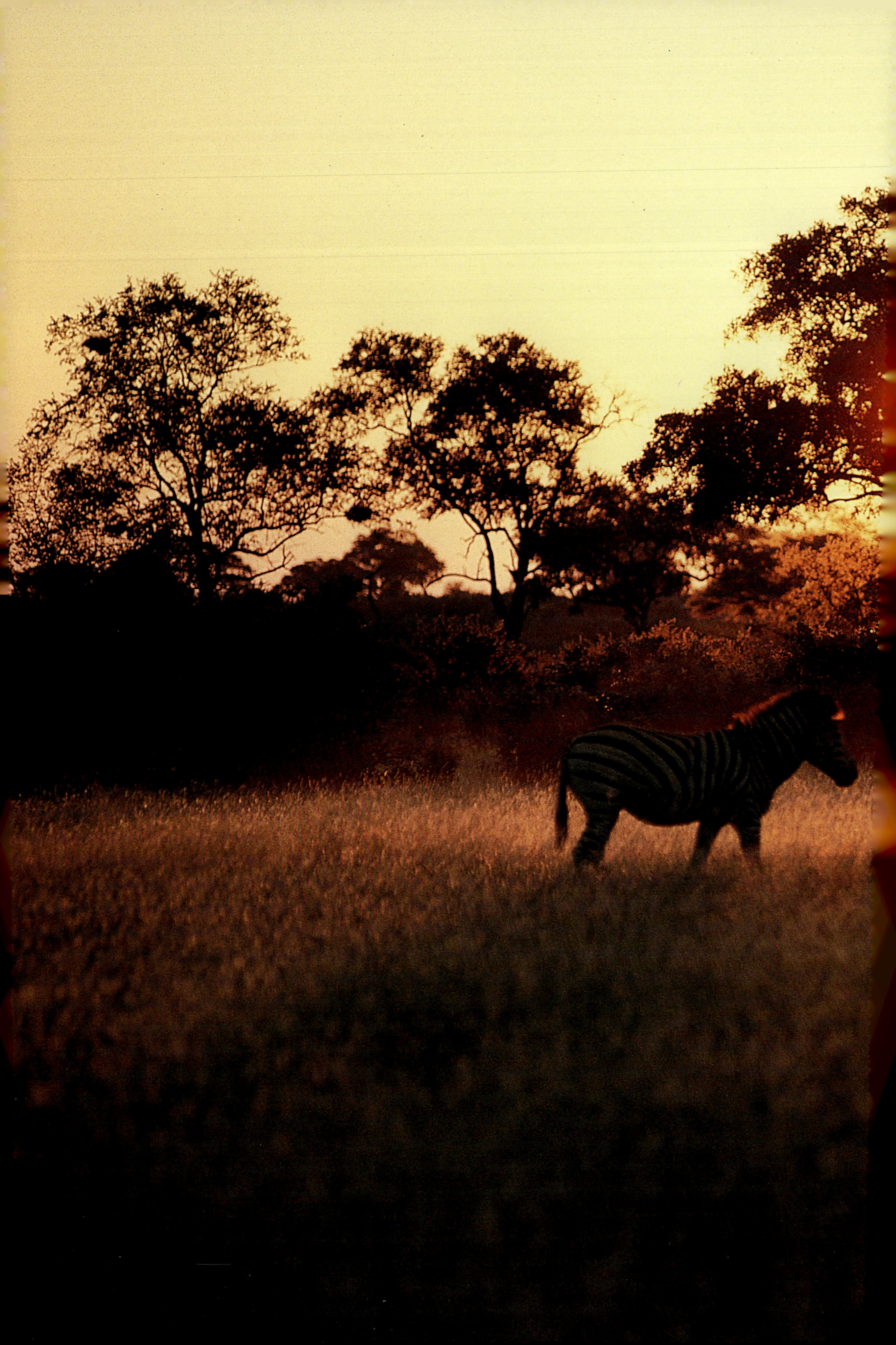

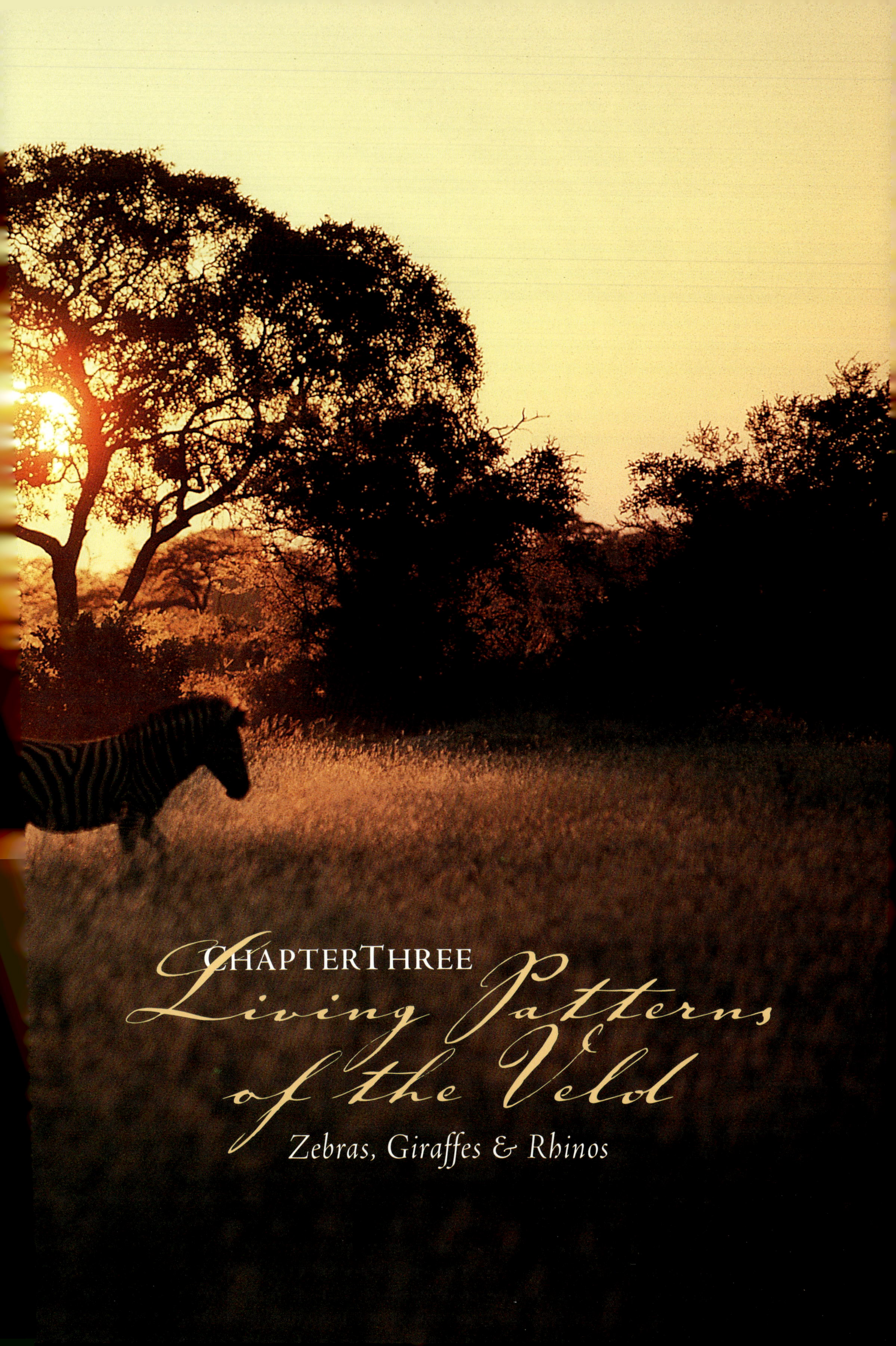

# Chapter Three

# Living Patterns of the Veld

## Zebras, Giraffes & Rhinos

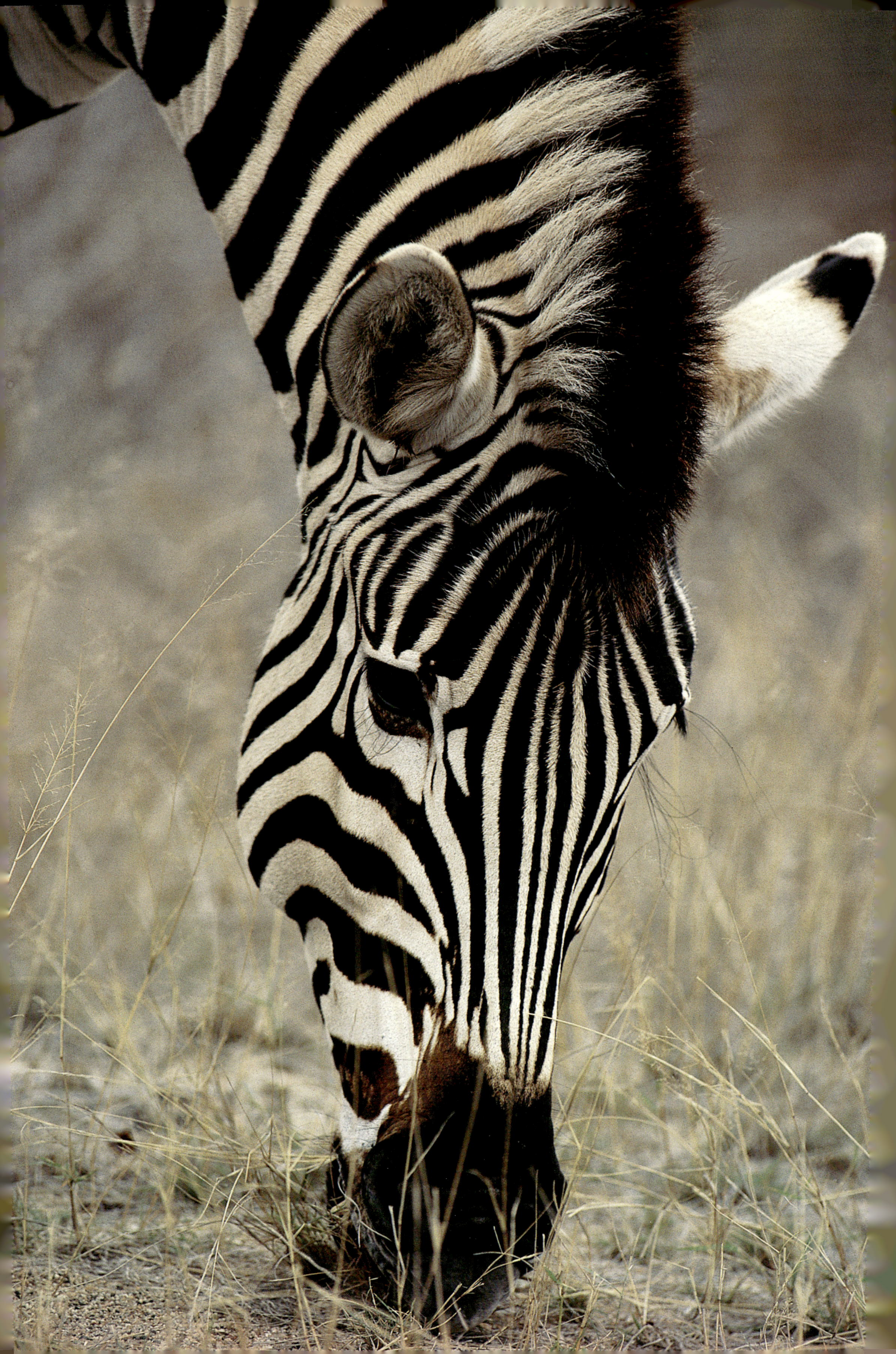

*Stripes, spots and folds. An assortment of fabrics? No. These words describe the hides of three distinctive animals; zebra, giraffe and rhinoceros. All three are well known for their unique appearance, and they make wonderful photographic subjects.*

The ancient Romans called them horse-tigers. A children's song tells of a horse wearing striped pajamas. Chances are, the first word one recalls for the letter "Z" is – zebra.

An equine famous for its striped coat, the zebra is a common sight on the African plains. There are several species and subspecies of zebra, each with its own distinctive stripe pattern. In addition, each individual has a unique arrangement of stripes; like fingerprints, no two are alike. Many believe the ancestor of all equines was striped. Studies of zebra are inconclusive as to the reason for striping. Some have suggested camouflage, but their habitat is not conducive to blending in except in dappled light. Others point to visual confusion for predators, but individual zebras are regularly singled out and killed. One extensive study in optics proposed that the stripes are visual stimuli which have a calming effect associated with grooming rituals learned in the months after birth.

PRECEDING PAGES.
*We waited for the zebras to move into position just as the sun was setting behind them.*

FACING PAGE
*Zebras are often quite nervous and it's very difficult to get close. This stallion allowed me to approach near to him without showing any fright.*

FOLLOWING PAGES.
*My impression when I saw this zebra is that he was running for the sheer joy of it.*

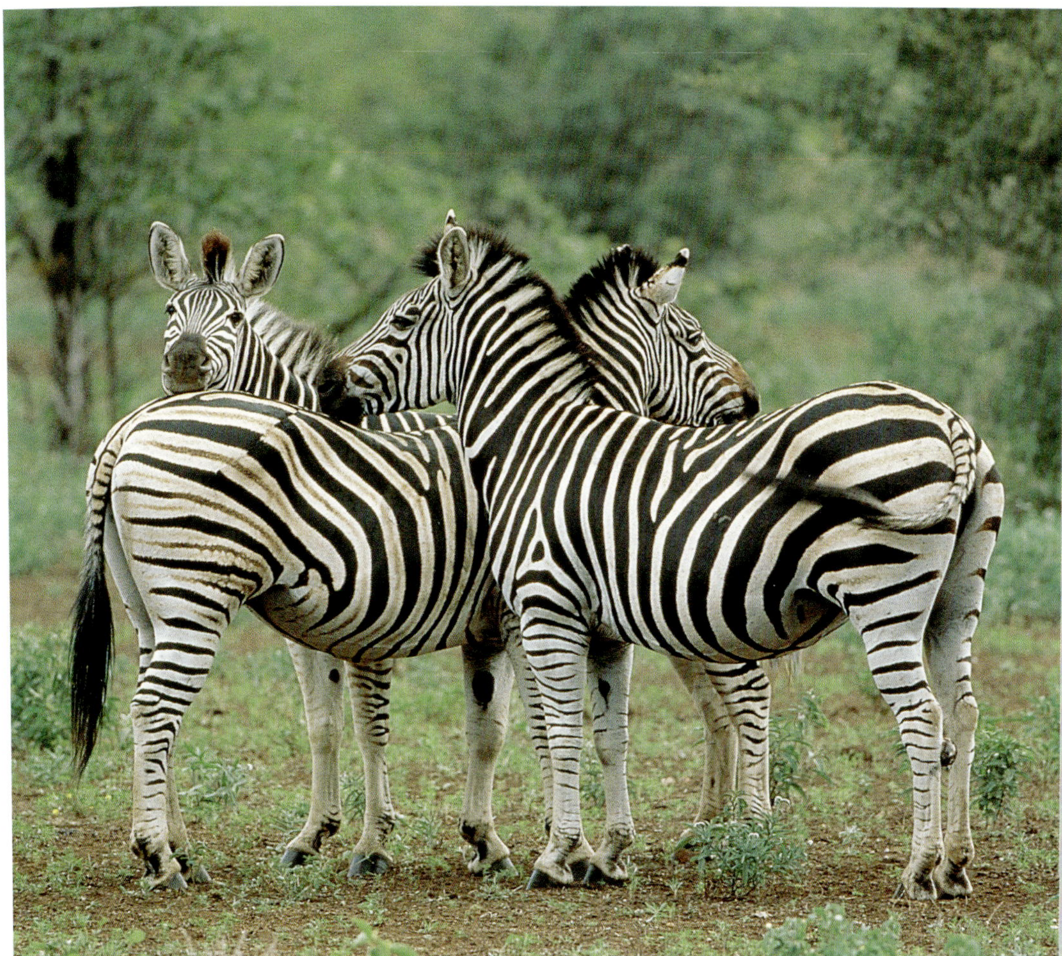

*By posturing themselves in this position, zebras can maintain a vigil against predators that may approach from any direction*

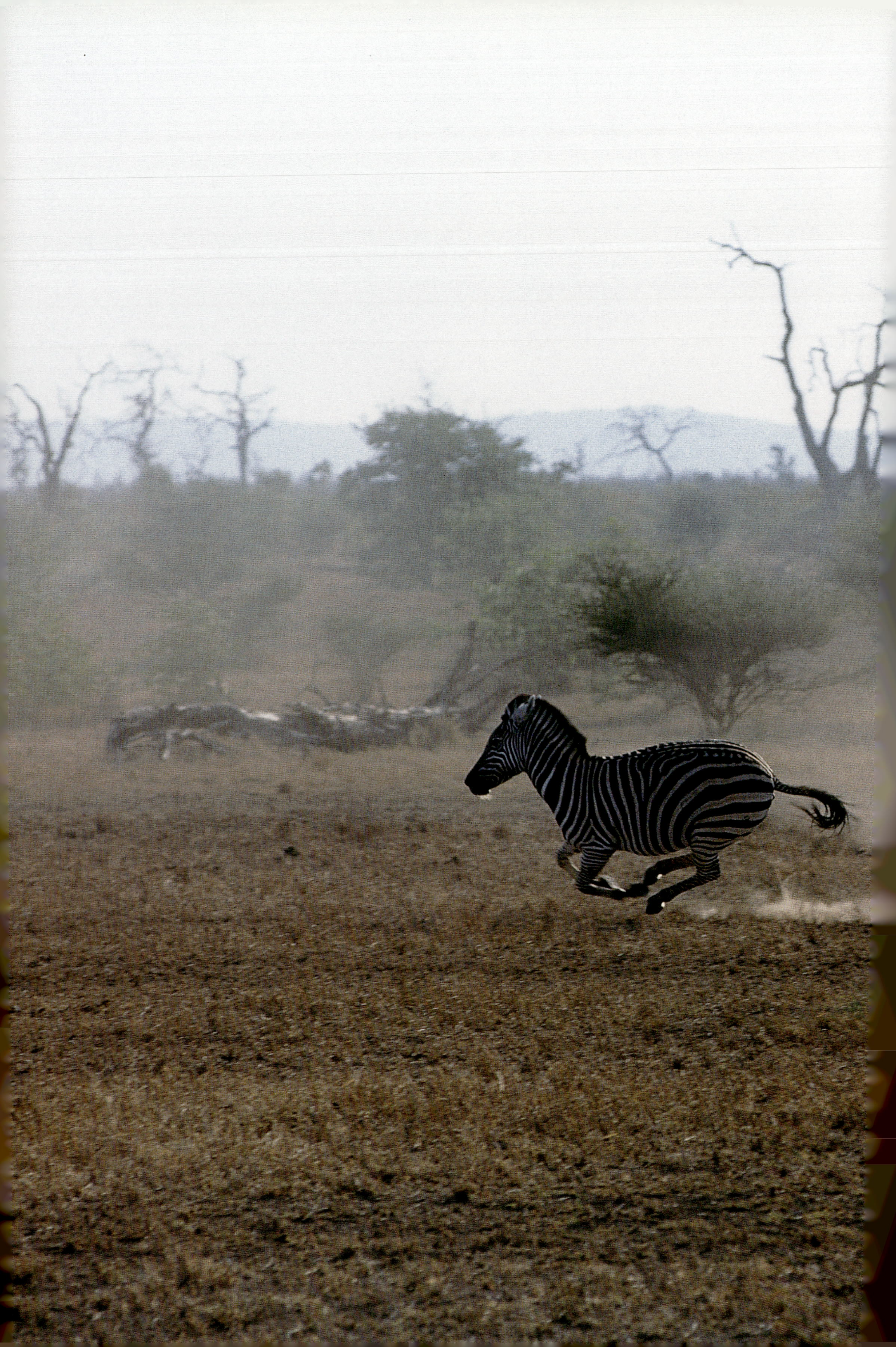

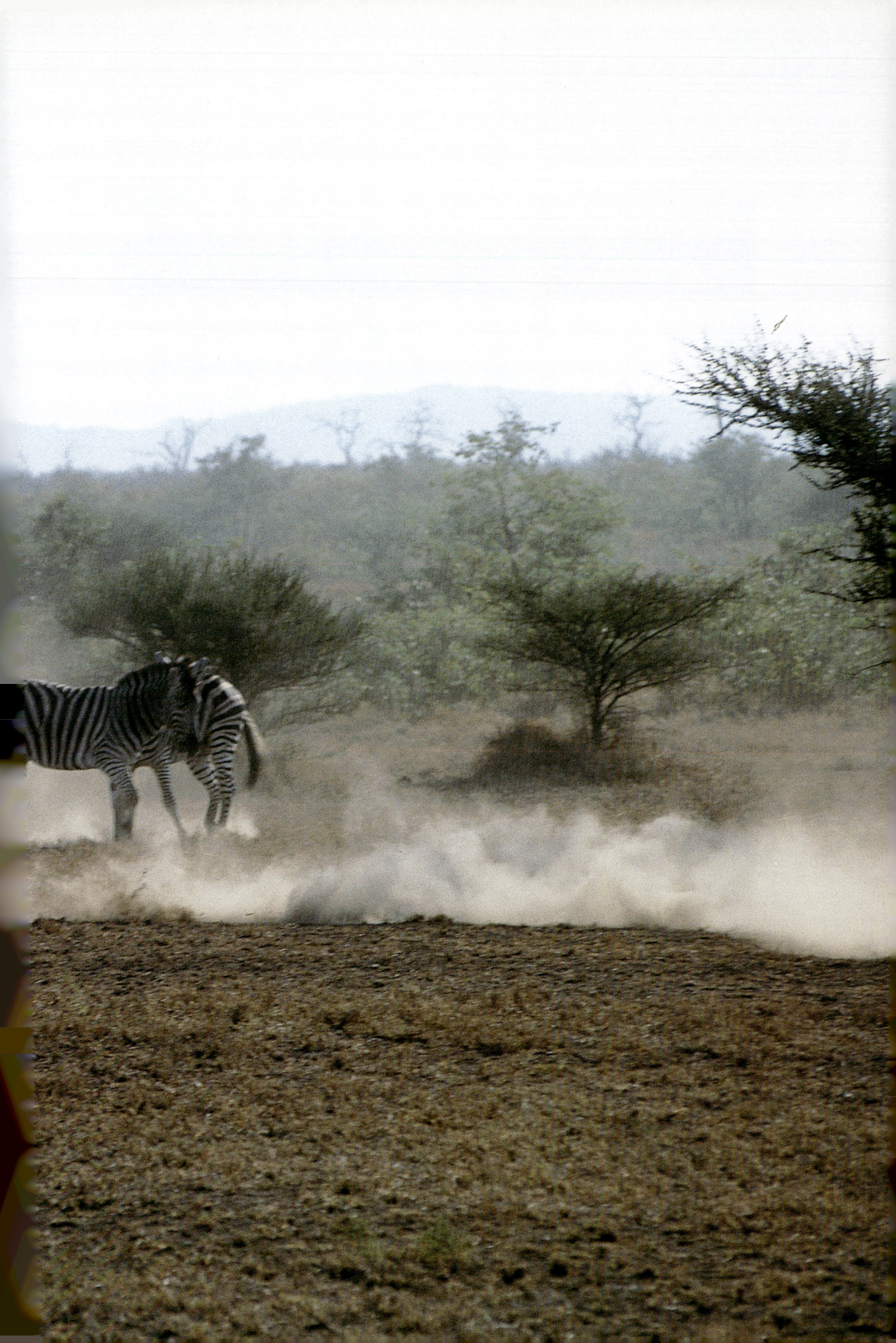

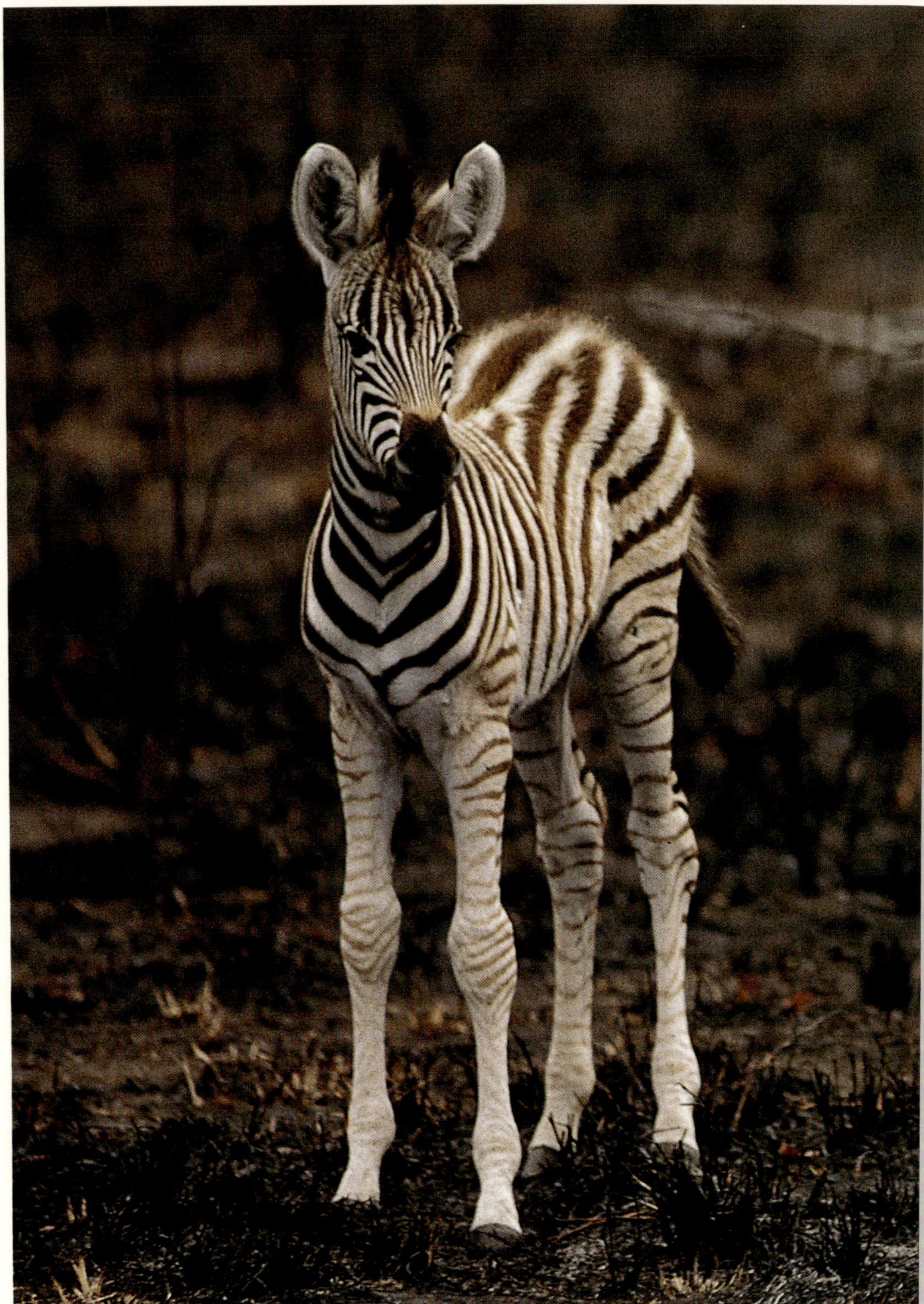

*The zebra foal looks abandoned in this burnt part of the bushveld.*
*However mom is close by and sweet grasses will soon sprout in this seemingly destroyed area.*

Facing Page

*If this zebra mare were human, she surely would be a high-fashion model or starlet.*

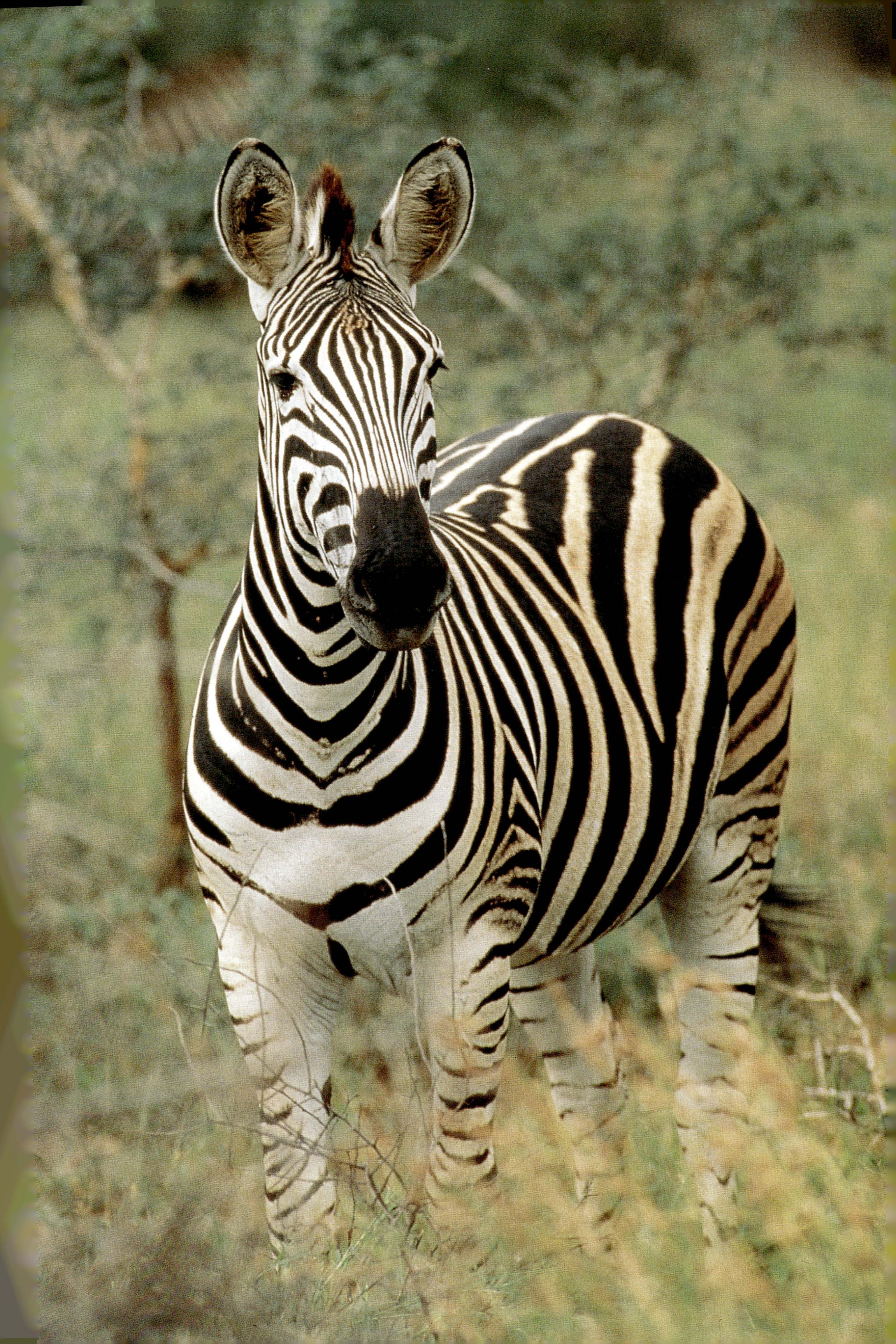

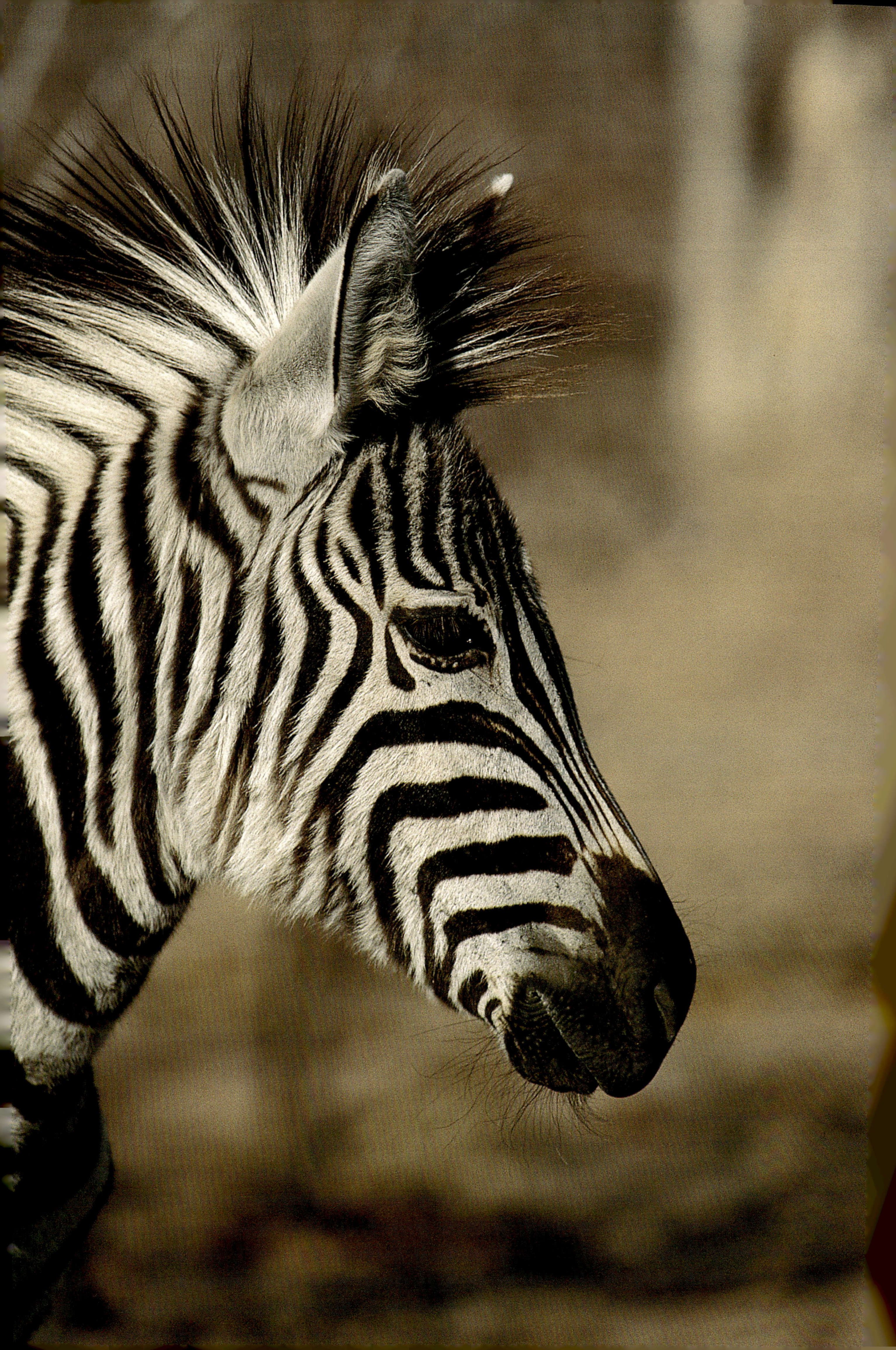

The Burchell's zebra, common to Southern Africa, is white with black stripes and with greyish *"shadow stripes"* in between. Short in comparison to other equines, an average size Burchell's zebra is fifty to fifty-five inches tall at the shoulder and weighs around five-hundred to six-hundred pounds.

The constantly vigilant zebras have acute senses which alert them to danger. Their ears twist in different directions to hear telltale sounds. Their keen sense of smell picks up the slightest odors of predators. Good peripheral vision can catch glimpses of stalking lions or hyenas. When standing side by side in opposite directions, two zebras' field of view is 360 degrees. Zebras can sense when predators are hungry or not and will graze peacefully nearby while the would-be hunters relax between meals. But, when the predators need to eat – zebras must run to survive! They have strong muscles and long legs which enable them to gallop in great strides up to thirty-five miles per hour. Large lungs help with endurance for running long distances without fatigue. When being chased, zebras run a zigzag course which tires their pursuers. Those unfortunates, who lose the race and are attacked, will kick and bite in defense; efforts which not always save their lives.

FACING PAGE

*A zebra foal looks as if he is sporting a spike haircut or, if you prefer, wearing a Roman helmet.*

BELOW

*I like to think of this as a posed school picture. Albeit, a very small school.*

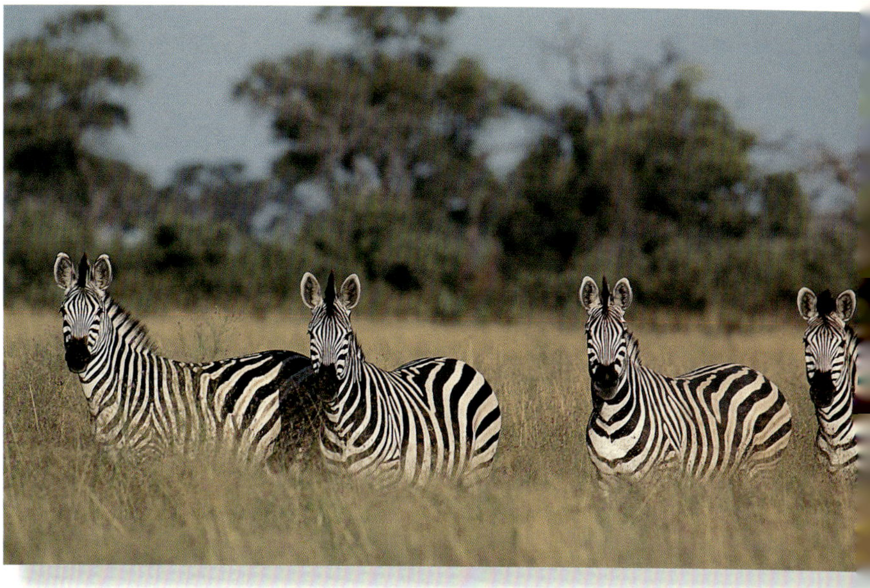

Zebras are social animals. A herd is made up of several family groups which consist of a stallion, a mare or two, and foals. Zebras will breed anytime of year, but most foals are born between October and March. Eighty-five percent are born at the peak time between December and January. Gestation lasts three-hundred sixty to three-hundred ninety days, after which one foal is born to a mare. At birth, the brown and white striped foal weighs sixty to seventy pounds and measures a height of three feet, most of which is legs. Because it is extremely vulnerable right after birth, a foal is on its feet within fifteen minutes. With legs nearly as long as its mother's, it is able to run with the herd within one hour. In a few short weeks the young colt (male) or filly (female) is grazing with its family and is weaned at between eight and thirteen months. Family members keep track of each other with noisy vocalizations. They recognize the braying and barking *kwa-ha-ha* sound of kin and of individuals considered to be special 'friends.'

On the vast savannas where they live, zebras graze in early morning and late evening. They eat grasses, mostly, but occasionally will munch the leaves of shrubs, roots, bark and fruit. The fibrous diet grinds down their teeth but they never wear out, for the molars continue to grow even into old age. As they graze, they need to stay within four or five miles of water. The changing seasons prompt movement to where water and good grazing are available. Wherever they roam, zebras often mingle with herds of other animals – wildebeest and giraffe. Together, their varied patterns create a feast for the eyes.

High above the patchwork of living patterns that graze the savanna stretches the spotted profile of the world's tallest land mammal. Giraffe was not named for its spectacular height, however, but for its movement. The Arabic word, *xirapha,* means' the one that walks very swiftly.' With its long legs matching the length of its six foot neck, the giraffe is capable of fifteen foot strides on feet the size of dinner plates. The walking style is unusual. The giraffe picks up both legs on one side at a time, stepping forward, which gives it a graceful looking gait. A male reaches sixteen to seventeen feet tall; a female, fourteen to fifteen feet tall. Giraffes also possess a handsome six foot tasseled tail. Like zebras, giraffes' coat patterns vary from region to region and no two individuals share identical spot placement.

*An African Story*

*This tale relates that after God made the other animals, He had parts left over from His camels and leopards. So, He created a creature aptly called a camel-leopard. Early Greeks and Romans also had such stories. And so, this tall spotted creature's official scientific name is Giraffa camelopardolis – 'the fast walking camel leopard.'*

Giraffes live in openly spaced herds of from a few to hundreds of individuals. The composition of the herd changes constantly since these aloof animals tend to wander off and join other groups at will. Giraffes are not territorial, but the bulls quickly learn their place in the herd hierarchy. They reinforce their standing by participation in ritual fighting called 'necking.' This dominance display is carried out by two opponents which stand side by side, necks intertwined. Each will push against the other in a contest much like barroom arm wrestling matches. Biting and head butting sometimes occur as well. Little harm is done, however, and the contests serve to strengthen relationships.

FACING PAGE

*These two young giraffes in Namibia's Etosha National Park were most gracious and curious. I ran out of film while I was photographing them; but they did not budge an inch. After reloading I finished shooting and we all looked at each other for several minutes. Then they moved off.*

The wooded grassland and open woodland habitat supply giraffe's favorite food-leaves of trees – especially acacia. They also enjoy vines, herbs, flowers, seed pods and fruits. With those long necks they can browse a six and a half foot band of foliage which is beyond the reach of other animals. Only elephants can also obtain food at that level. Males and females each have a different browsing stance. Males reach up to the highest branches and use their eighteen inch tongues to gather leaves. Females prefer to bend their necks and select the lower branches and shrubs. From afar, it is easy to tell the gender of feeding giraffes.

At dawn, dusk and on bright moonlit nights, they can be found browsing – consuming up to seventy-five pounds of food per day. They can cover hundreds of miles in a few months, browsing just a few mouthfuls from each tree as they pass. Fortunately, giraffes do not need to drink water very often. Deriving most of their water from their leafy diet saves them from having to assume the awkward spread-legged pose necessary for leaning over to drink. Water holes can be dangerous places. Lions lurk nearby just waiting for a giraffe in such a vulnerable position – unable to stand quickly.

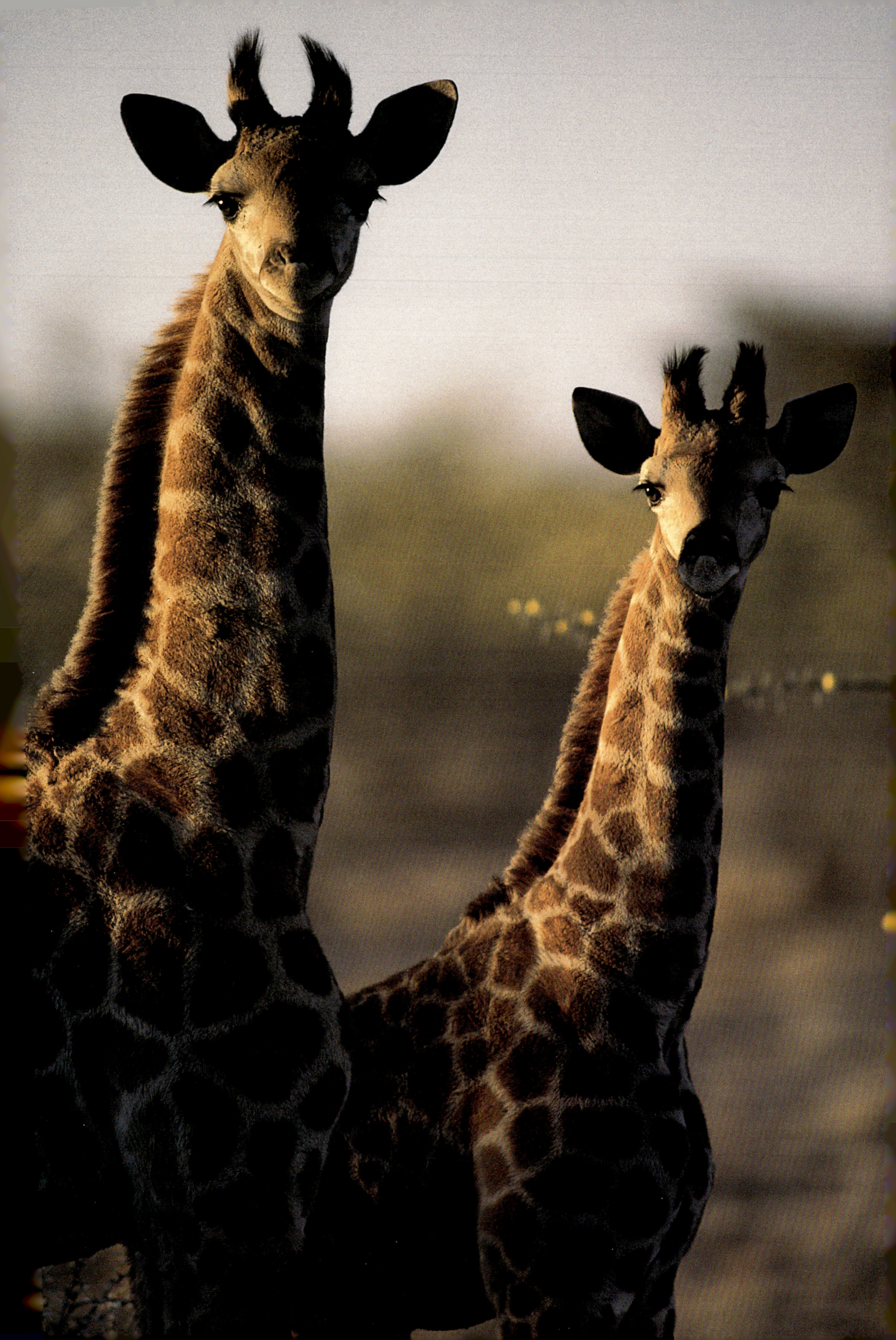

They remain watchful at all times, using their sharp senses of hearing and smell to alert them to danger. Their best sense, though, is sight. From their towering height giraffes can survey their surroundings and keep track of each other for miles. Giraffes do take time to rest, but they take short naps which amount to only five to thirty minutes of sleep a day. Though seldom heard, fearful giraffes do vocalize with grunts and snorts. Calves emit bleating sounds when alarmed.

When not browsing, bull giraffes spend much time looking for females in heat. In fact they prefer seeking mates to eating. When they do find receptive cows, they readily mate. Breeding can take place year round. Males begin at eight years old; females at five. When mated, gestation lasts about four-hundred fifty-seven days. At the end of that period cows will migrate to traditional calving grounds. There, several cows will give birth to single calves. At birth, a calf will weigh over two-hundred pounds and be over six feet tall. Like zebras the newborns are up on their feet within fifteen minutes and able to walk with their mothers after one hour. In the following months they grow rapidly, several inches per month. Their height doubles within two years. Mother giraffes keep their calves in nurseries looked over by a single female. This allows the mothers to browse at a distance during the day for much needed nutrition in order to nurse the rapidly growing offspring. After a year the calves are weaned, though they begin to browse much sooner.

FACING PAGE

*One can readily see how tall a giraffe is when compared to a zebra – an animal the size of a small horse.*

Usually considered gentle giants, giraffes are capable of defending themselves and their calves with deadly kicks. Kicks powerful enough to kill a lion. If attacked, their spotted hides are tough enough to withstand, for some time, the bites of lions, leopards and hyenas. Young calves, however, are often taken

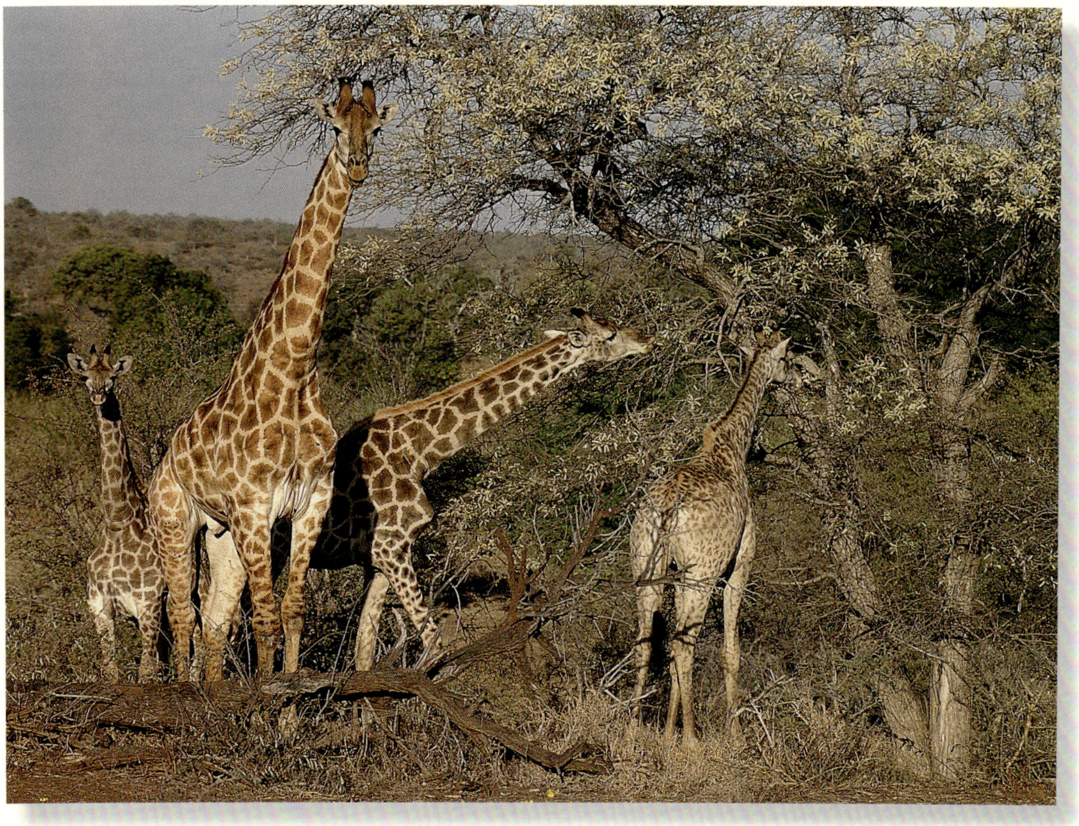

*Although giraffes do not form close bonds with each other it appears that way when they feed from the same tree.*

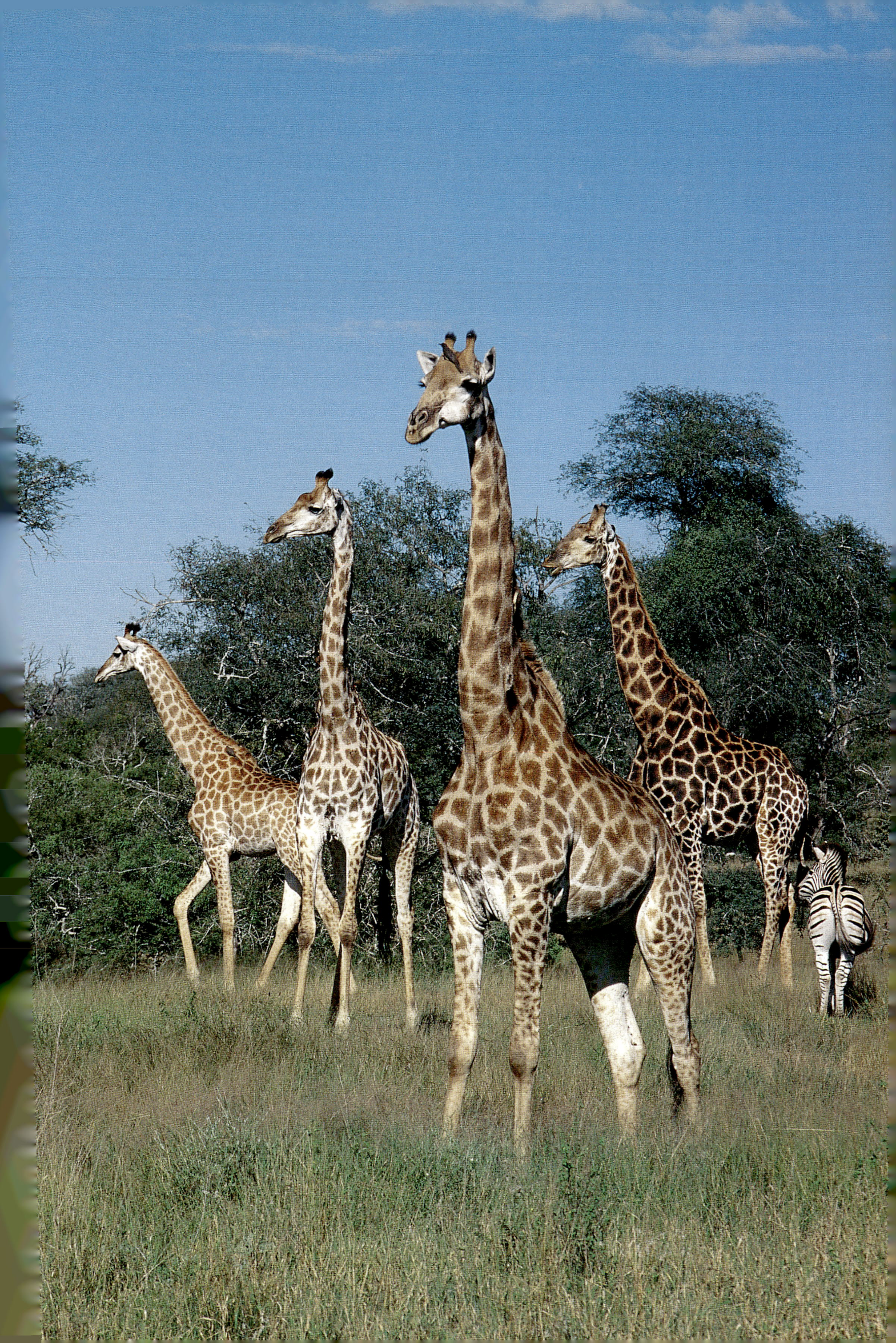

by these predators. Fifty to seventy-five percent of calves are lost to lions and hyenas. Adults are less often hunted successfully.

FACING PAGE
*Barely a half day old, a wobbly, newborn giraffe is still able to keep up with his mother.*

Although life can be perilous on the plains, the giraffe does have an ally. It comes in the unlikely form of a small bird called an oxpecker. The colorful little bird can often be seen riding on the head and hide of giraffe, a hitchhiker with a purpose. Theirs is a symbiotic relationship, each doing a favor for the other. Bushman legend tells of the origin of their friendship:

## *African Fable*

*In early times, there was once a terrible brush fire. All the animals fled for their lives. Oxpecker had a nest in a tall tree which held her young flightless chicks. She pleaded with the fleeing animals to carry her chicks to safety, but none would stop to help. Soon Giraffe came running and once again Oxpecker called out for rescue. Giraffe was kind and sympathetic. Even with the fire roaring closer, he stopped long enough to scoop up the little family and carried it to a safe place. Mother Oxpecker was grateful and asked what she could do in return. Giraffe didn't have to think long. He asked her if she would relieve him of the parasite ticks that bothered him so. She cheerfully agreed .... and from that time on Oxpecker has been a constant companion to Giraffe, pecking away tasty ticks.*

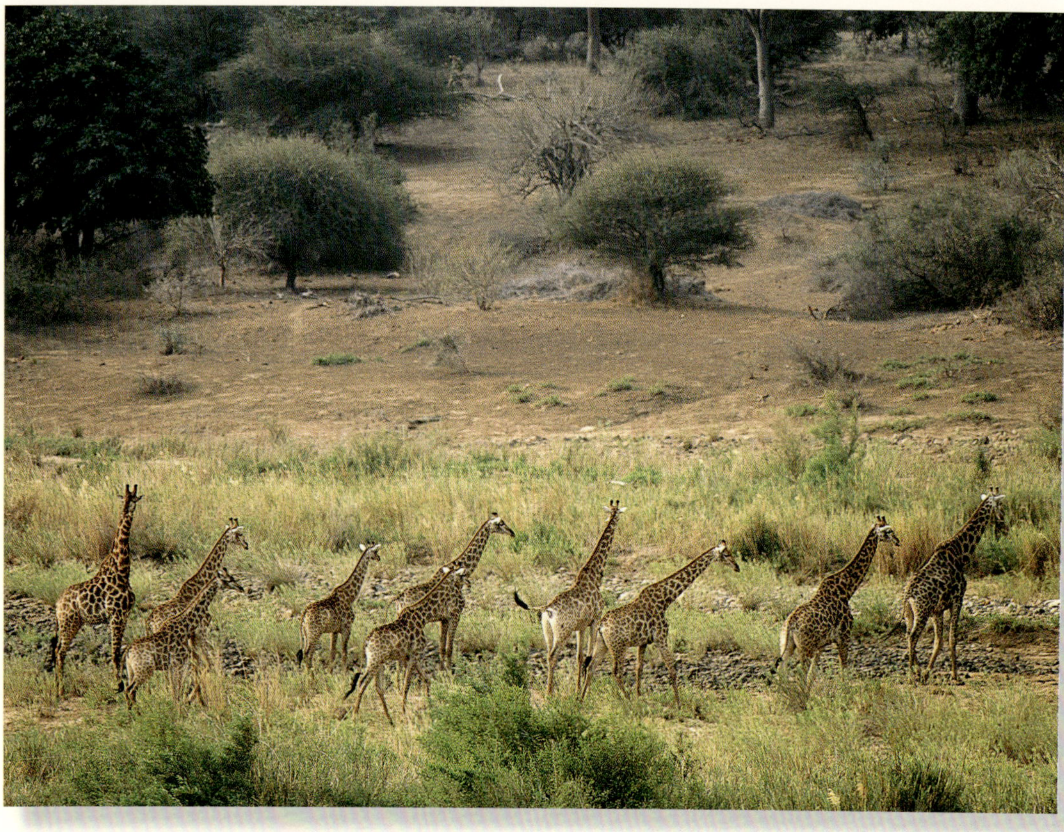

*The Letaba riverbed in the Kruger Park offers a perfect setting for this ambling herd of giraffe.*

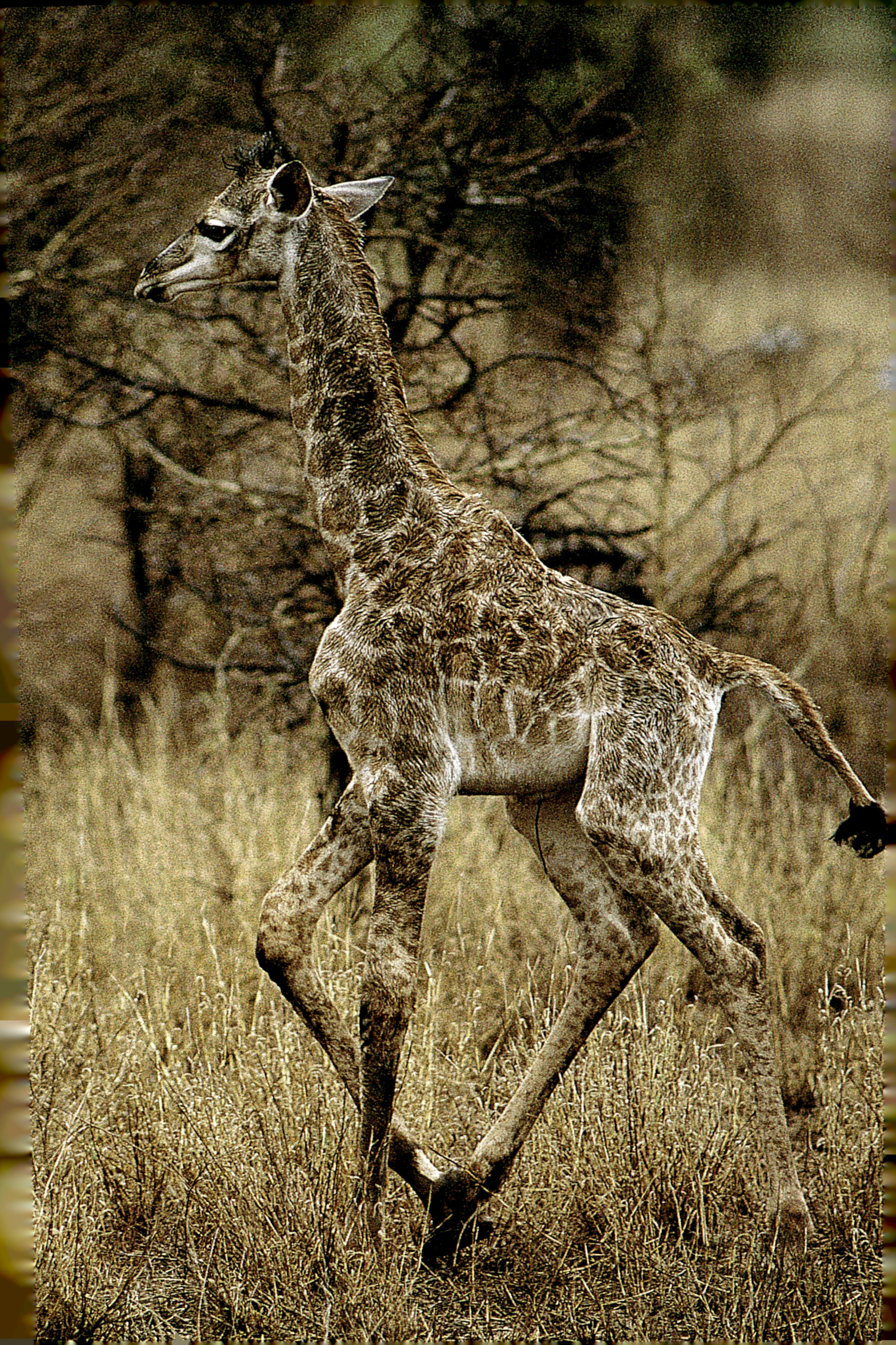

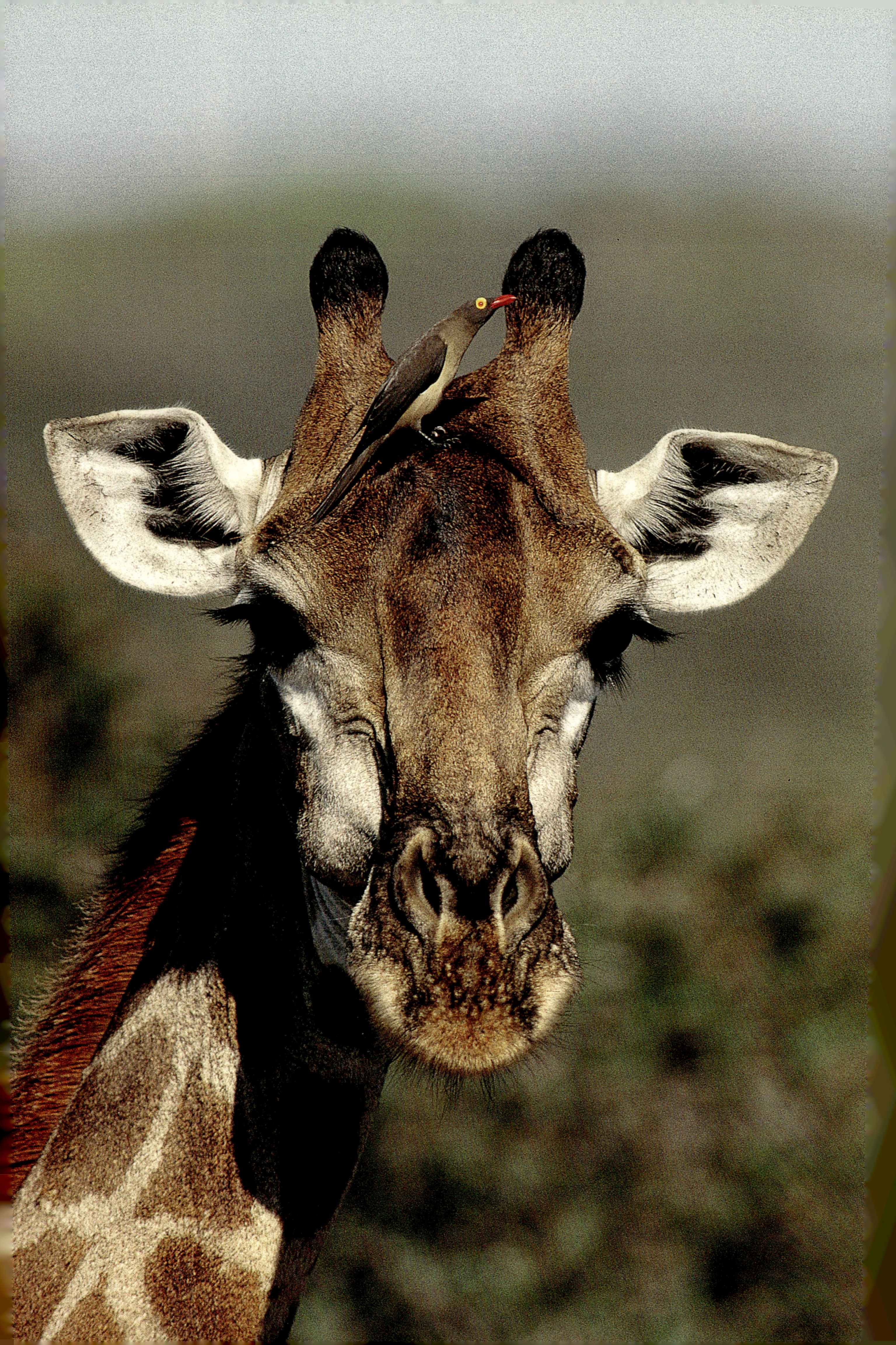

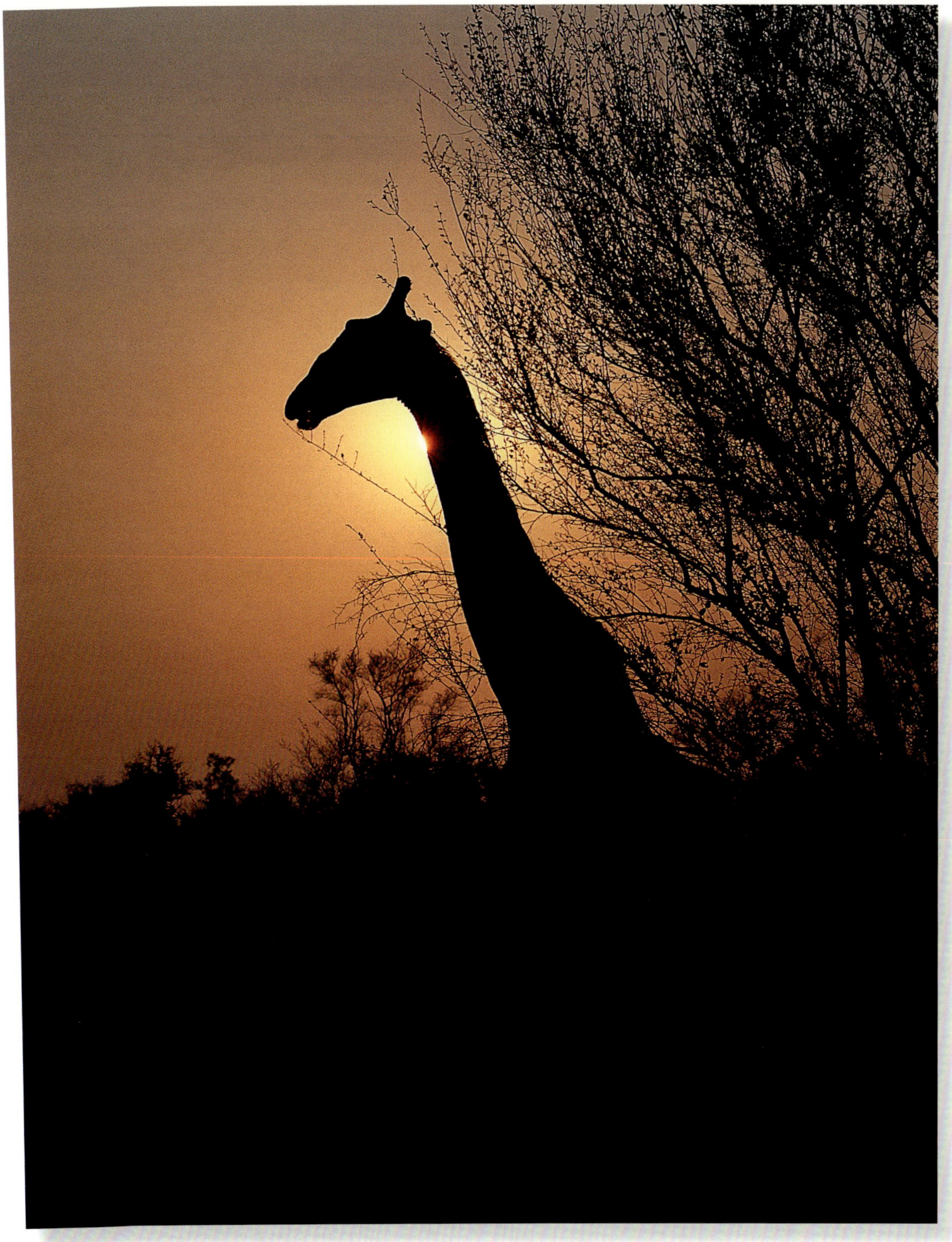

*The end of the day often brings a pleasant surprise, such as a giraffe in silhouette.*

FACING PAGE

*I had just a few seconds to get this shot but the giraffe and oxpecker got into an excellent pose in front of the camera.*

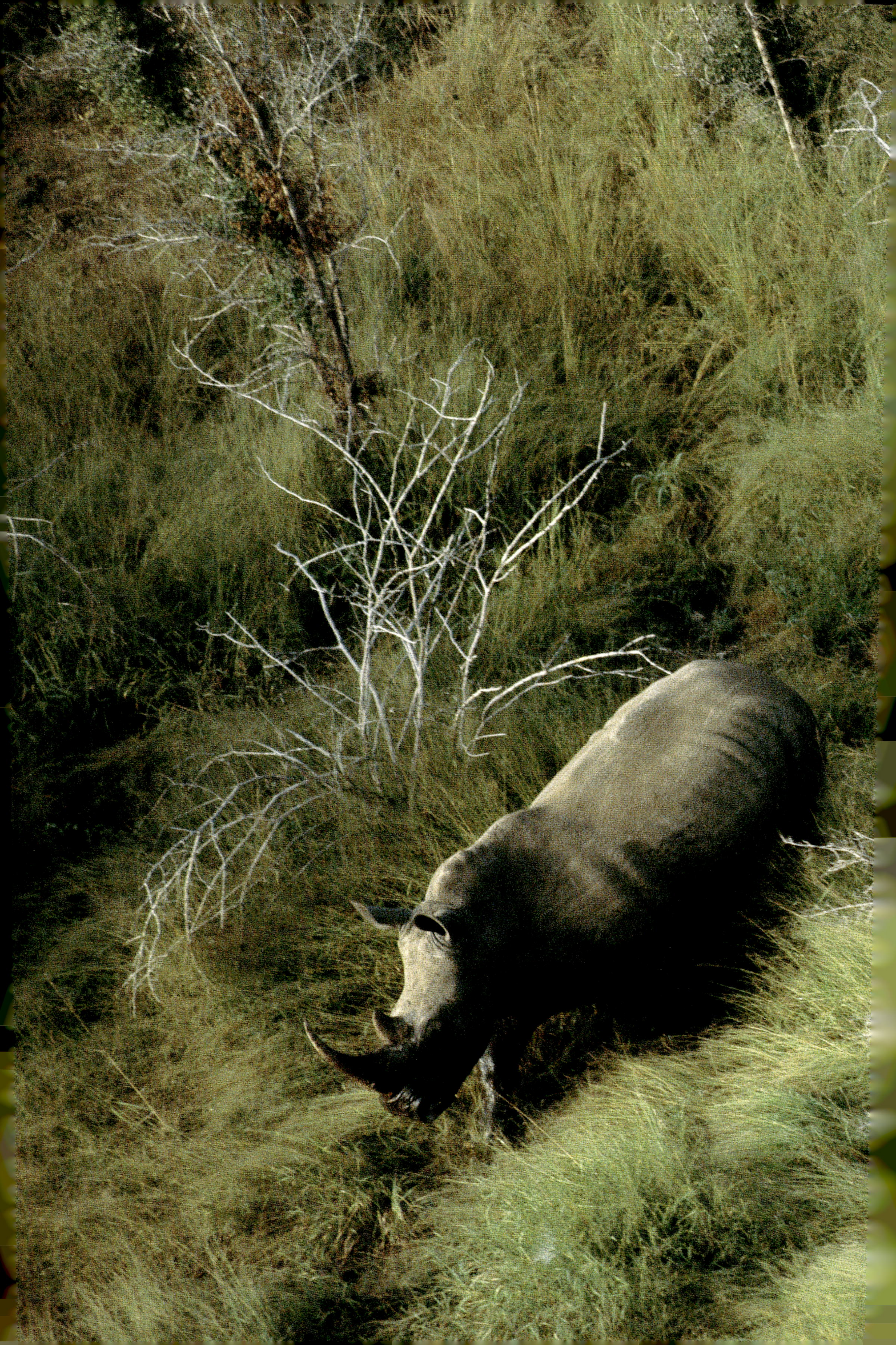

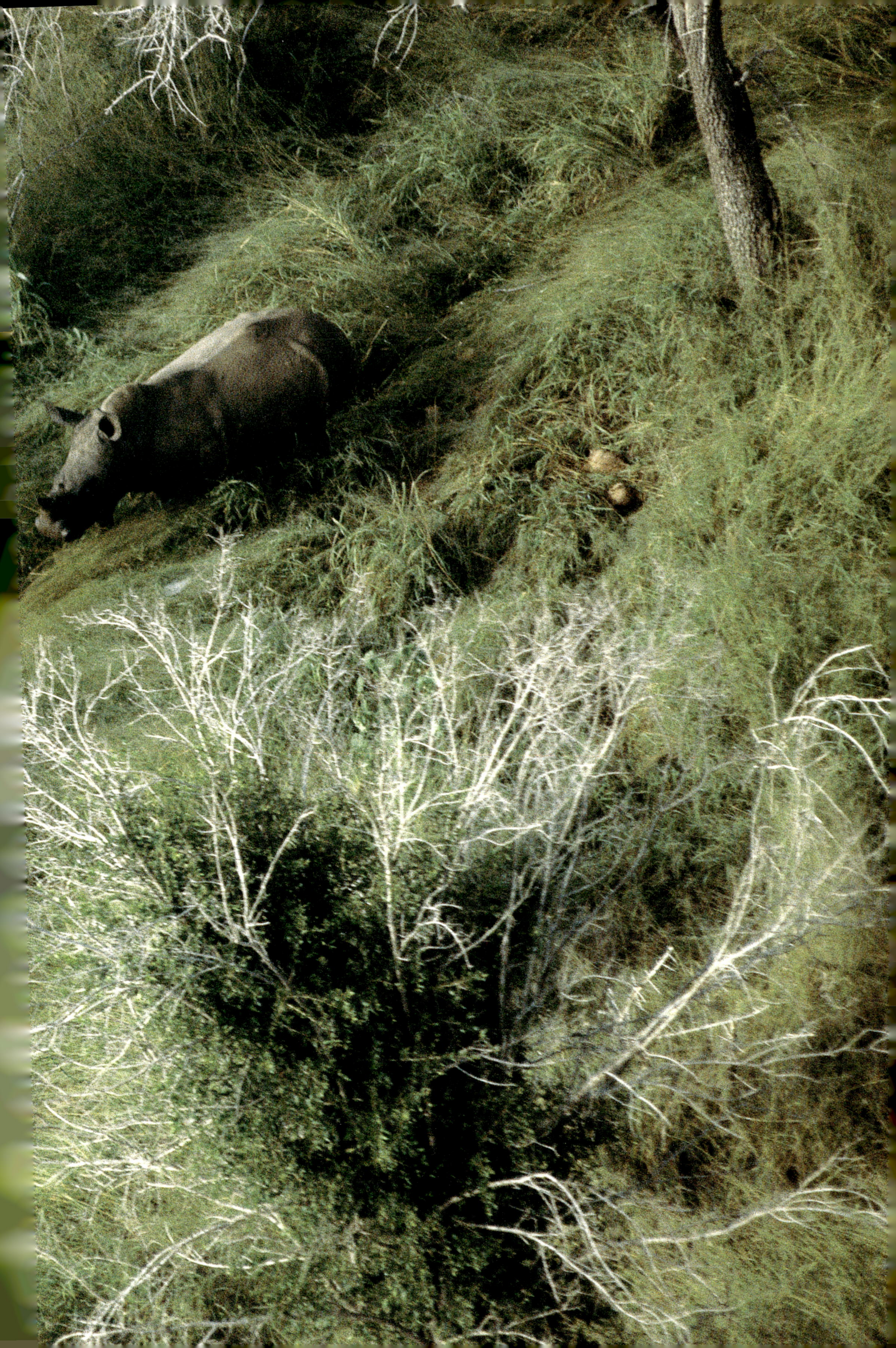

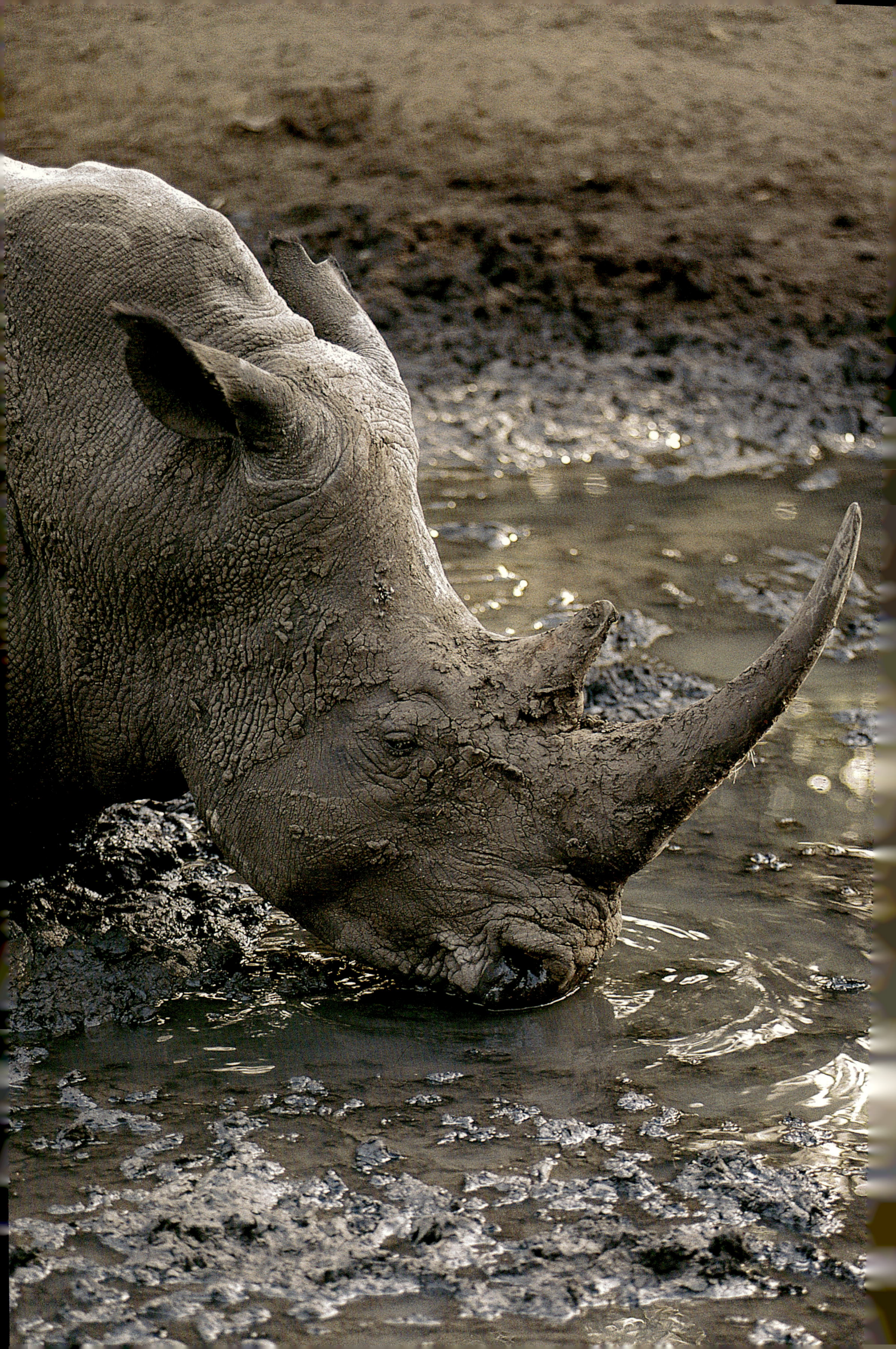

Looking earthward from the towering giraffe, on the muddy flats by a water hole, can be found a tank-like creature that loves nothing better than wallowing in gooey mud. Rhinoceros. With jutting facial horns and rough, corrugated hide, the rhinoceros conjures up images of ancient denizens of eons past. Named rhino (nose) - ceros (horn), this giant beast is left over from prehistoric ancestors that lived forty to two million years ago. Today, there remain only five species.

The White rhinoceros, indelibly recorded in prehistoric rock art, once occurred from Morocco to South Africa. Today it inhabits only National parks in Zaire and South Africa where it is a protected endangered species. Six feet tall at the shoulder and up to forty-nine hundred pounds, it is the largest of all rhinos. Its head alone weighs hundreds of pounds. The massive head is supported by strong muscles and ligaments composing a hump on its

PRECEDING PAGES

*A rhino cow and calf were moving through bushveld in the southern part of the Kurger Park.*

FACING PAGE

*Notice the caked mud on this huge rhino. They all wallow in the mud to protect their hides from biting insects.*

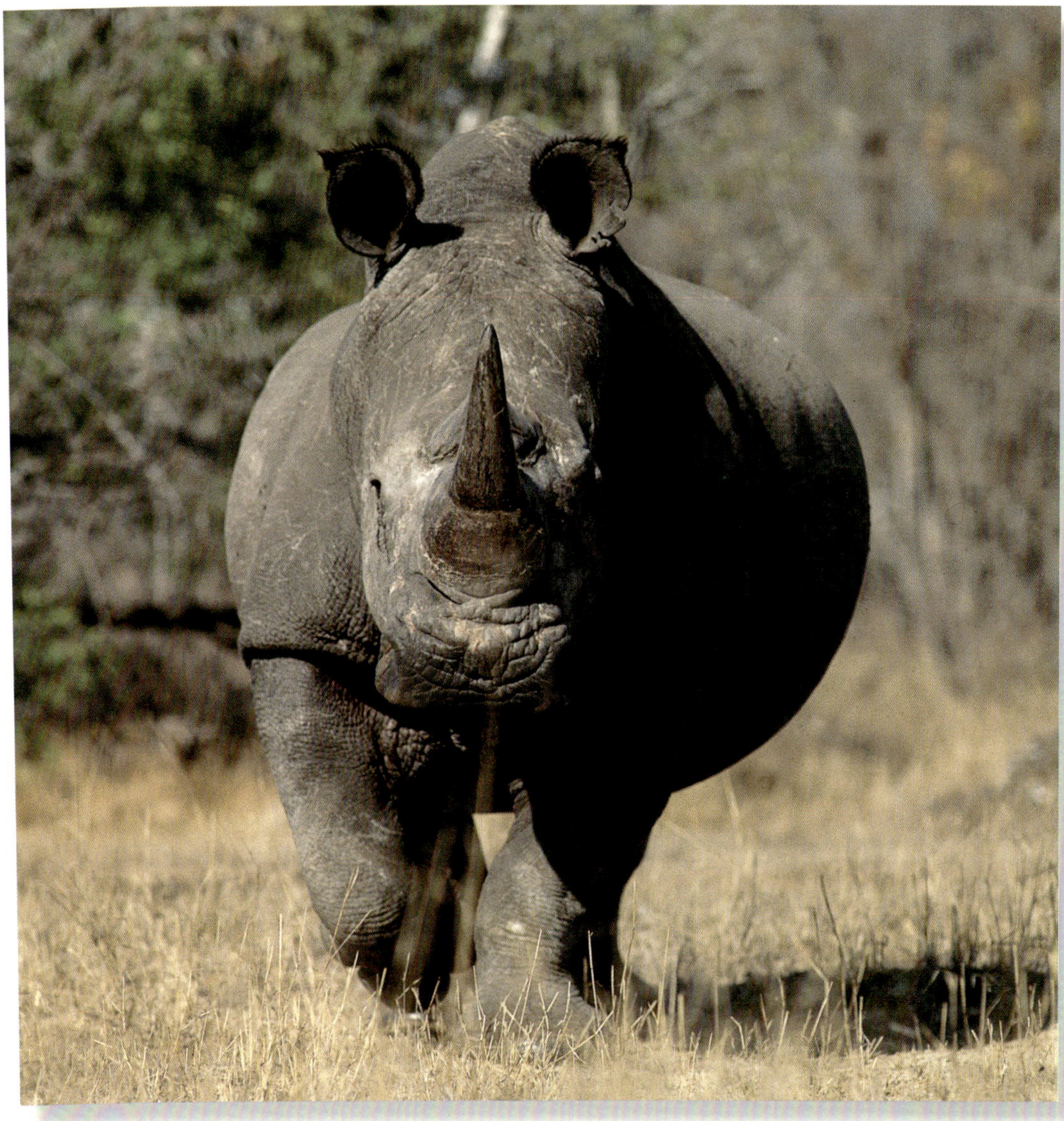

*While on foot in the company of a ranger, I asked myself, "What am I doing here?," as this two-ton rhino came trotting toward us.*

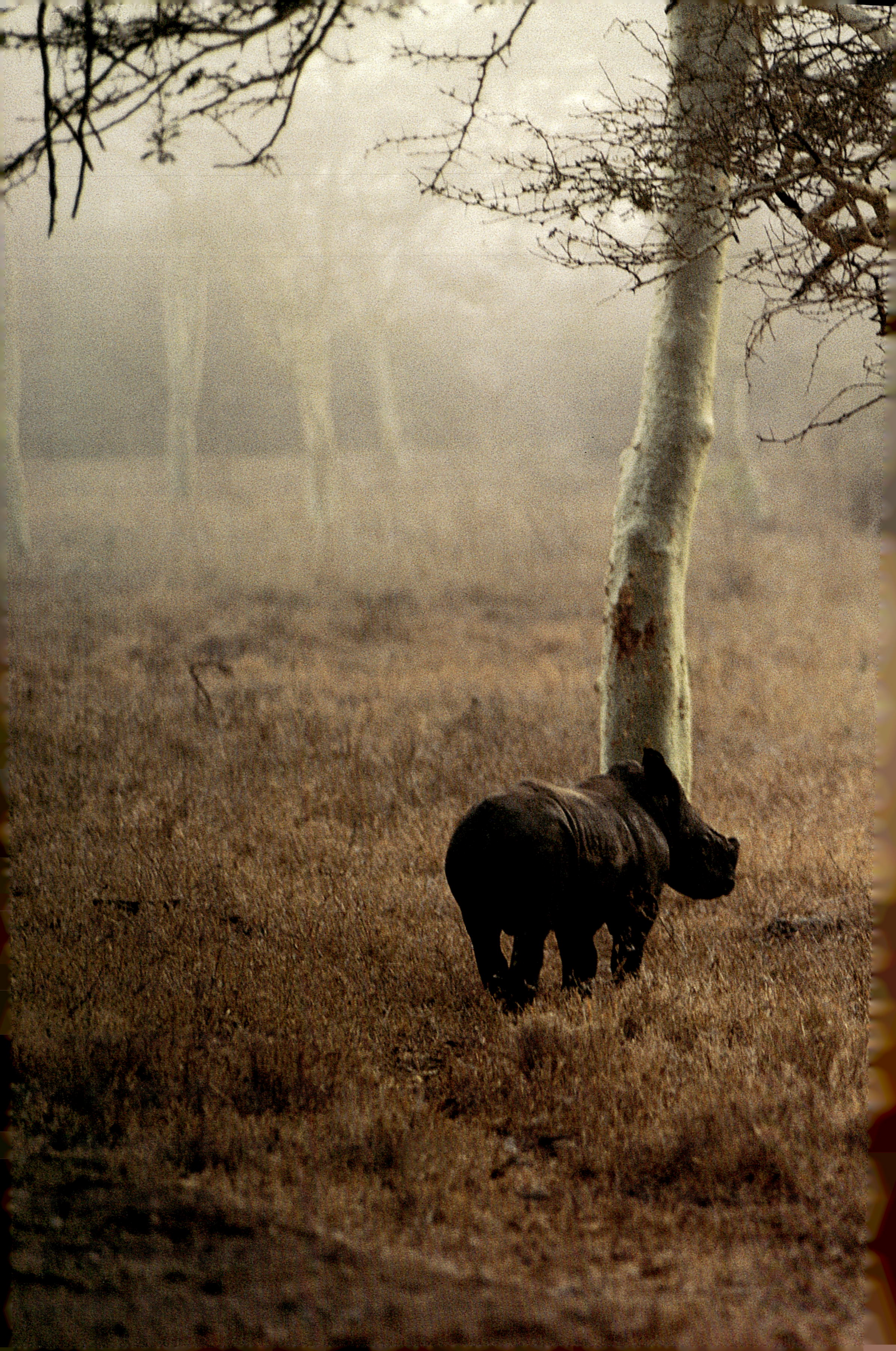

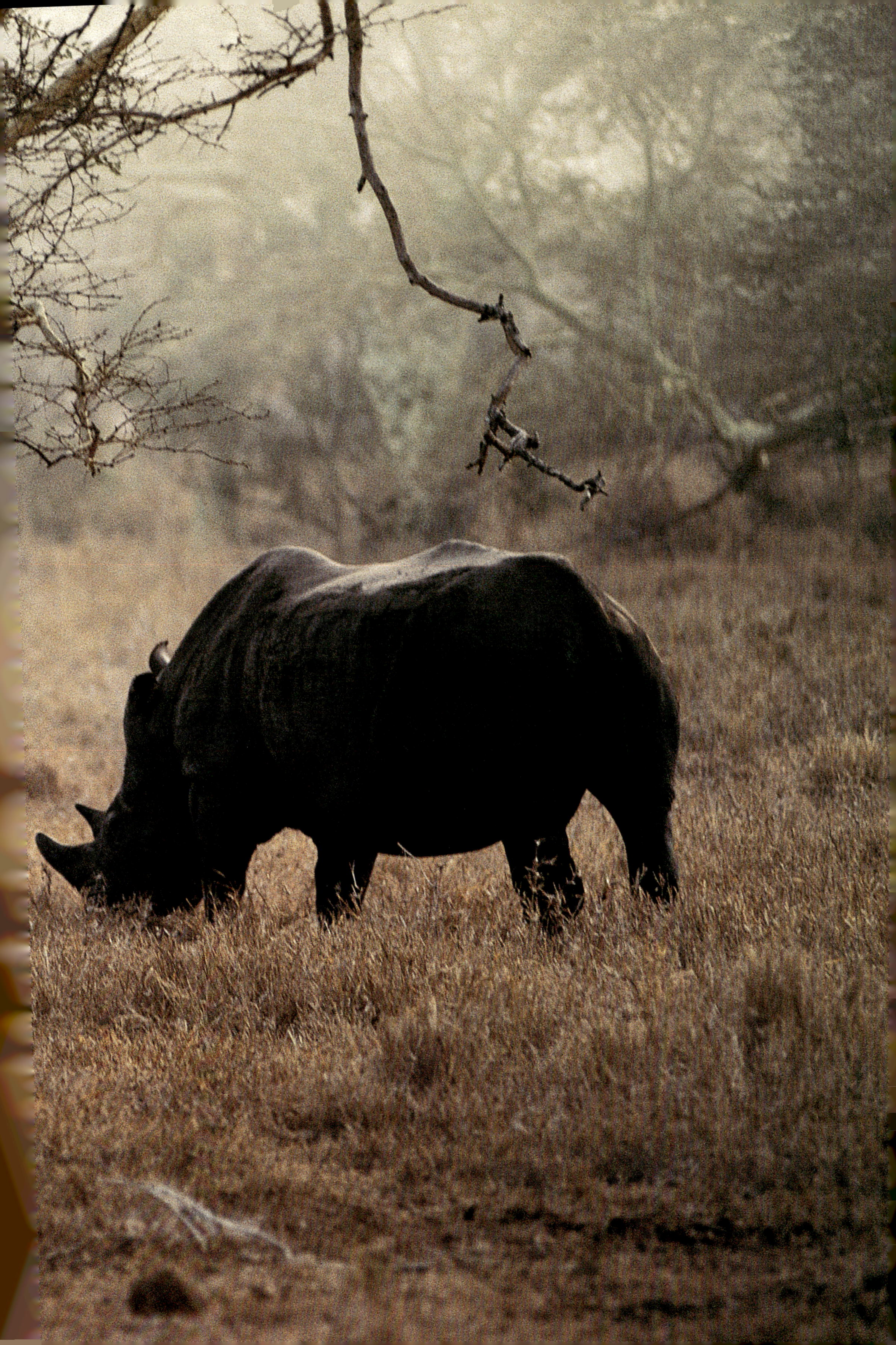

back. The White rhinoceros is one of three species sporting two nose horns. The front horn is the longest – a record length horn was five and one half feet. Different from other horned animals, the rhinoceros horn is composed of tightly packed strands of keratin, the substance of which fingernails are made. The horns are used for plowing the ground to find mineral salts and also as weapons.

The primitive urge to fight remains in the genes of rhinoceros. Males fight each other for territory. The battles can be lengthy and bloody – ending in injury or retreat of one combatant. Males will fight with females as well, during courtship.

Mating behavior, conducted year round, is a primitive affair as well. The main stimulus is scent marking. Bulls breed at around ten years old, cows at from six to eight years. After a sixteen month gestation, a single calf arrives. Like other plains animals, the calf is up and mobile within one hour. The pinkish-grey calf tends to stay a few steps ahead of its mother as they move. A White rhinoceros cow makes a good parent. She is a formidable defender against any danger. The calf will stay with its mother for two or three years.

PRECEDING PAGES

*Early morning fog created a dramatic mood in this fever tree forest for a rhino cow and calf.*

Mostly solitary animals, they sometimes gather in small herds. Females will greet one another with nose touching. Socializing takes the form of wallowing in mud together. Mud cools the animals in summer and serves as an effective insect repellent.

The mud covering is also thought to be the source for the different rhinoceros names. Mud in different regions has different colors, hence, the Black rhinoceros and the White rhinoceros. In reality, both have thick folded grey skin. Another popular explanation for the White rhino's name is that the early Dutch word for *wide* sounds like *white* in English. Wide refers to the lip shape of the White rhino which

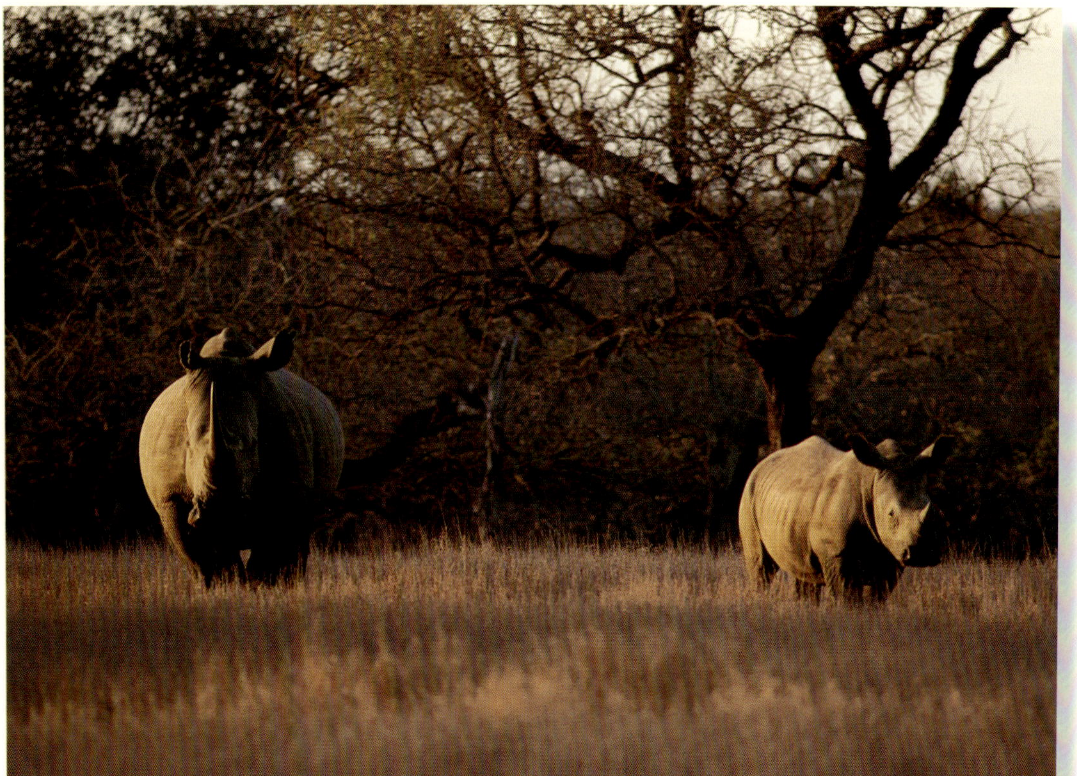

*We saw this rhino and her calf in Mkuze. I like to think of these reserves as little jewels because they add so much richness to photo opportunities.*

is wide and square. The lip is adapted for grazing. The rhino clips short savanna grass between a hard lower lip and the sensitive square upper lip.

The required habitat for a rhinoceros must include grassy plains for eating, and water, for drinking and to produce the ever popular mud wallows. It also needs brush cover at times.

It is thought that the rhino's ancestors lived in closely wooded areas where everything it needed was close by. That may explain why vision is its poorest sense. Like Mr. Magoo, it must be very close in order to see something. And, a rhinoceros doesn't take chances that the blurry image before it might be unfriendly. It will sometimes charge. Individuals are even known to charge trees and rocks at times.

To compensate for poor vision, the sense of smell is acute. It is used to find mates and to identify intruders that may invade its territory. Any upwind interloper can be sensed and the rhino may charge – hopefully just as a warning. Any human in the way of a powerful charging rhino is in peril. Fourty-nine hundred pounds of solidly muscled rhinoceros can attain a speed up to thirty five miles per hour. When it feels threatened it may emit noisy vocalizations in the form of snorts, roars, squeals, shrieks, honks and puffing sounds.

Generally, though, the White rhino is non-aggressive compared to other species. It would rather run away than confront an intruder. When danger is perceived by a group of rhinos, they will form a star shaped circle with rumps together and facing out in different directions; a formidable defense for any but man, rhino's most dangerous enemy.

His horn has long been valuable to Asian medicine and for men's dagger handles in the country of Yemen. Tragically rhinoceros are killed to obtain the horns. Hopefully, education and control of poaching can save these magnificent remnants of earth's distant past from extinction.

*The day had begun early with my good friend, Ian Milne, who was guiding me while we searched for rhinos. We were in the south of the Kruger Park in an area known to have a good share of rhinos. Ian, knowing well the habits of these sometimes shy, sometimes aggressive 'gentle giants', was looking in all the right places, but the rhinos were being quite elusive. We were driving carefully down a fire break road when we suddenly heard an extremely loud, somewhat menacing, grunt. Just ahead of us we spotted this huge and magnificent animal.*

*He turned from us and disappeared into the bush and away from the road. Being familiar with the lay of the land, Ian said that we should follow on foot because after a short distance, the brush would open into low grassland. We carefully made our way through the acacia shrubs and soon came to a rise in the land from which Ian could see the rhino grazing, about one-hundred to one-hundred twenty-five feet distant. The wind was in our favor so he could not pick up our scent, but he probably heard us. Since we were in his territory, he began a lively trot toward us to investigate. I thought to myself, "Why am I doing this? That animal, weighing over two tons, is coming at ME." Fortunately I regained my composure, steadied myself on my knees and began 'shooting.' About that time the wind shifted and he picked up our scent and stopped. This 'gentle giant' was about thirty-five feet from us. My shutter was opening and closing in sync with my heart. Within a minute or two he began to graze and apparently lost interest. After a few minutes we quietly made our way back to the road and our vehicle. Having this experience and getting some pictures was certainly worth the entire day.*

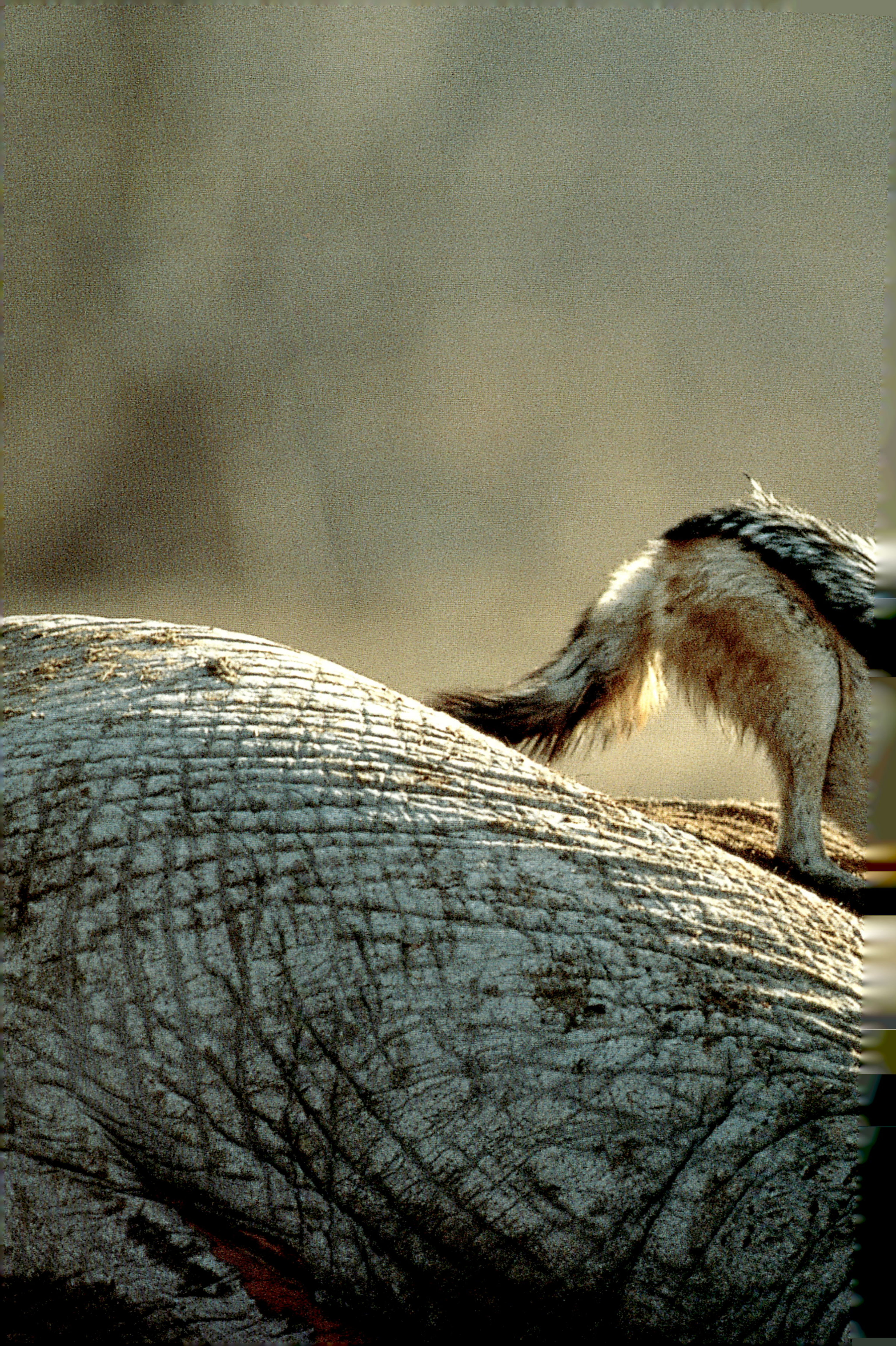

# CHAPTER FOUR

# *Voices of the Night*

*Wild Dogs, Black-backed Jackals & Hyenas*

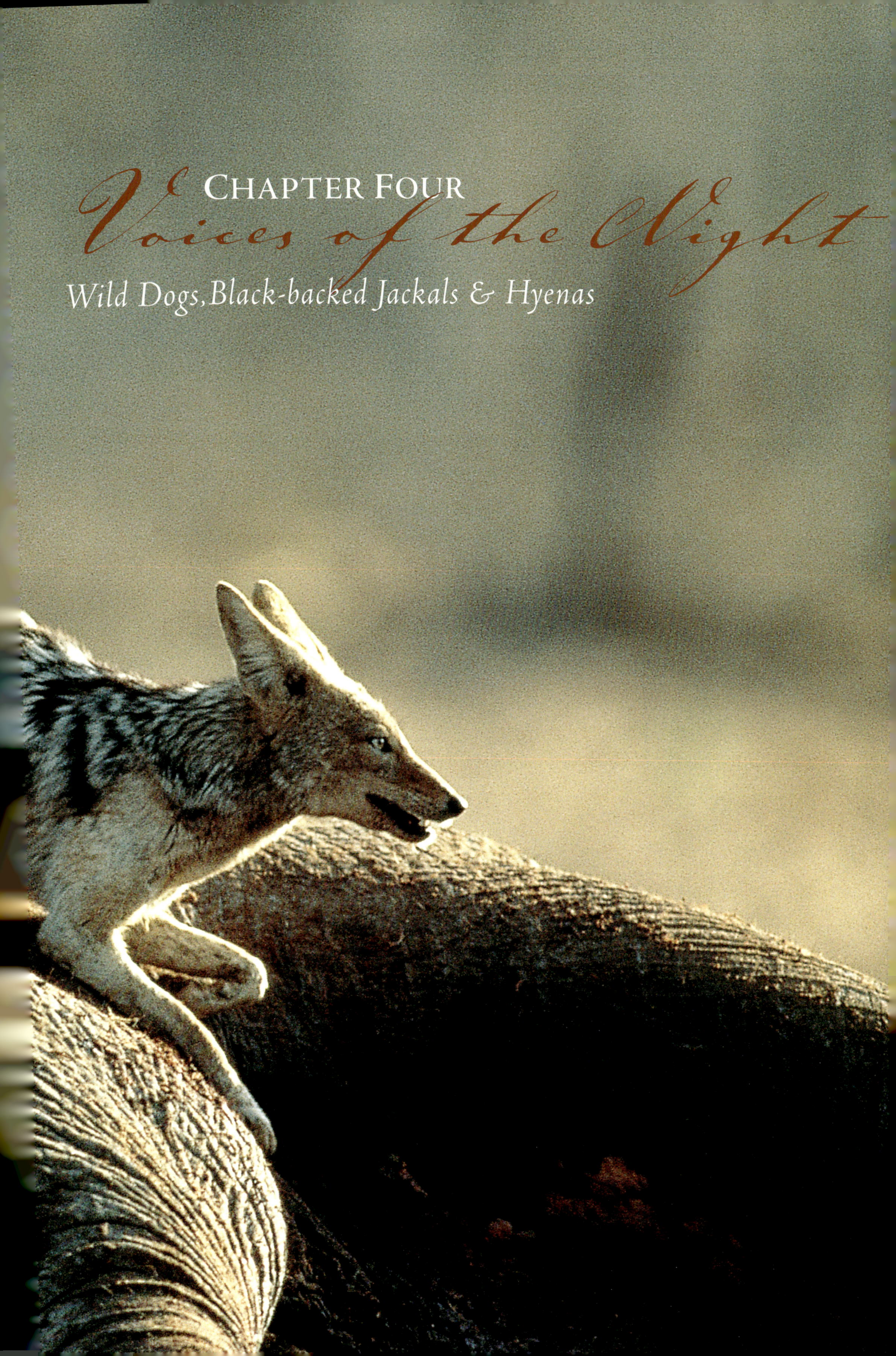

When the sun pulls its last rays of light below the horizon of the bushveld, pale lunar beams reveal only shadowy images and silhouettes. Where the eye fails, the ear becomes the instrument attuned to the magnified sounds of the African night: from a distance, bell-like hoots of Wild dogs; the keening *nyaaa ya ya ya ya* of the Black-backed jackal; a Spotted hyena's rising whoops – calling to its clan. These vociferous carnivores, with equally voracious appetites, roam the plains searching for unsuspecting prey to satisfy their hunger.

PRECEDING PAGES
*The jackal is being ever so cautious while scavenging on this elephant carcass because there are lions nearby.*

BELOW
*Wild Dogs are an endangered species in many areas of the world and are vulnerable everywhere.*

The strictly meat-eating African Wild dogs hunt as a pack for antelope and impala, and sometimes zebra, wildebeest and small mammals. Running at thirty miles per hour, the pack will chase its prey for up to three miles, tiring it, so that a kill is easily made. Wild dogs kill 85 percent of animals they chase. They hunt in the shadowy hours of dusk and dawn and on moonlit nights, crisscrossing up to thirty miles of their range; a range that might total more than five hundred square miles. Although mostly found in savanna and arid zones, they *have* been seen in mountainous regions. Several packs may have overlapping ranges and if they meet, the larger pack will chase off the smaller. Although they are sometimes seen in packs of up to twenty, the average for this highly endangered species is seven to eight adults.

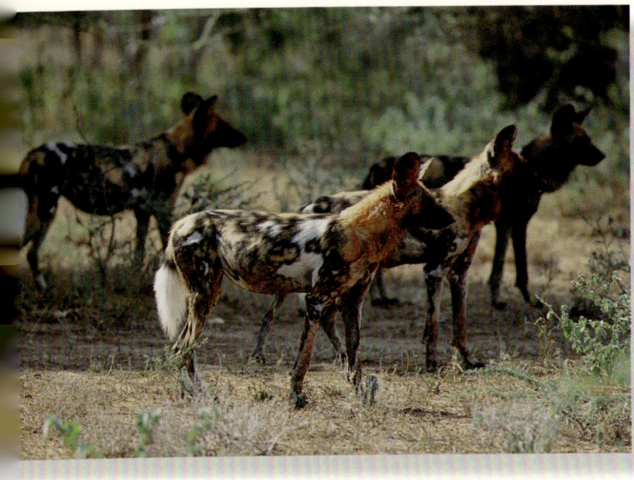

Wild dogs are long-legged, big-eared canines with tri-colored fur. The species name 'pictus' means *painted*, referring to the blotched black, white and yellow coat. They stand around thirty inches at the shoulder and weigh forty-four to sixty pounds. Their best sense is sight, which assists them in hunting the open spaces of their preferred habitat.

Wild dogs are highly sociable. To prepare for a hunt, they excite each other by active movements; running, playing and emitting bird-like twittering sounds. After a kill, the twittering resumes, accompanied by submissive behavior identified by retracted lips and toothy grins, flattened ears and lowered forequarters. In this way they compete for food rather than by fighting. Pups that show submissive behavior are allowed to eat first.

Many aspects of their social system set them apart from other gregarious mammals. Whereas lions, baboons and elephants have groups of related females, packs of Wild dogs are composed of related males. Males of those other species leave their natal family to breed. The opposite is true for Wild dogs. Females are the aggressive sex – with more fighting among themselves. The males tend to live longer.

Each pack has a dominant pair which breeds once a year – at a time when food is abundant. Blind and helpless, pups are born in the safety of a den.

Each weighs about fourteen ounces. Their eyes open at three weeks. At ten weeks, they are weaned and the entire pack provides regurgitated food for the pups until they are fourteen months old.

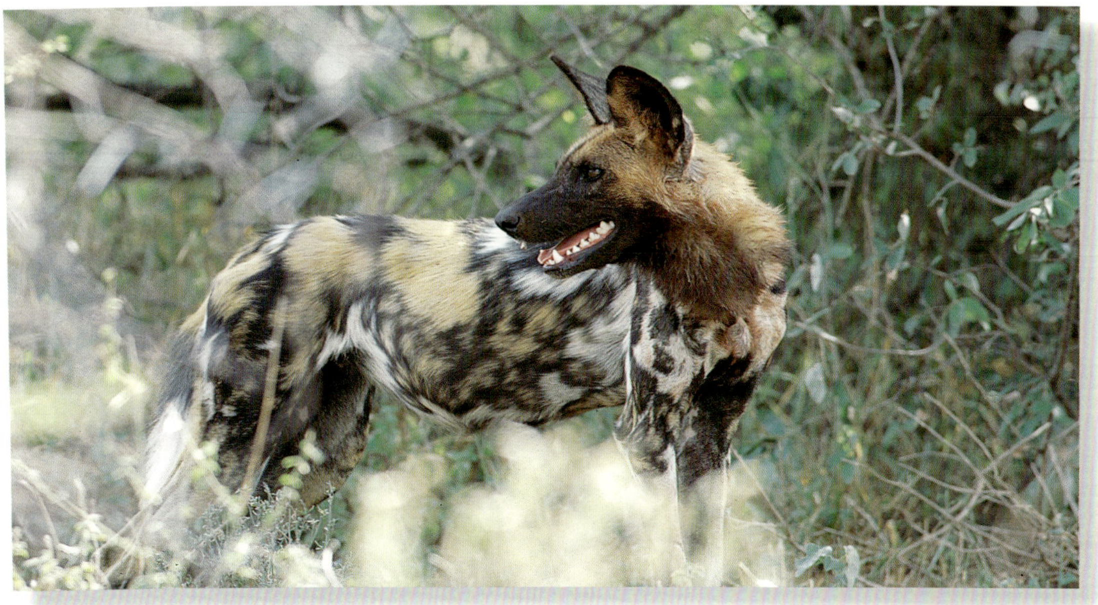

*In a drastically short time, Wild dogs have suffered more loss of sustainable habitat than any other carnivore.*

If other females have pups, fights break out over them. The pups become victims of these fights and are often injured or killed. As pups mature, they play together and begin their vocalizations early. They whine and bark and are even capable of ultrasonic calls. When mature, the female pups leave the pack together to find unrelated males with which to form a new pack. Males stay with their first family for their entire ten years of life.

Wild dogs have little to fear from other animals. They retreat only from the 'king of beasts.'

The Black-backed jackal has other enemies to fear. Leopards and eagles will prey upon young pups. Smaller than Wild dogs, they average fifteen to twenty inches high and weigh only about twenty-four pounds. Their signature feature ia a black saddle interspersed with silver markings. The appearance also earns them the name Silver-backed jackal.

Unlike Wild dogs, Black-backed jackals are omnivores. They consume a varied diet including rodents, reptiles, birds, fruit, and invertebrates such as grasshoppers, beetles and termites. They can sometimes take small antelope calves as well. They have little need for water. Eighty percent of their food is gained by hunting and the rest is scavenged. Competition with lions and hyenas at a kill is common. They avoid challenge by these large predators by being quick – able to nimbly snatch morsels before they are noticed.

Black-backed jackals hunt by night and by day in a varied habitat, from acacia woodland to arid regions and savanna. When at rest they prefer borrowed burrows and rock crevices.

Family groups consist of a male and female pair which are mated for life, and often, a pup from an older litter which serves as a helper when new pups arrive. Beginning at one or two years of age, Black-backed jackals breed annually in the dry season. After sixty to sixty-five days gestation, three or four pups are born. A helper assists both parents in caring for the pups. It will bring food to nursing mothers and will watch pups so that the parents can go hunting. A helper will guard the pups and alert them

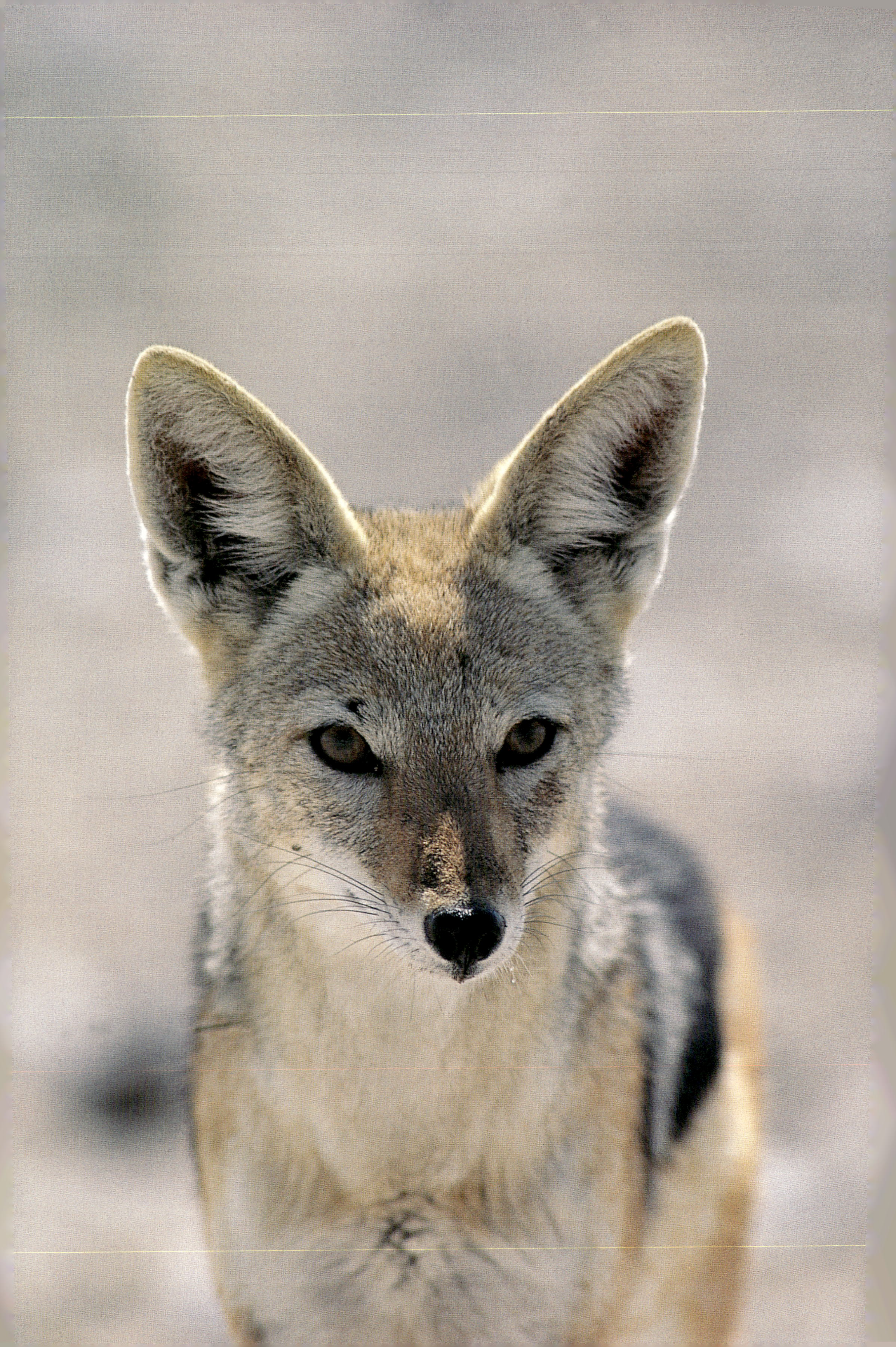

to danger, when after fourteen weeks, they leave their den to explore and learn. Helpers have been known to chase and bite intruders such as hyenas. By six months of age, the pups are skilled at hunting and leave the territory at merely eight months old. Black-backed jackals' territories are small, averaging one square mile. Pairs will scent-mark the territory together. When one of a pair dies, (by accident or old age at eight or nine years), and only one is left to scent-mark, others will quickly detect the void, due to a highly developed sense of smell. The territory is then vulnerable to take-over.

FACING PAGE
*Black-backed Jackals are among the handsomest creatures in all of Africa.*

BELOW
*The Nossob riverbed in the Kalahari Desert at sunrise is the setting for this jackal's wake-up call. Even animals like to get a few extra yawns in the early morning.*

Like other canids, Black-backed jackals vocalizations are varied. The characteristic *nyaa ya ya ya ya* punctuates the night. A shrill chattering means the animal is cornered or stressed, They communicate with a sharp yelping sequence for which one African tribe named them *bweya!*

But, neither canine above matches the variety of calls voiced by the Spotted hyena. This vaguely bear-like animal has eleven ways to vocally express itself. The loud series of whoops that carry eerily through the still night air are calls to gather the clan at a kill. Fast whooping and grunting, yelling, whining, growling and even ultrasonic calls are likely conversations over a feast of carrion. Clearly audible, too, is the famous giggling sound, a high cackle that earned them the name Laughing hyena.

Although dog-like in appearance, it is not a member of the canine family. The Spotted hyena belongs to the second smallest family of carnivores which also include two other hyenas, the aardwolf, civets and mongooses. It has a massive head and forequarters which slope down to shorter, smaller hindquarters. At birth its coat is black/brown. As it matures the fur changes to light red-brown or yellow-brown with dark oval spots. To complete the look, it sports rounded ears, a short bristling mane which grows in reverse, and a ten to twelve inch tail with brush-like black hairs at the tip. Males and females look so much alike that it is difficult to tell them apart.

All of the Spotted hyena's senses are acute. Nocturnal hunting requires keen night vision. Good hearing allows for long distance communication and the ability to listen for other predators at a kill. A clan of hyenas is swift to intervene when easy food is available and will compete with Wild dogs, jackals and

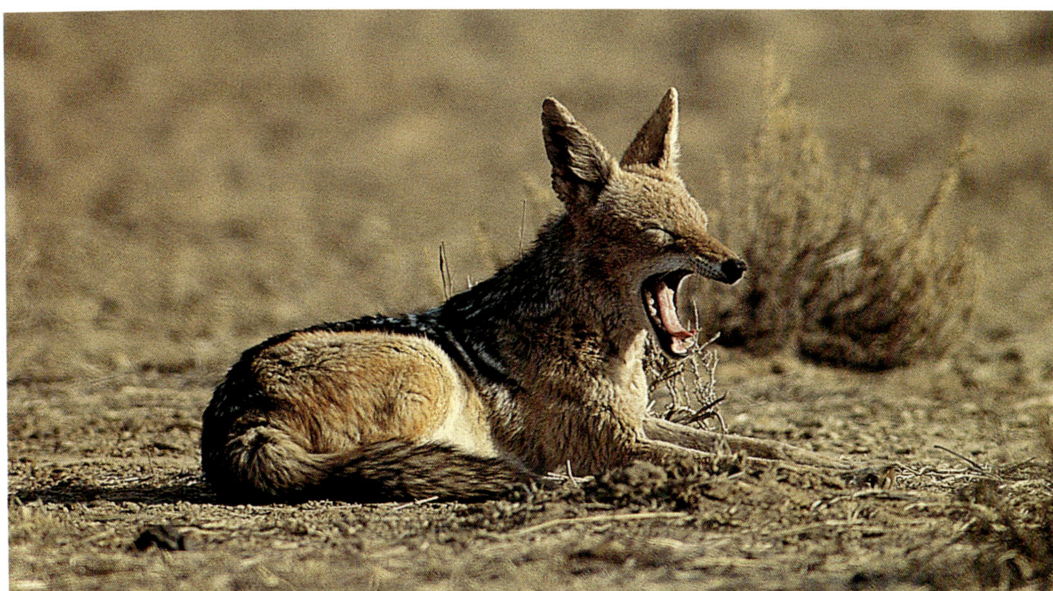

vultures, leopards, cheetahs and lions. They even mob lions to take over a kill. When they do, they can eat very fast. With powerful bone-crushing jaws, a party of ten or fifteen hyenas can tear apart and consume an entire Zebra in fifteen minutes. An efficient digestive system can handle bone, horn and even teeth within twenty-four hours.

Most carnivorous of all hyenas, the Spotted Hyena prefers medium size prey; zebra, wildebeest, gazelle and gemsbok. It also eats birds, fish, reptiles, snails and fruit. And to wash it all down, a hyena needs a lot of water. Traveling long distances to find it is common. While the Spotted hyena is an opportunistic scavenger, sixty-five to seventy percent of its meat is hunted. Anything remaining from a kill is often cached in water or mud. The hyena's excellent sense of smell enables it to easily find leftovers when hungry.

BELOW

*Mother hyena seems to be saying, "Don't come any closer!" Who would argue with a creature with so formidable a set of teeth and powerful jaws?*

FACING PAGE

*Although not fully grown, the hyena pup has all the markings of an adult.*

Smell also plays a large part in the Spotted hyena's social life. Both male and female have scent glands which contain a paste-like substance that they deposit on plants, especially in the border regions of their territory. Often mistakenly thought to be solitary beasts, since individuals often hunt alone, they are actually very gregarious animals with a complex social system. Females are dominant in groups called clans. An average clan has four to eighteen adults, but sometimes there are up to one hundred members.

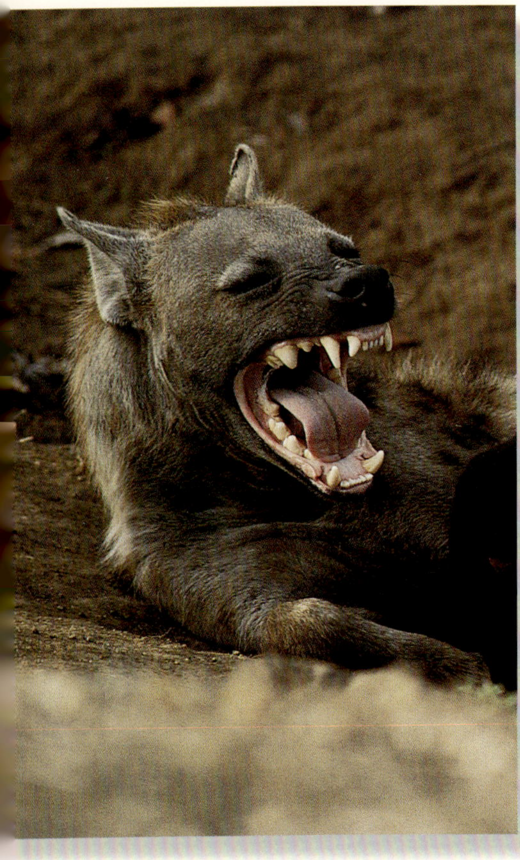

As well as their repertoire of calls and unique scent marking, they communicate with mutual licking and sniffing rituals when they meet. They can be aggressive with one another but will often show submissive behavior by crouching and grimacing. They also enjoy playing together with bones and sticks.

Spotted hyenas breed at any time of year with the peak in late summer when bovid and zebra young are abundant. Two cubs are born to a mother and are raised in a communal den with pups of other mothers. Unlike canines, they are born with eyes open and a full set of milk teeth. They are voracious eaters, able to nurse for extended periods time. Only hours after birth female cubs begin to fight for dominance using adult fighting techniques such as neck biting and shaking. Same-sex sibling rivalry claims 25 percent of cubs in the first month. Adult males do not play a parental role and are not allowed near the den. Cubs accompany mothers to feeding sites starting at from three to five months old. Even so, they are known to throw tantrums when their mother attempts to wean them. By one and a half years old, a cub is competent at hunting on its own.

In the National parks, only large lions pose any threat to the Spotted hyena. If pursued, the hyena will run in its strange loping gait, glancing back periodically and laughing its high cackling call, off into the sunset – a real-life cartoon simile.

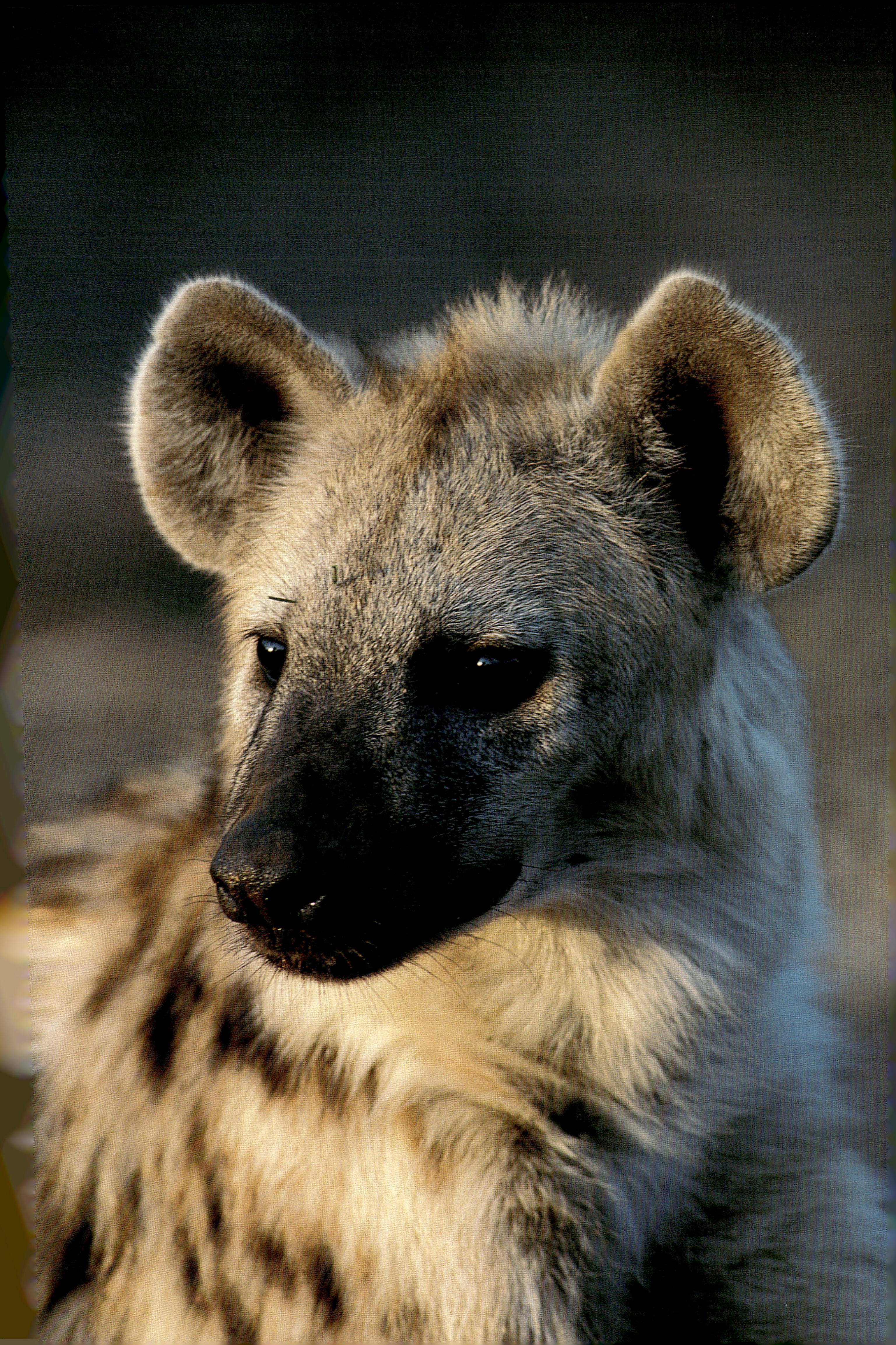

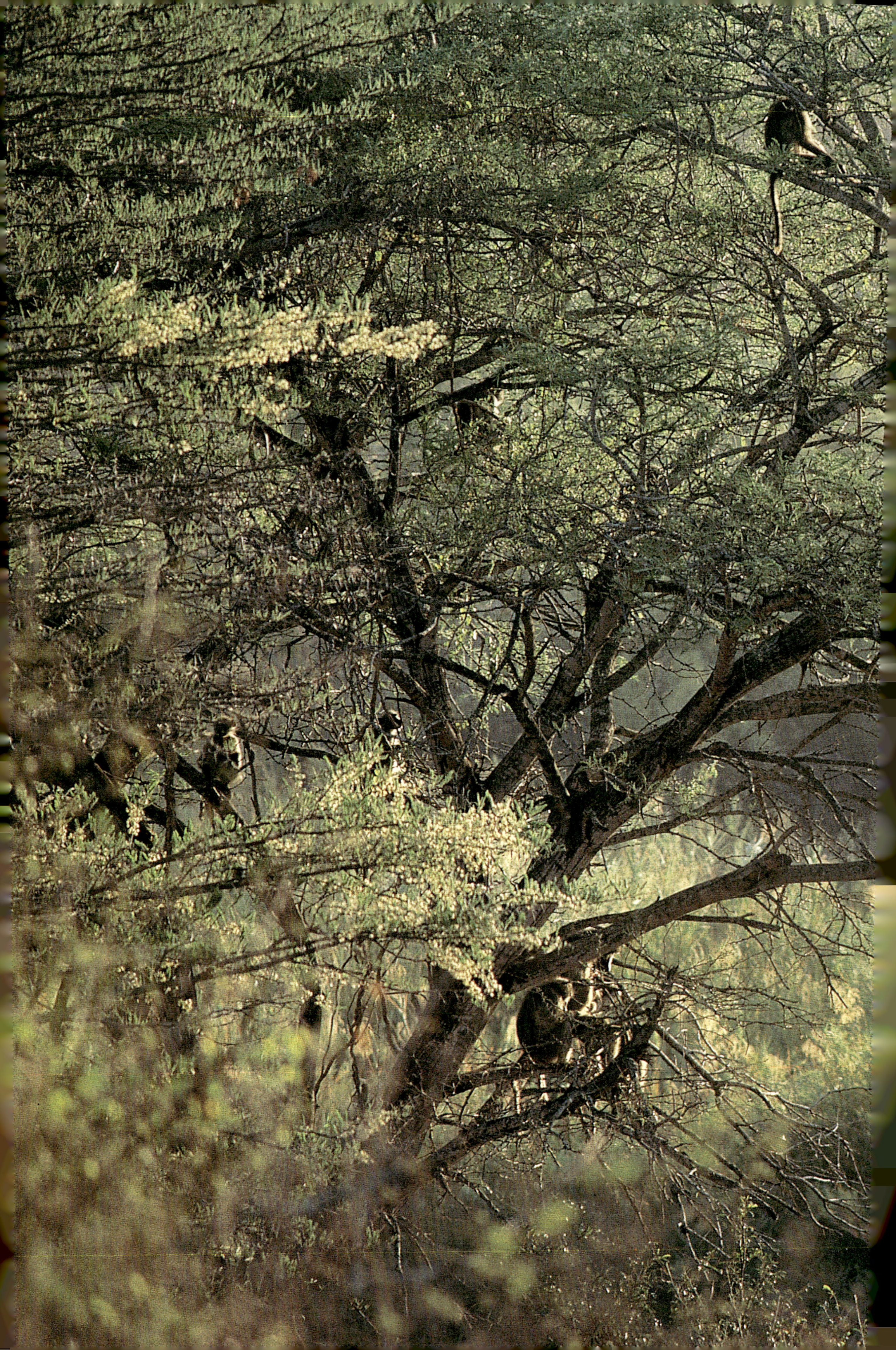

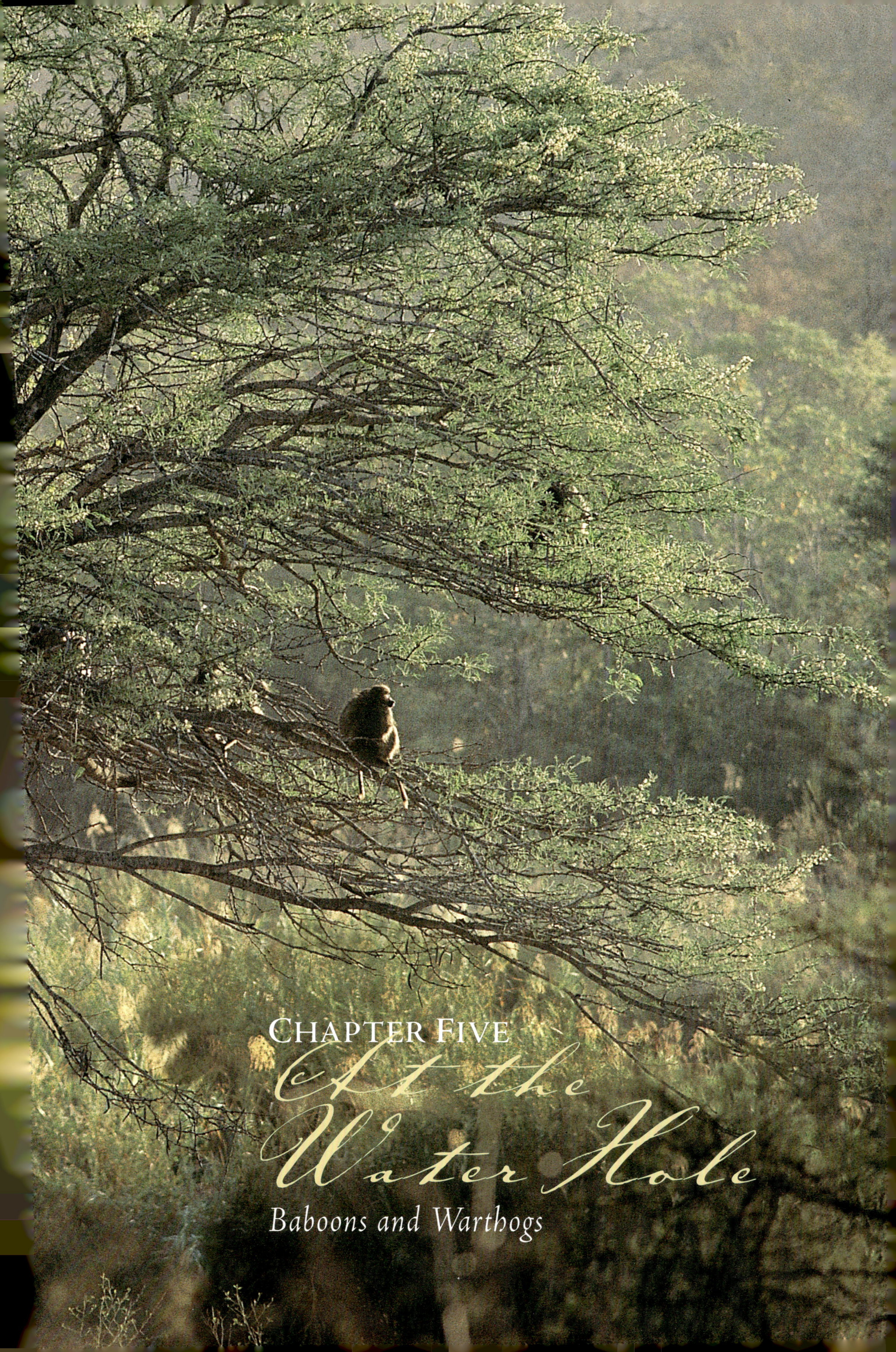

# Chapter Five

# At the Water Hole

## *Baboons and Warthogs*

When the sun beats down and the heat rises up from the savanna, those drawn daily by a powerful thirst gather at the local water hole. Among the regulars are troops of Chacma baboons, bellied up to the brink, sipping long refreshing drinks. On the banks, sounders of Common warthogs wallow luxuriantly in the soft mud – their thirsts already quenched by the cool aqueous beverage. Species as diverse as baboons and warthogs may be, these curious characters not only share a love of water, but other traits as well.

Baboons are omnivorous, but primarily eat plants. Grasses, leaves, and fruit are their common foods. For variety, they also eat flowers, seeds, twigs, mushrooms and other plants. In the dry season, they dig up nutritious rhizomes, tubers and roots; a habit shared by the Common warthog.

The dry season also drives baboons to sample invertebrates such as grasshoppers, spiders, scorpions and shellfish. Lizards, turtles, frogs, fish and eggs are on the menu as well. Adult males will, on occasion, hunt small rodents, hares – and even young antelopes.

PRECEDING PAGES

*A troop of baboons in an umbrella thorn tree is enjoying the warmth of the early morning sun.*

FACING PAGE

*A large male baboon is armed with knife-edged canines that are about 2 1/2 inches long.*

FOLLOWING PAGE

*Mother was sitting in the grass sunning herself after a chilly night. Her baby was playing nearby, but as soon as he saw me, he popped right into the safety of his mother's arms.*

One of four subspecies, the Chacma baboon is common to the savanna regions of Southern Africa. It is a monkey with a dog-shaped head covered with short, coarse, greenish-brown fur and darker lower limbs. The head has a partially naked face with close set eyes, and a projecting muzzle which contains powerful jaws and cheek pouches for storing food temporarily. About twenty-nine inches from head to calloused rump, it weighs from fifty-nine to ninety-seven pounds. The baboon uses all of its senses well. Sight includes color vision. The sense of smell is implicated in social status and courtship. Hearing is important for many social and survival encounters.

Chacma baboons have a complex social structure. Family units called troops consist of thirty to forty members on average. Most are female and half are immature. A female will often have several favorite males. When in heat, she will mate with one of them. The ensuing pregnancy lasts six months. After birth, the baby clings securely to its mother. Within two months, baby is able to climb well, and at six to twelve months it can frequently be seen riding astride its mother's back as she moves. Babies begin to play together at three to four months, wrestling and chasing each other. Gradually they become more adventurous, and by eighteen months are independent. Daughters inherit rank from their mothers and females stay together in their natal troop; while males migrate periodically from troop to troop.

Body language, facial expressions and vocalizations are important means of communications for baboons. Troop members greet one another with nose to nose contact, kissing, sniffing, and touching. Lip smacking, teeth chattering and grunting sounds are exchanged. Grooming is an important activity. They use fingers and lips to remove parasites and debris from each other's bodies.

A high ranking baboon will have a look of confidence, be casual and relaxed, hold its tail high and walk with a swagger. A baboon of low rank looks anxious and nervous with a low curving tail, hunched back and lowered head.

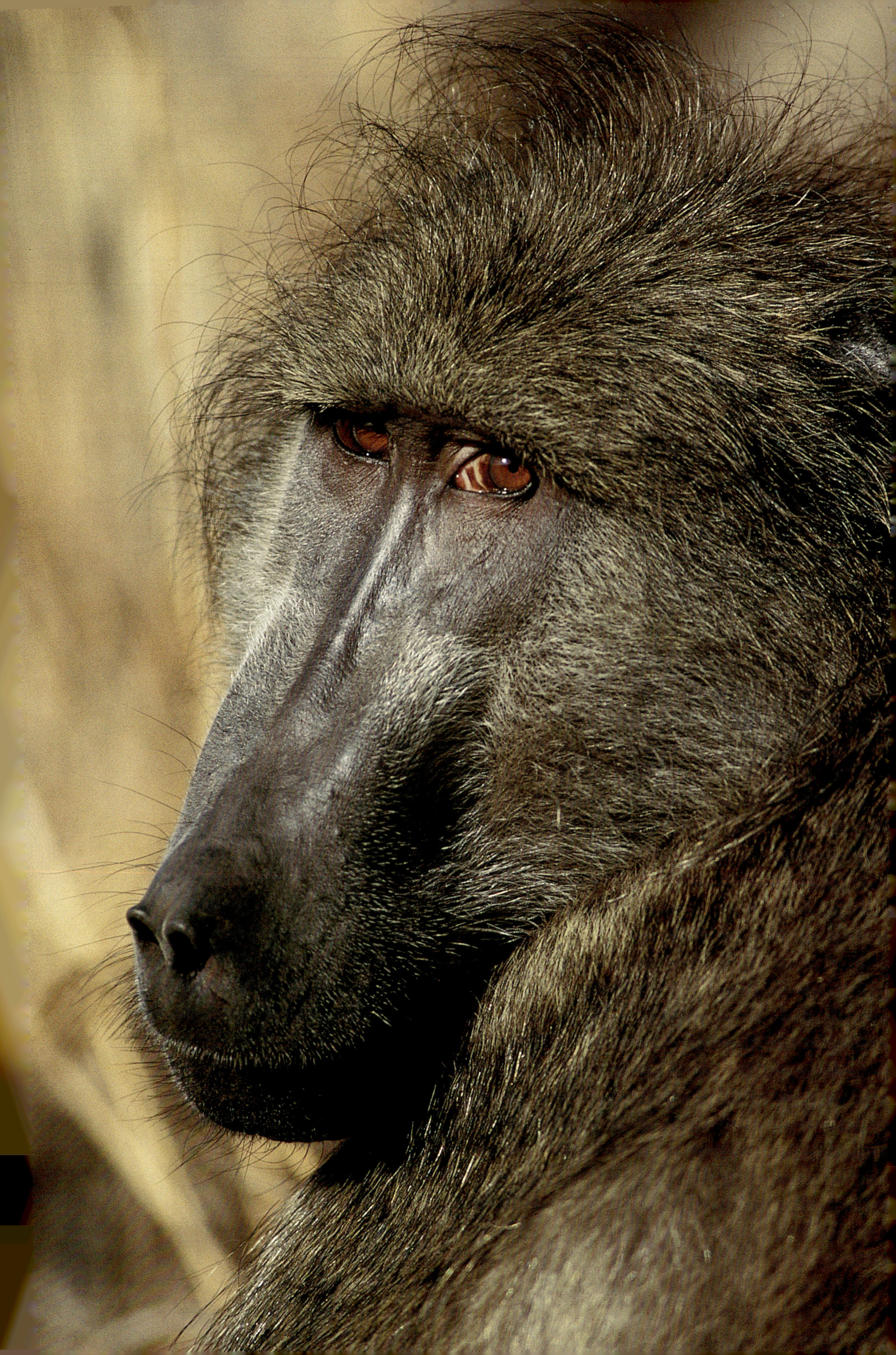

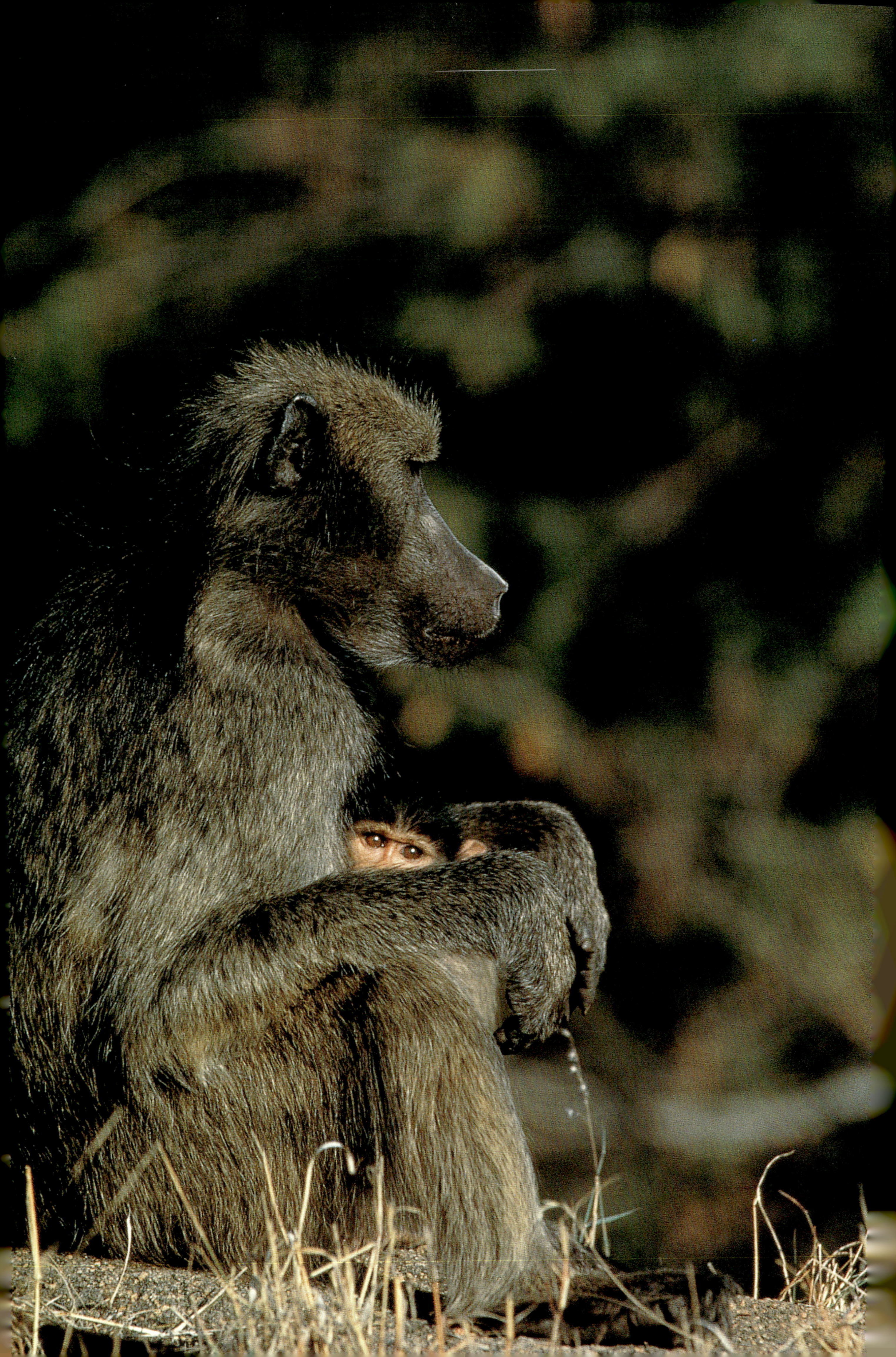

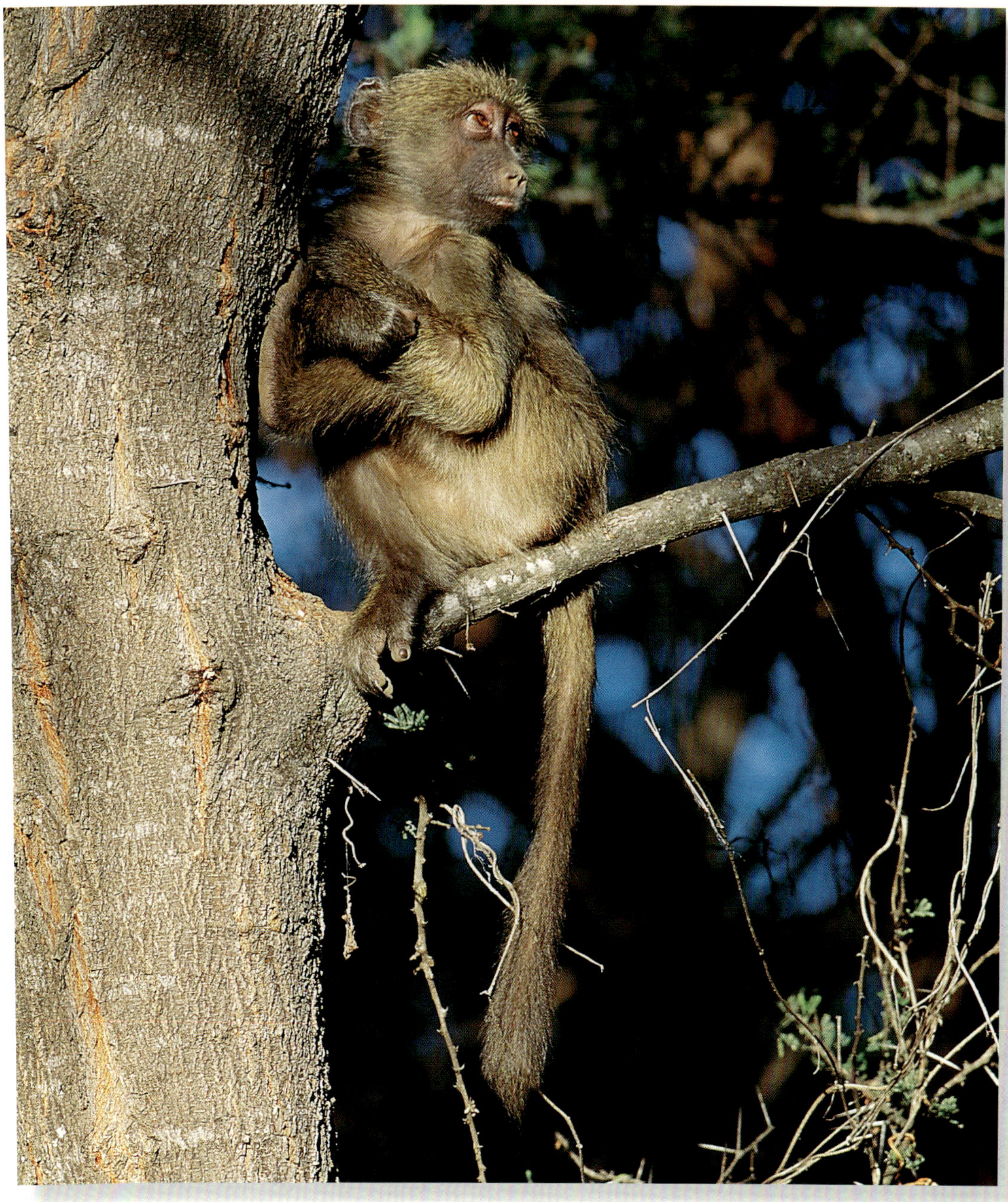

*The expression on this young baboon shows a bit of concern. He possibly sees the approach of a larger and more dominant baboon.*

The bare face easily shows subtle expressions that are used to show aggression, submission, and fear. There are also expressions for curiosity, affection and invitation to play. A baboon will stare down a rival, and bare its teeth to threaten; or use an intimidating yawn with raised eyebrows, flattened ears and aggressive gestures. To show submission, a threatened individual will glance away from a stare and pretend to be preoccupied with grooming or some object – as though thinking, "If I act like I don't notice, perhaps he'll go away." A fearful baboon will grimace with lips turned up and teeth clenched while squealing and screeching in alarm. If predators are lurking about, the response may include head bobbing and pushups that escalates into branch shaking and jumping around. Screaming often follows.

The baboon's repertoire of calls covers every activity. While foraging or on the move, they keep in touch with soft grunts. If lost, whistles and barks reconnect the troop. An encounter with a rival will initiate a rhythmic – *uh, uh, uh, uh* – from a male, which can build into a grunting roar. A shrill bark will alert the troop to the presence of a predator.

Eagles and jackals can take baboon infants. If attacked, males will join together in a coordinated effort against large cats or hyenas. At night leopards are a threat, but most often the baboons are safe sleeping in their beds, high up in trees or on cliffs.

FACING PAGE

*Is this the face only a mother could love? Despite the warts and elongated face, this is a wonderful animal to observe and photograph. After a while they almost become cute.*

Just after dawn they rise from their roosts – first grooming one another and socializing, then setting off to forage. They travel an average of three to four miles per day to find nourishment, and of course, to visit their favorite water hole.

And there, as usual, wallowing happily in the gooey mud, is the family of Common warthogs. Not just a recreational activity, wallowing protects the warthog's skin from biting insects. Beneath the earthy frosting breathes a sturdy beast noted for its ill-favored visage. Weighing one-hundred ten to two-hundred forty-five pounds and twenty-four inches tall, the sturdy body supports a long massive head with small eyes and a flattened nose. Two pair of facial warts indicate a boar while a sow will have only one pair. Curving out from the upper lip are two tusks which are actually enlarged canine teeth. The warthog's coat is a sparse covering of black and white bristles over grey skin. Longer hair forms cheek whiskers and a brownish mane. The tail ends in a brush of stiff bristles.

The warthog has scent glands near its eyes which are used to mark trees and other objects. The secretions are often visible on a male's face as dark tear-like stains. The sense of smell is well developed. It is used to detect the presence of others, females in heat, and to find food.

By day, family groups graze in the wet season on savanna grasses, their preferred food. When the earth becomes dry and grasses wither, their flexible noses sniff out the buried rhizomes, tubers and bulbs below the surface. Contrary to popular belief, warthogs do not dig with their long tusks, rather, they root with the tough upper nose edge. And, while foraging they frequently shuffle along on bent, calloused knees. For dessert they enjoy fallen fruit.

Warthogs live in a family group called a sounder, commonly consisting of one or two sows and their piglets. Adult males are present only at breeding time. Usually they are solitary or live in a group of bachelors. Sounder members are affectionate with one another. Like baboons, they greet one another with nose to nose touching. They help each other scratch those hard to reach places with mutual grooming and enjoy close contact when sleeping. They reside in an underground den often previously excavated by aardvarks. The den offers protection from heat and predators and serves as a nursery for newborn piglets.

Warthogs breed in the dry season and farrow about five months later as the rainy season begins. They are the only hoofed mammal to produce multiple young with an average of two to five piglets forming a litter. The hairless piglets weigh only eighteen to thirty-two ounces at birth. They stay safely underground for six or seven weeks, but occasionally leave the burrow for short periods. After a couple of months they join their mother to graze – following her in orderly single file fashion. Piglets enjoy playing; running, tail chasing and tossing objects. Activity also involves sparring and aggressive behav-

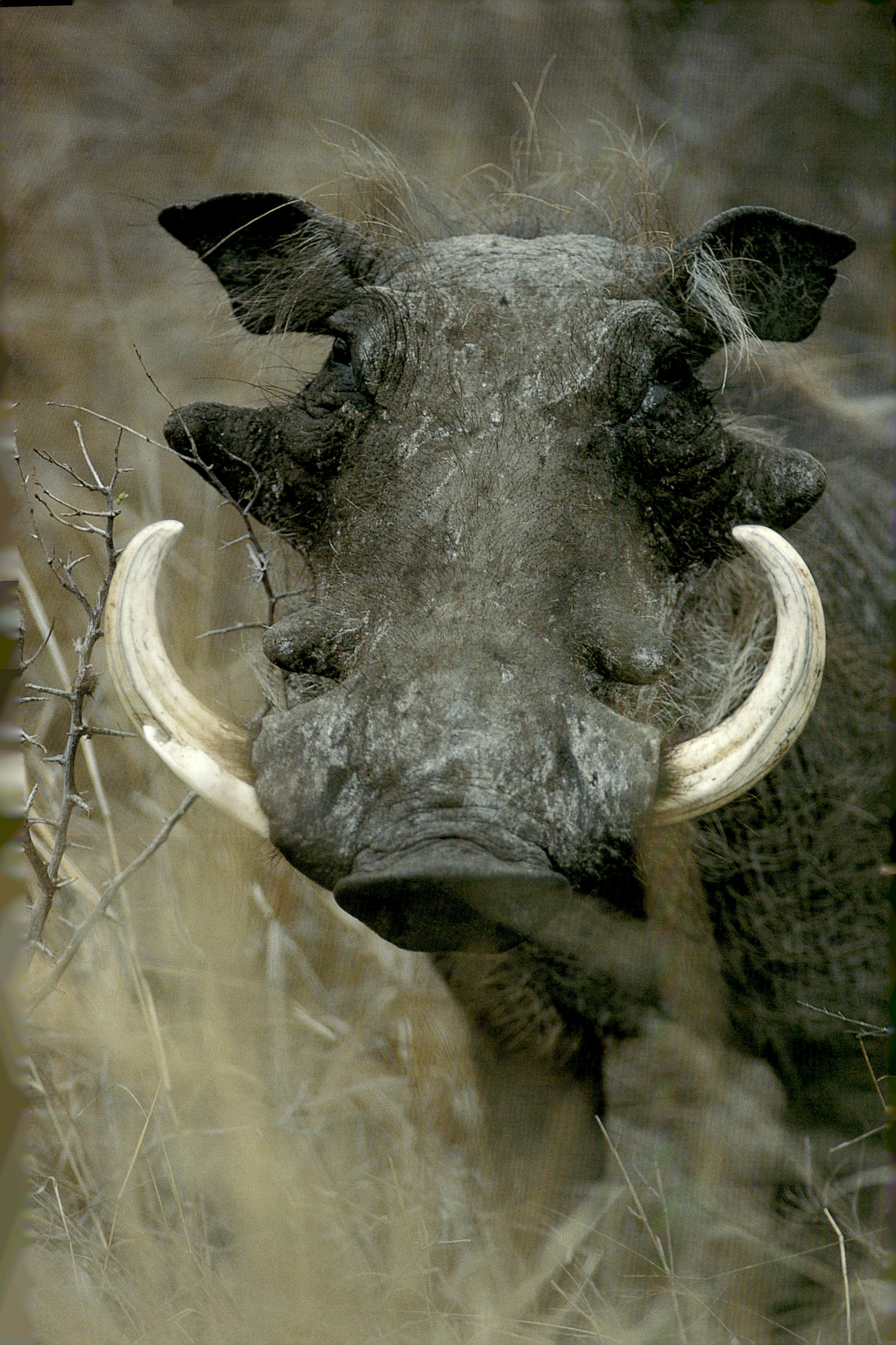

ior that establishes rank at a young age.

As they mature, boars continue to grow larger and larger. They maintain their status with dominance displays. Grunting and growling and grinding of teeth accompanies broadside strutting, which can escalate into snout boxing and tusk gnashing with an opponent. The losing party retreats with a show of submission; first by backing off, then by running away squealing.

FACING PAGE

*A warthog and mud wallowing is almost synonymous. Like all wallowers, they protect themselves from biting insects with a good frosting of mud.*

BELOW

*The color of warthogs depends upon the color of the mud in which they have been wallowing.*

Warthogs run, too, from hungry predators with a taste for pork; wild dogs, hyenas, and big cats. First, a warning grunt alerts the sounder, which bristles in alarm. Piglets may initially freeze or hide. Some adults may charge the predator, but most often they flee in a noisy gallop, squealing and snorting, tails held erect – headed for the safety of the den. Piglets scurry in first, followed closely by the adults. The last adult quickly turns around and enters, tail first, blocking the entrance. Wedged there, it bravely confronts the pursuer with warty face and formidable tusks, protecting the family within. They will live another day, to root for tasty bulbs and to wallow in the cool mud by the water hole.

## *How the Warthog became so ugly.*

*Warthog was once a very handsome beast, but vanity and rudeness were his ugly traits. He often insulted the other animals which reacted by chasing him away. He would always run for the safety of his burrow. One day, while grazing, Warthog teased and insulted Lion. As usual he was quickly pursued, high-tailed and squealing, to his den. This day, however, in Warthog's absence, Porcupine had slipped into the cozy burrow for a midday nap. Upon his hasty arrival, Warthog piled headlong into the sharp spines of the sleeping Porcupine. Before he knew what had happened, he reversed and dashed out as quickly as he had come, squealing in pain from a face full of quills. Ignoring angry Lion who waited outside, he bolted past, frantically searching for relief. Having alienated all the other animals, none would help him, so he was left to suffer alone. The quills caused his skin to fester and left bumps and warts, permanently disfiguring his face. The ordeal served to make him humble, and one by one, the other animals accepted him. But Warthog remained wary, and to this day he enters his den backwards..... just in case a sleeping porcupine awaits.*

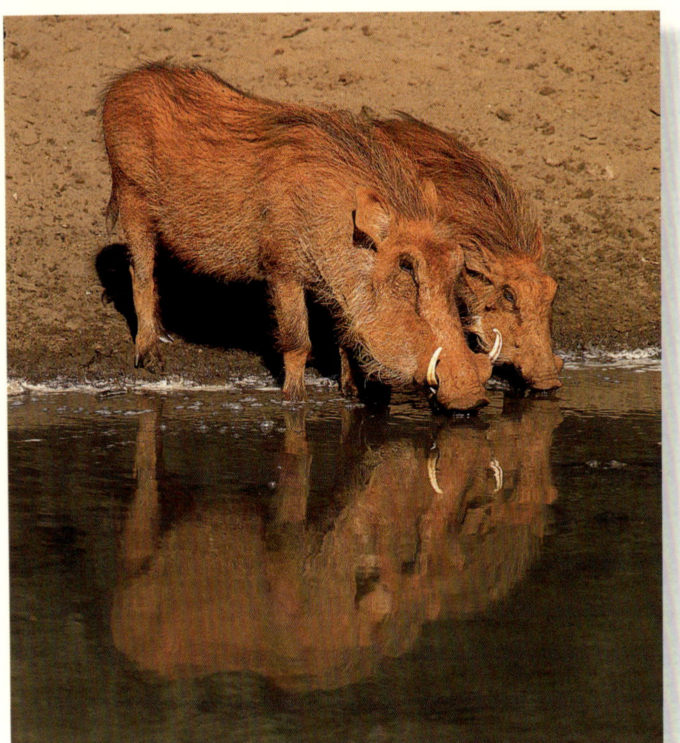

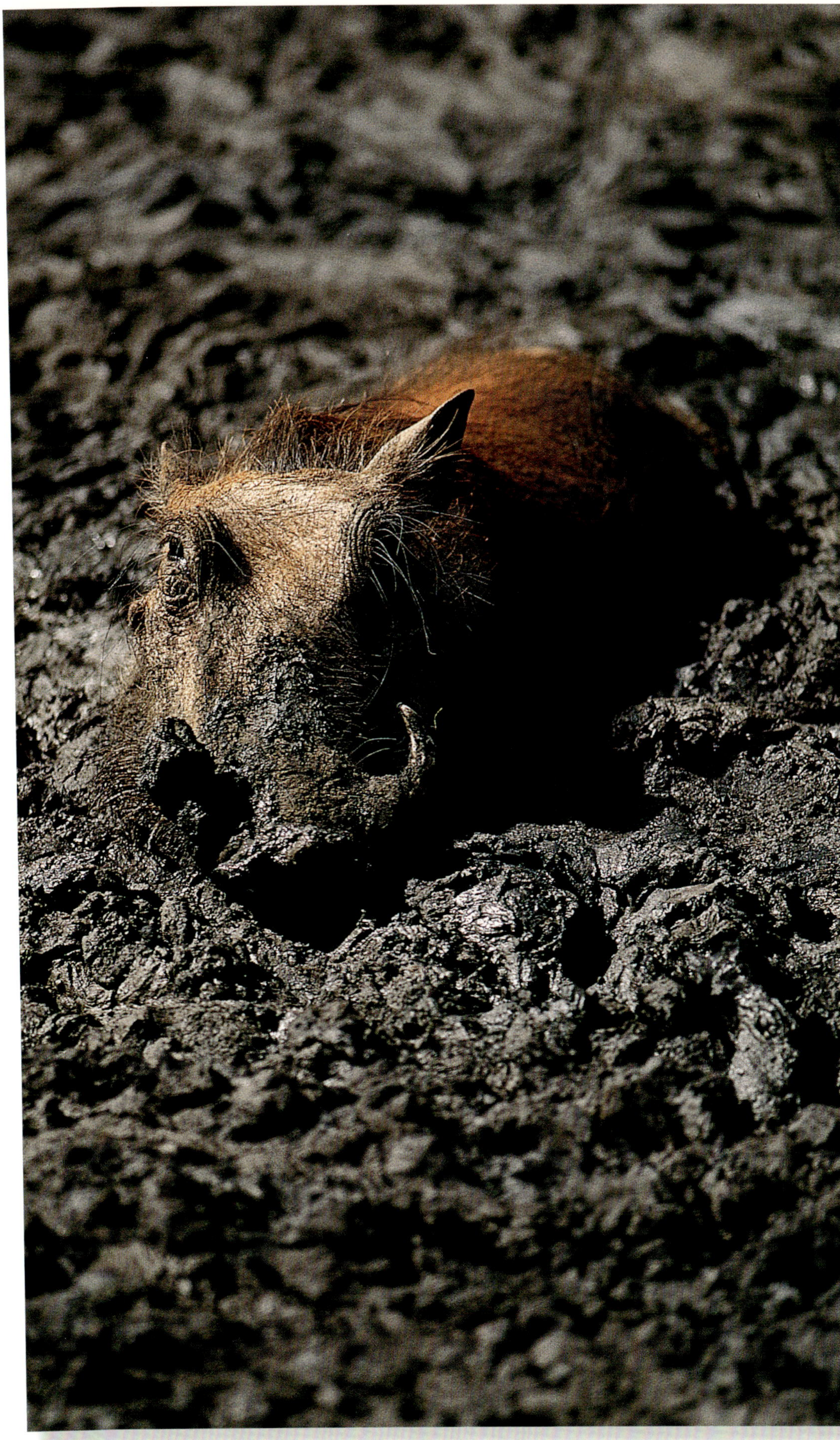

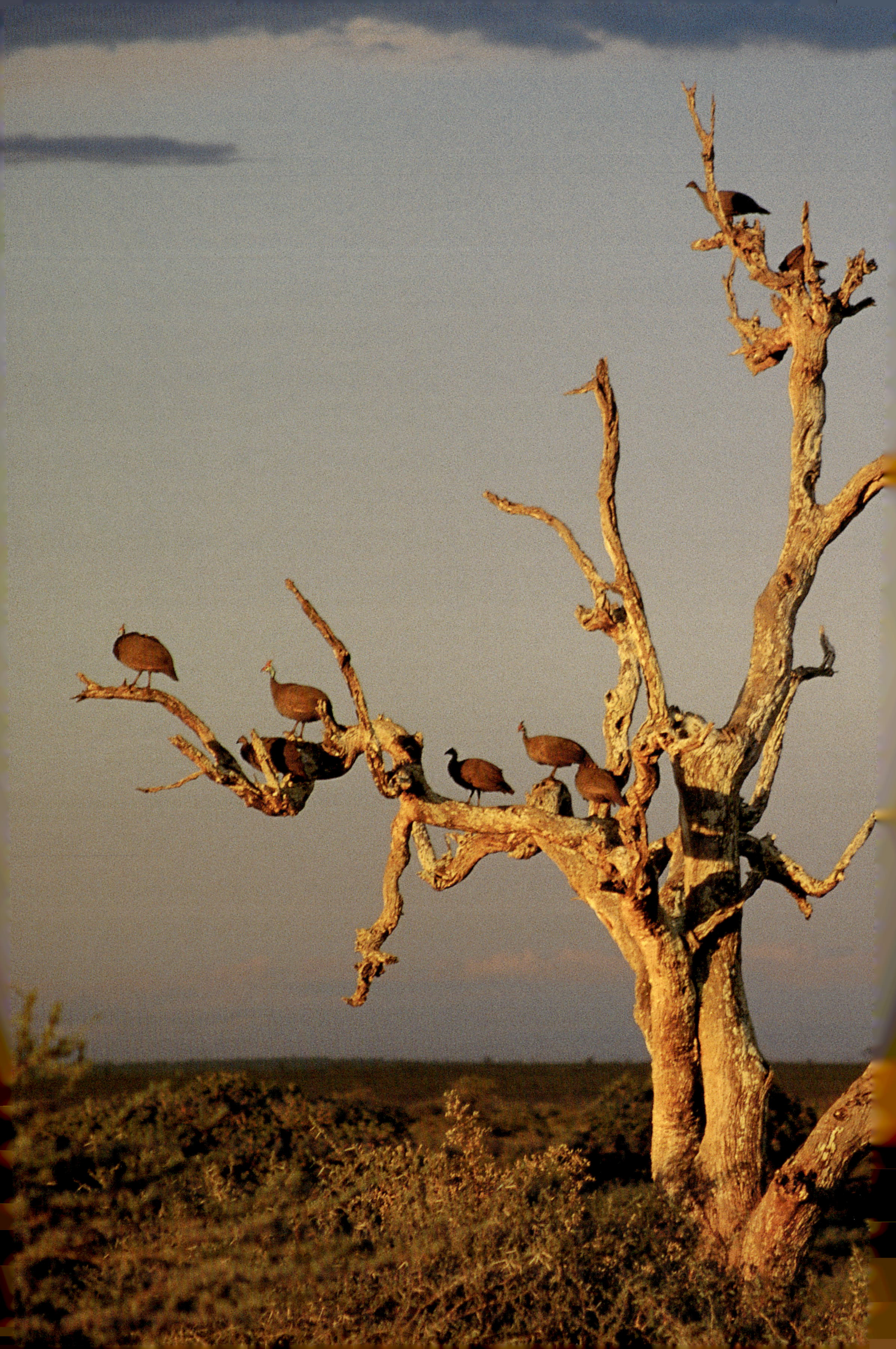

# Chapter Six
# Walking Birds
## Guinea Fowl, Bustards, Hornbills, Secretarys and Ostriches

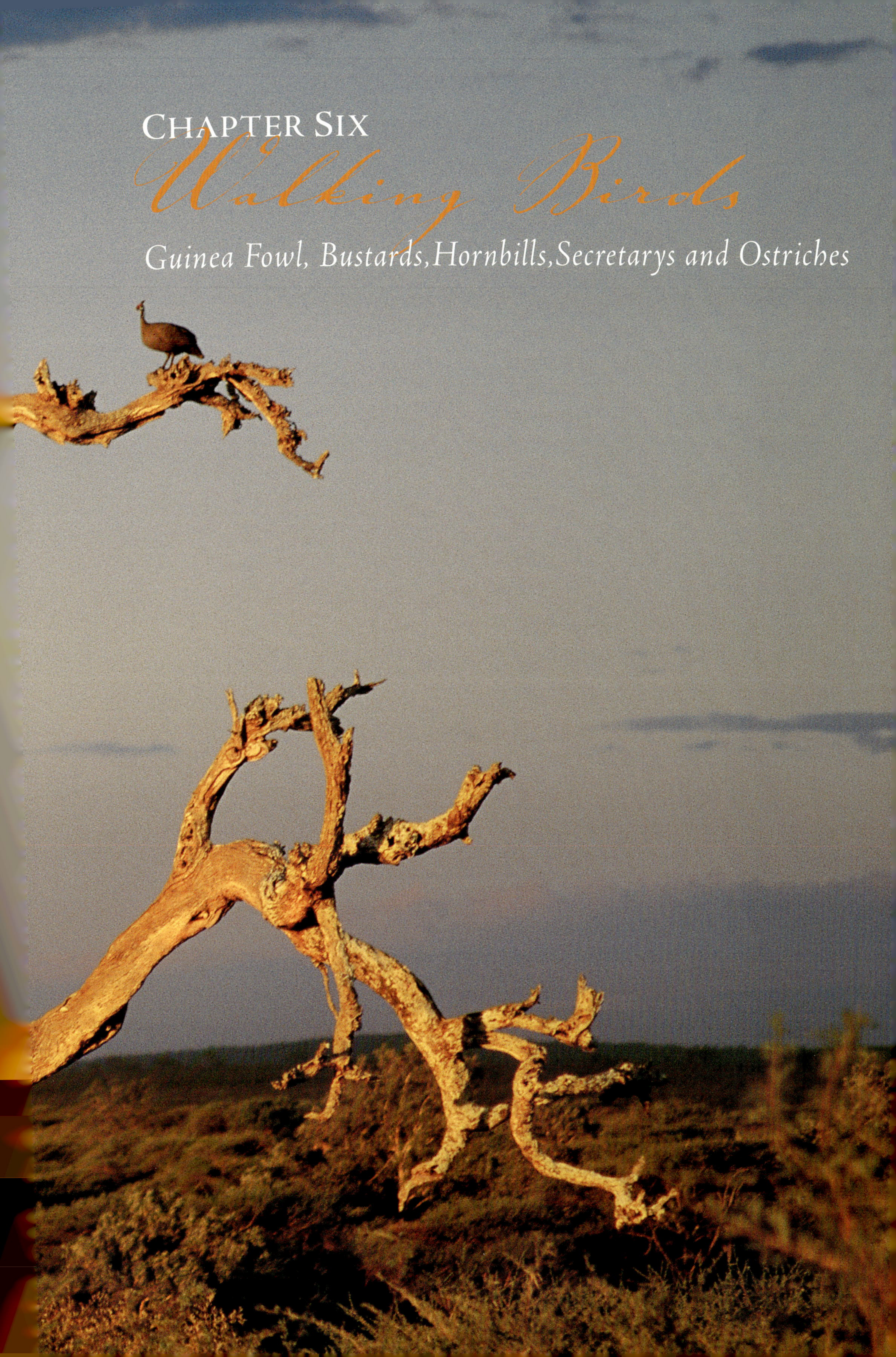

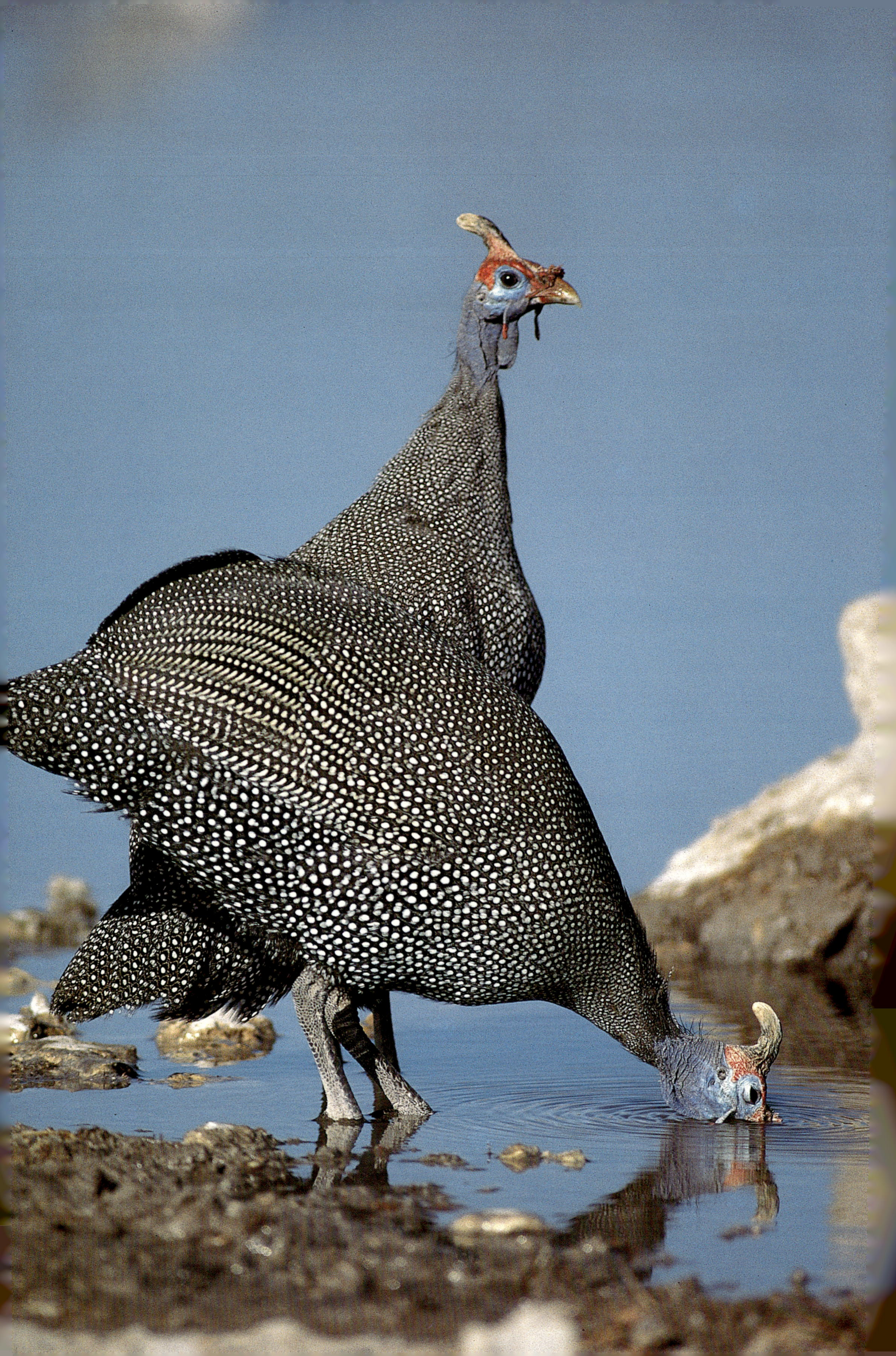

Early evening at the water hole. Shadows outline a crisscrossed pattern of assorted footprints, tell-tale signs of the myriad animals that came and went from the refreshing pool. The warthogs linger there soothing their itchy hides. Soon, materializing from an opening in the bushveld, a parade of Helmeted Guinea Fowl arrive on the opposite shore, marching in single file. They come to sip their daily drink, and to snack on any parasites the warthogs may have lost in their wallowing.

These walking birds are common and highly visible with their naked blue necks and red caps topped with the horny helmet for which they are named. The head, being unfeathered, can regulate brain temperature. About the size of a chicken, they have rounded bodies covered with grey feathers flecked with white. Longish black legs and feet are adapted for walking and scratching the ground for favored foods. Guinea Fowl forage for seeds, bulbs and berries, insects, millipedes and snails with their strong arched bills

These birds are quite gregarious, found in flocks of several hundred. Moderately vocal, the males call *ker bek, ker bek, ker bek, krrrr,* to the females which answer, *buck whéét, buck whéét, buck whéét.* At breeding time Guinea Fowl separate into pairs and call to each other, *pittoo, pittoo, pittoo.* Males will engage each other in ritualized chasing. Females enjoy this sport – watching for males that endure the longest, which are then chosen for mates. Nests are but a scrape in the ground, lined with a bit of grass and concealed in dense cover. Hens share a nest, depositing from six to nineteen eggs together. After hatching, the young, called keets, are cared for by both males and females. Keets are precocial, meaning they are highly developed when they hatch; covered with down, able to walk and to feed independently, like ducks. By their fourth week they are fully fledged.

Being small, terrestrial birds, Guinea Fowl are wary, watching out for jackals, baboons and snakes.

PRECEDING PAGE

*Although this impressive tree has lost its leaves and bark, it remains a sentinel overlooking the bushveld. Guinea Fowl add life to an otherwise barren tree.*

FACING PAGE

*These two Helmeted Guinea Fowl were getting a much needed drink at one of the water holes in the Etosha National Park.*

BELOW

*Helmeted Guinea Fowl are very gregarious and they're seen in large and small flocks.*

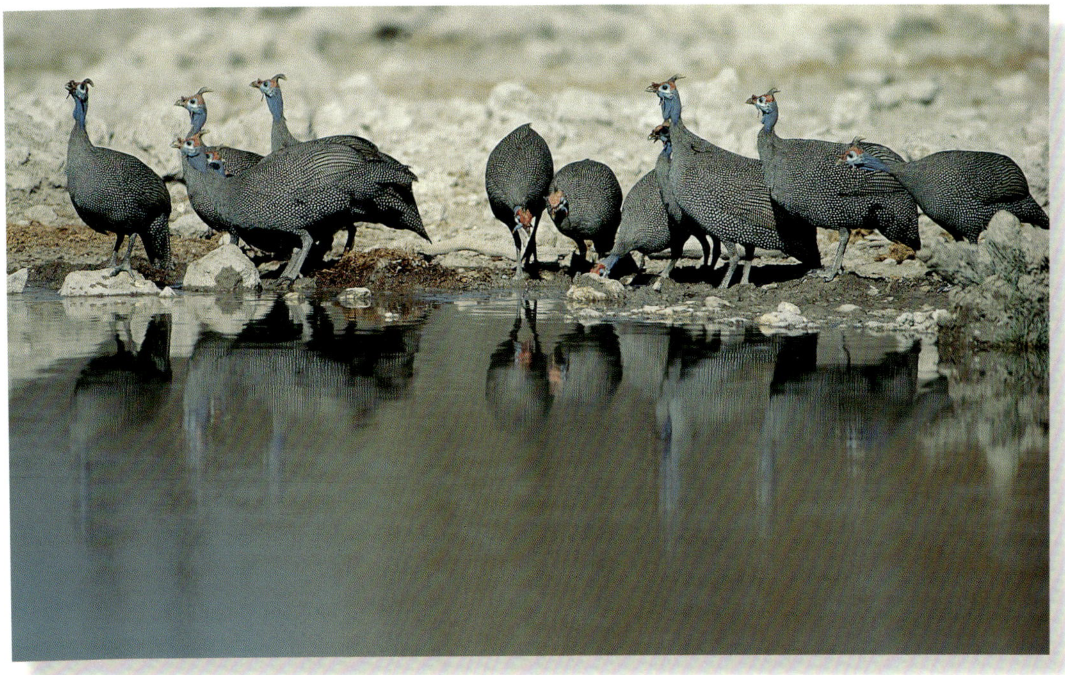

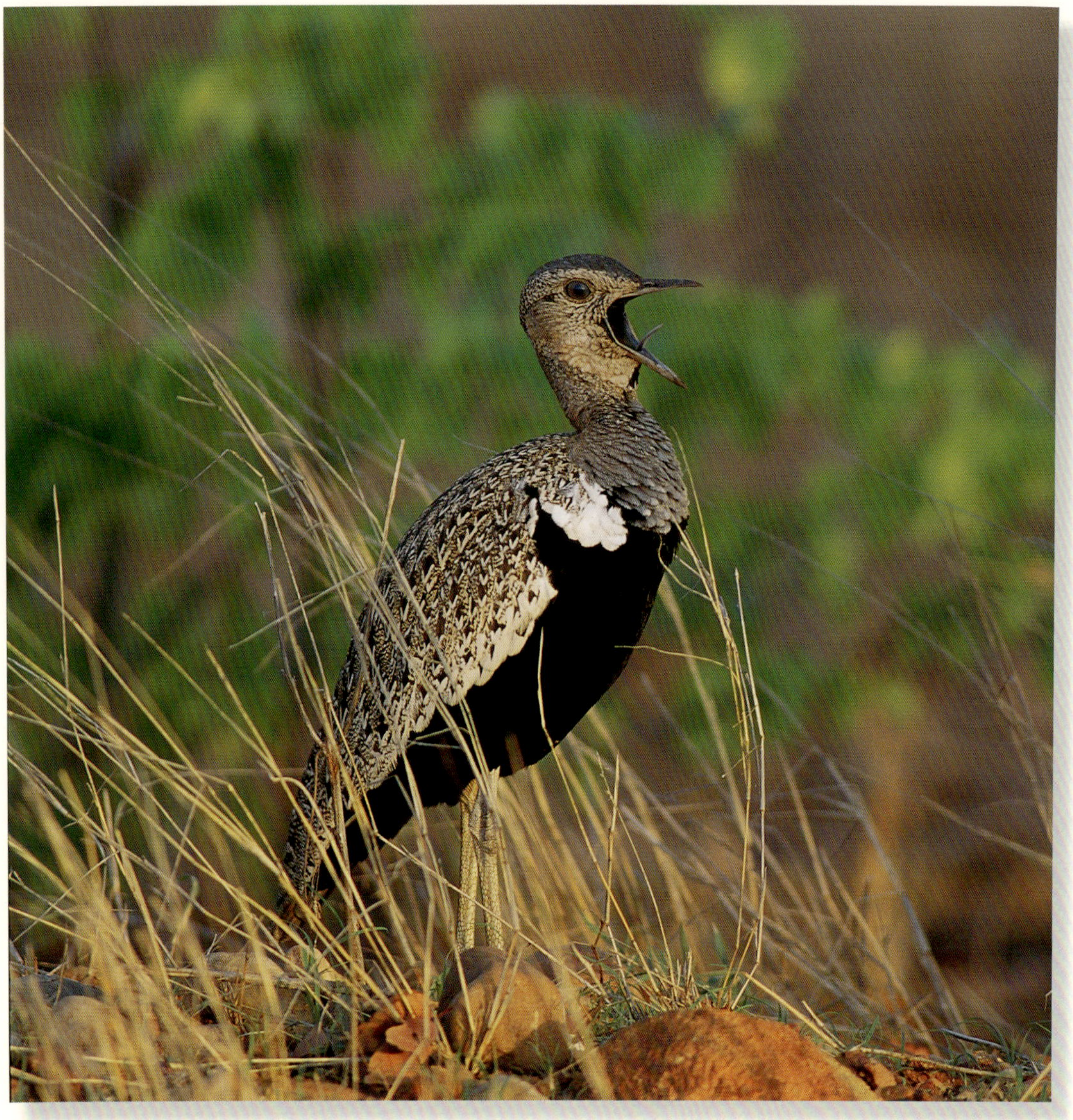

*A male Redcrested Korhaan was calling for a mate with a croaking sound, ever increasing in volume.*

To escape an attack, they *are* able to fly to the safety of tree branches. It is there, as well, that they roost at night to sleep, after an evening at the water hole, of course.

Another family of ground dwelling birds called bustards, inhabits the savanna and arid plains. Bustard, loosely translated, means 'the bird that walks.' They occur in many parts of the world, but two-thirds are found in Africa. These nervous birds have various colorations that keep them well camouflaged. Most are patterned in combinations of grey, chestnut, black, white and buff.

Bustards forage slowly, meandering through grasslands, intent on finding a wide variety of foods. Their omnivorous diet includes shoots, flowers, seeds and berries, as well as invertebrates, reptiles, young birds, eggs and small mammals. Like Guinea Fowl they nest in a bare scrape on the ground. One to six eggs are commonly laid. When alarmed, bustards are easily disturbed and will leave the nest. The eggs are then subject to predation. This behavior makes bustards especially vulnerable to the growing problem of habitat exploitation.

At thirty-six pounds, the Kori Bustard is the largest bustard in the southern region. It is common only in game reserves, occurring as single individuals or in groups. When breeding, pairs are evident. The male has a distinctive courtship display. He balloons out a throat patch covered with bristling feathers. His head is drawn back, with a raised crest, while his tail is held erect and wings droop low. All the while he calls to his mate with a deep seductive *wum, wum, wum, wummmm.*

A medley of clicks and shrill whistles identify another bustard, the Red-crested Korhaan. Its calls are familiar sounds in summer on the grasslands. This elusive bird will stand in the tall grasses, unmoving, its feathery camouflage successfully concealing it from predators. The signature red crest is seen only in a courtship display which is quite impressive. Pairs call in duet – *wuk, wuk, wuk, wuk,* which rises

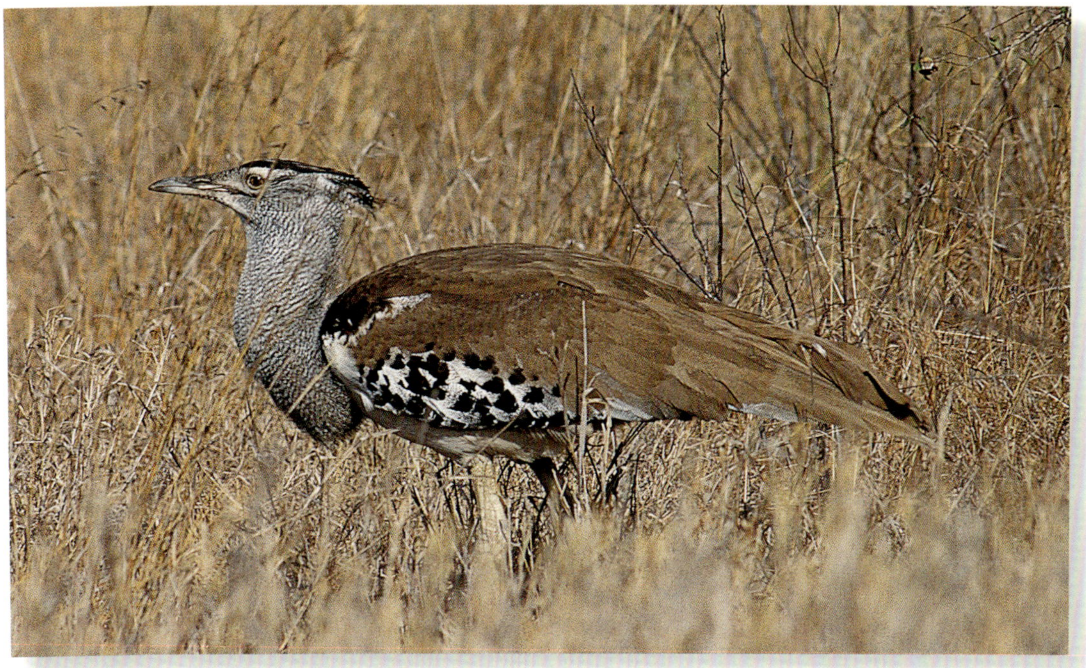

*The Kori Bustard, forages on the ground and is a very large bird, weighing about 36 pounds.*

to a crescendo – *wuka, wuka, wuka, wuka.* Then the male, seemingly out of sheer exhilaration will leap into vertical flight, do a backward somersault and then recklessly plummet back to earth, only to pull up at the last second and gracefully glide to a safe landing. He then struts stiffly toward his intended – red crest upright and showing brightly, apparently proud of himself.

Hornbills are a family of birds with features certainly worthy of praise. They are known for their large bills topped with horny casques. Bold colors enhance bills and skin patches, while plumage varies from species to species on a theme of black, white, grey and brown. The heavy bill requires that the first two neck vertebrae be fused together for strength. They have long necks and short legs, generally. Hornbills nest in natural cavities found in trees, rock faces and earth banks. Most species seal the entrance to their nests, leaving only a small slit through which to feed their chicks.

FOLLOWING PAGES
*Note the large eyelashes on this hornbill; they are actually feathers and not hair.*

The exception is the Southern Ground Hornbill, which prefers an open nest on top of a broken off

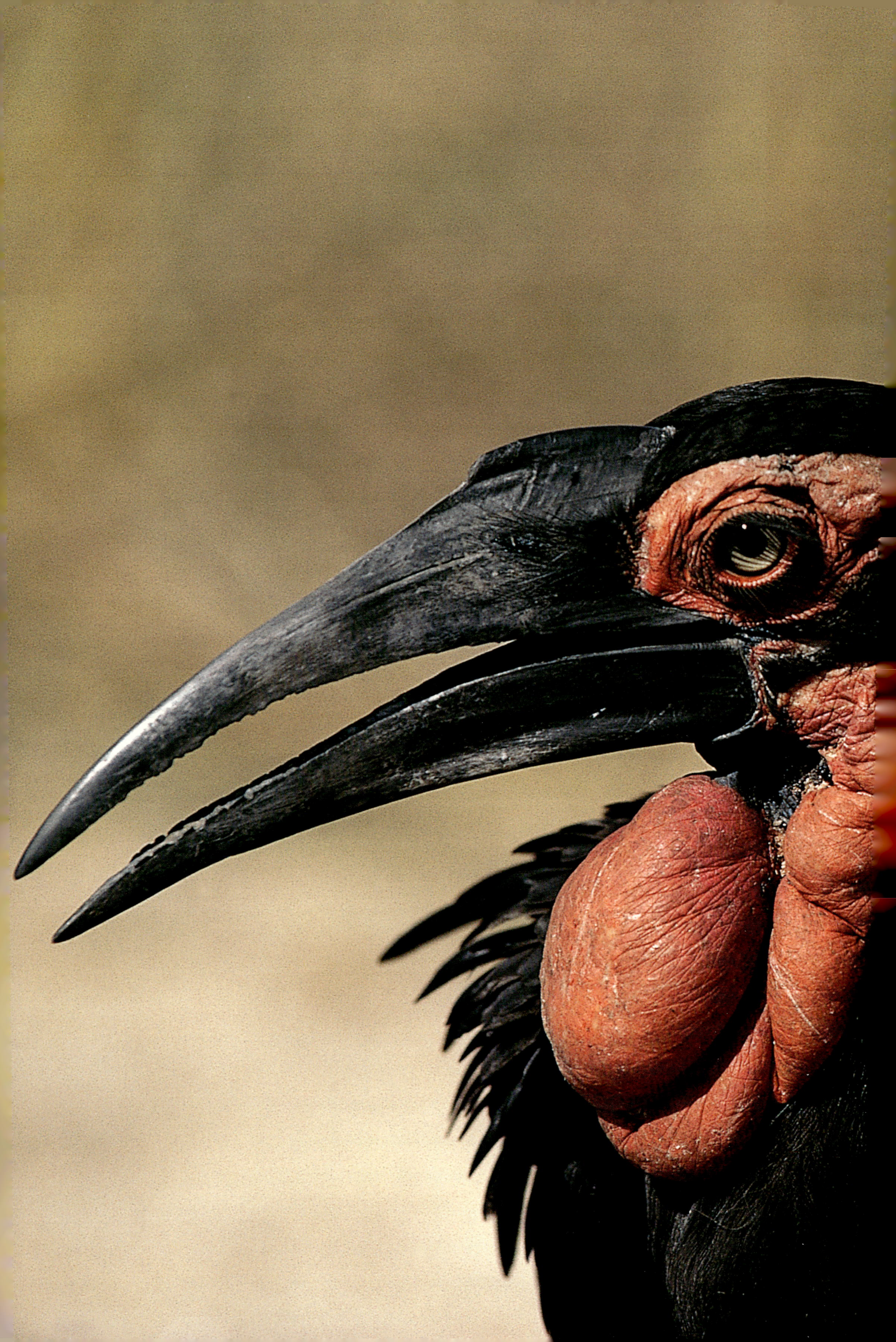

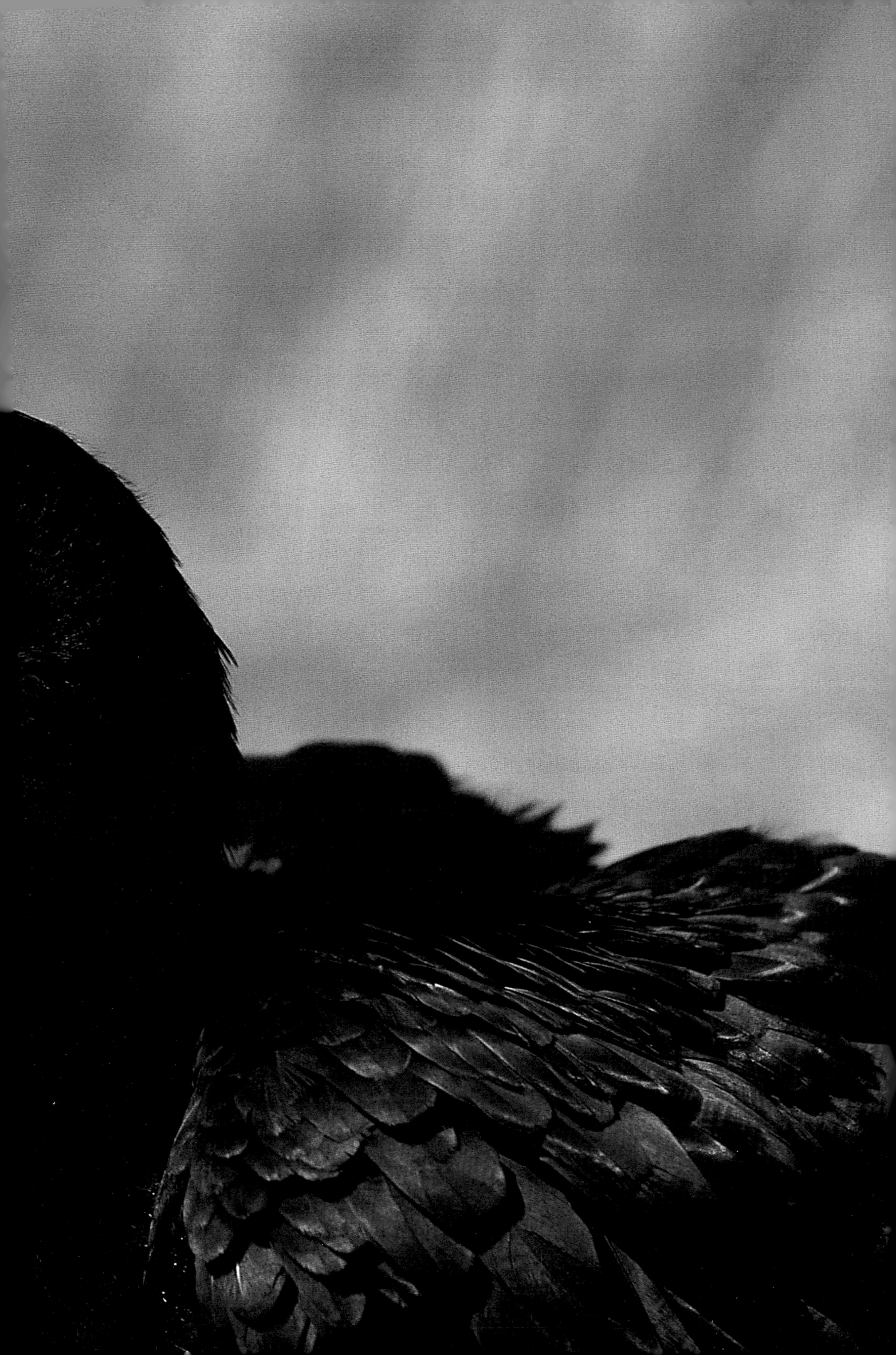

tree. This turkey-sized hornbill is chiefly black with a bright red throat patch that hangs as wattles. The female sports a bit of blue on her throat patch, while the immature has a yellow throat. Three front toes on the feet are partly fused. Ground Hornbills live in groups of up to eight. Strictly carnivorous, they hunt on the ground for insects, snails, frogs, reptiles, (including tortoises) and mammals up to the size of a hare. They communicate with elaborate wing displays and bring each other gifts of food. In early morning or late evening they call to one another. A booming *du, du, dududu* is answered *hu,hu, huhuhu.* When disturbed, Ground Hornbills take off in low flight, their wings making a whooshing sound. Flying is limited to escaping, or for reaching a roost or nest. Pairs produce only one chick at a time which will often remain with its parents for years. Sometimes mature male offspring take on the role of helper for young siblings.

The Yellowbilled Hornbill is one of the majority that seals its nest cavity. The female enters to lay her eggs and seals herself in, assisted by her mate. There she remains as the chicks mature, relying on her mate to bring food to the family. He deposits tasty morsels through the narrow slit in the seal. Sometimes congregating in small groups, the Yellowbilled Hornbills vocalize with a staccato *wuk, wuk, wuk, kok, kok,kok.*

One walking bird is the sole member of its family. The long-legged Secretary Bird is easily spotted as it strides slowly across the savanna. Looking like a cross between a raptor and a stork, this bird is considered a bird of prey. Popular belief holds that the Secretary's name is derived from the quill-like feathers projecting from behind its head which resemble old fashioned quill pens. Actually, Secretary is the anglicized form of the Arabic word *saqr-et-tair,* meaning 'hunter-bird.'

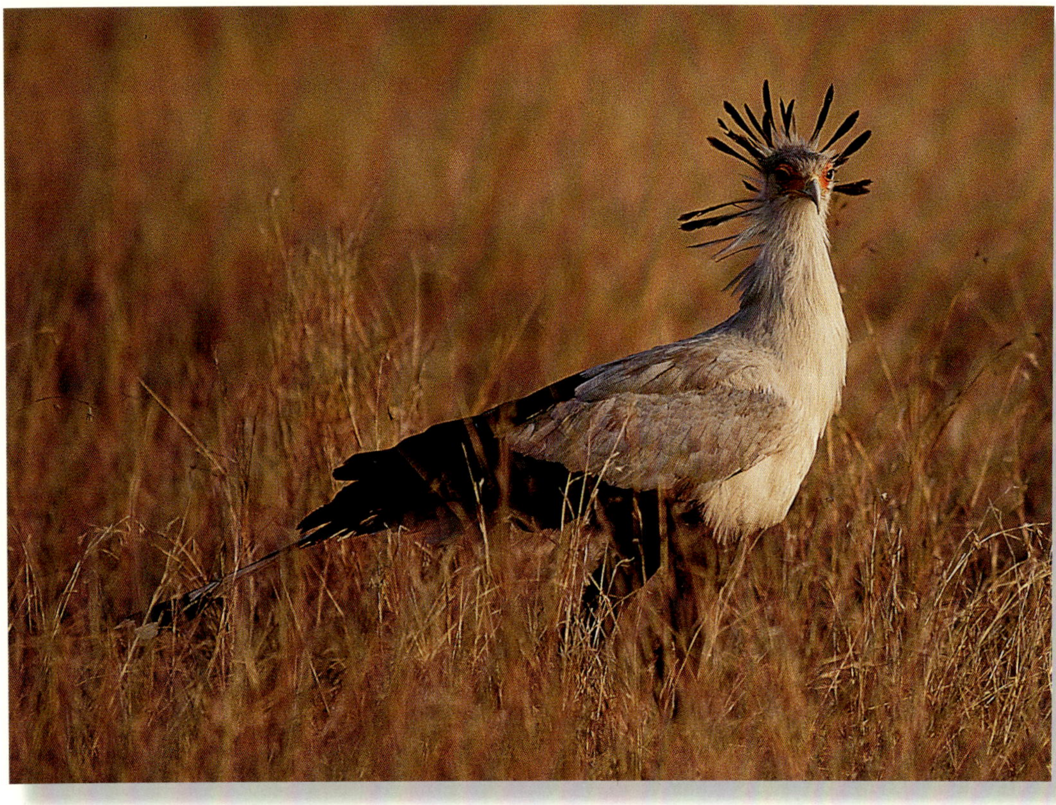

*When seeing the quills on the Secretarybird, it is easy to imagine a nineteenth century secretary using them to pen a letter.*

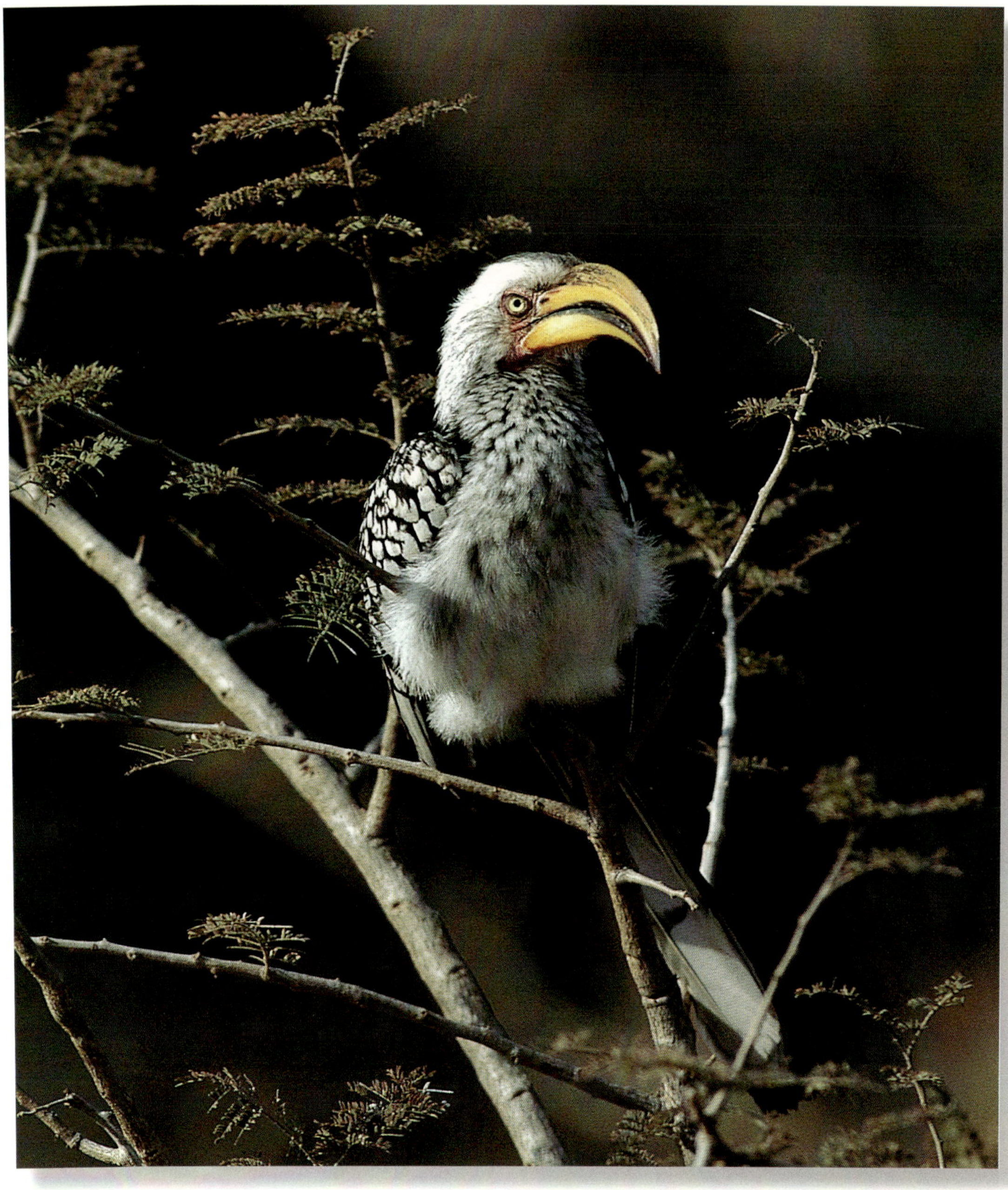

*While perched on a branch, this Yellowbilled Hornbill searched for insects.*

At about four feet, the Secretary Bird is very tall. Its pale grey mantle, black wing quills, rump and thighs, contrast with a colorful orange and yellow face. They are often seen in pairs or small groups. Large congregations can be found at water holes. Secretary Birds are usually silent but do occasionally vocalize with a frog-like croak.

They walk intently through grassy plains, searching for prey; insects, small amphibians, reptiles, young birds and eggs. Favorites are grasshoppers and rodents, which they usually catch with their bills. Often, they kill snakes for food by stomping on them with their strong feet. Most of their time is spent on the ground, but they do fly well and can soar in stork fashion. In the evenings the Secretary Birds' tall silhouettes decorate the branches of thorny bushes and acacia trees where they roost.

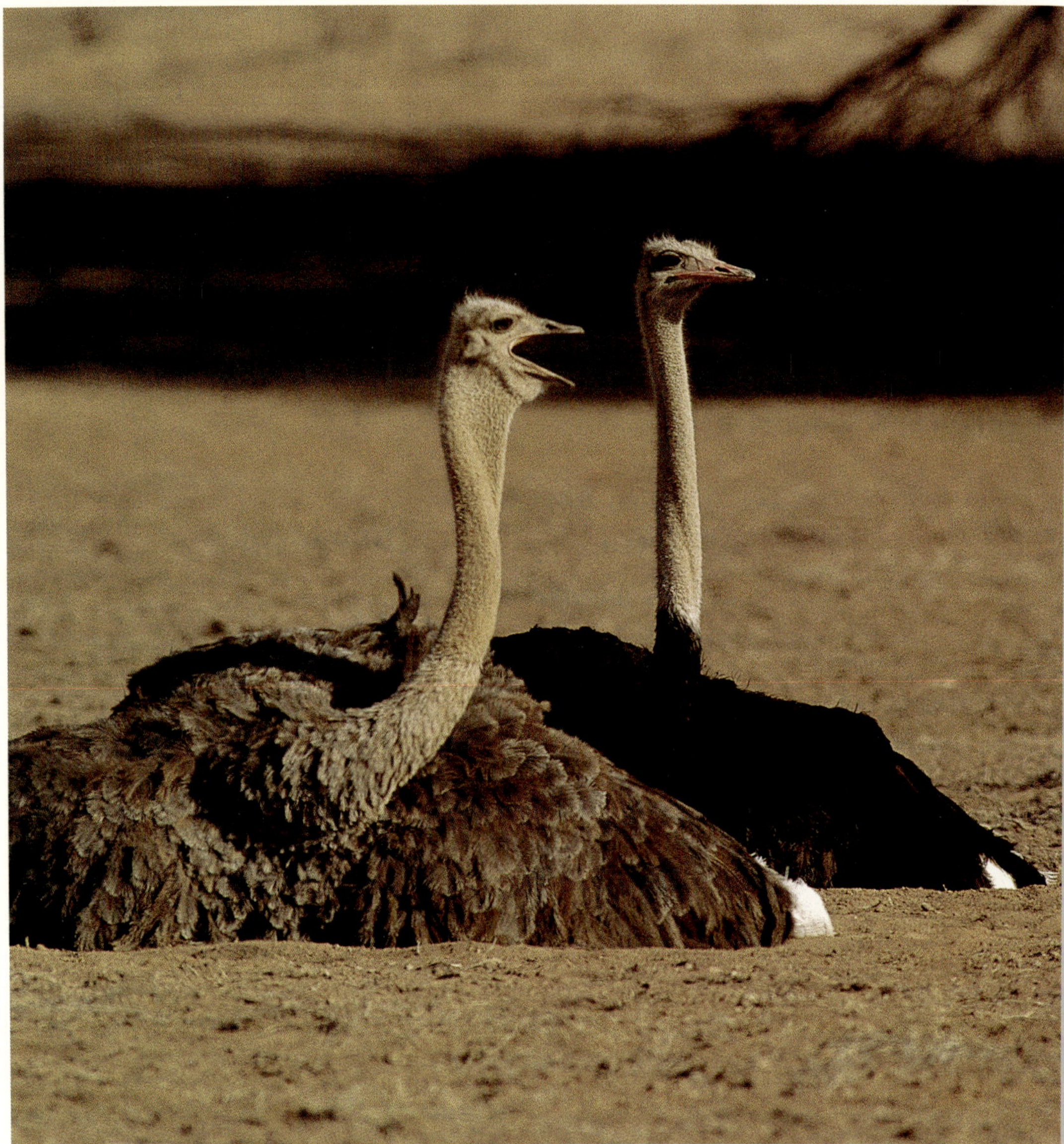

As tall as the Secretary Bird may be, it can't match the height of the ultimate walking bird – the ostrich. Holding the record for largest bird on earth, a male Ostrich towers over the bushveld at nearly seven feet tall, and weighs in at over two-hundred fifty pounds. His striking black and white plumage is highly visible at a distance. Brown and off-white feathers on juveniles and females allow for good camouflage. Long bare legs and two-toed feet enable swift running at speeds up to thirty miles per hour. The Ostrich is a member of a group of flightless birds called ratites. Ratite is from the Latin, *rata*, meaning 'raft', which describes the shape of the breast bone. The flat raft shaped bone is not structured for attachment of muscles needed for flight. And, Ostrich wing feathers are fluffy rather than barbed like those of flying birds. But the wings are not useless. At full running speed, the Ostrich uses its feather-duster wings like sails to change directions.

ABOVE
*Ostrich will often sit in the dry river beds of the Kalahari in order to take a dust bath.*

Physical resemblance to the camel, the long neck and enormous eyes framed with lengthy eyelashes, plus its ability to live in arid regions, have given rise to the name 'camel bird.' The long periscope-like neck and keen eyesight enable the Ostrich to see far – to watch for predators like wild dogs, hyenas and big cats. While running is the Ostrich's best defense, the strong legs also serve to deliver powerful kicks when necessary– blows that can cripple or kill even a lion. Burying its head in the sand is no defense, and is not, in fact, a true behavior.

Ostriches are found in pairs or in mixed groups of adults and juveniles. They walk tirelessly across the bushveld, foraging for succulent plants, grasses, berries and seeds. Small reptiles and insects are also to their liking. Ostriches have a graceful and lavish courtship ritual. The male squats down and alternately waves his wings. The female, in turn, lowers her neck and quivers her wings. They form a loose pair bond most of the time, but in dry years a male may have a harem of two or three females.

Hens will share a nest, a bare scrape in the sandy soil on the vast savanna. Each female lays three to eight large, thick-shelled eggs. Nests holding sixteen to twenty-three eggs are not uncommon. To the untrained eye, all the eggs look alike, but Ostriches recognize their own eggs. A dominant pair will guard the entire nest for the thirty-nine to fifty-three day incubation period, favoring their own eggs. Hatching produces well-developed precocial chicks. They must be closely watched by the adults and be protected from raptors and ground predators. Even with that, only about fifteen percent survive to adulthood, which can last forty years.

Modern theory holds that, eons ago, when extinction claimed the dinosaurs, ancient relatives of the Ostrich took their place. Perhaps that mighty legacy is echoed in the booming call of the male Ostrich. After nightfall, intermingled with voices of predator and prey alike, a deep leonine roar can be heard. Is it the king of beasts? Or, is it perhaps, an Ostrich?

BELOW

*The size of an ostrich becomes very evident when compared to these perched starlings.*

FOLLOWING PAGES

*Namibia's Etosha Pan is the setting for this lone ostrich. It is hard to believe that any animal can sustain life in such a forlorn looking area. However the pan is surrounded by enough food and water for animals to thrive on. During very heavy rains the pan actually fills to a depth of about three feet.*

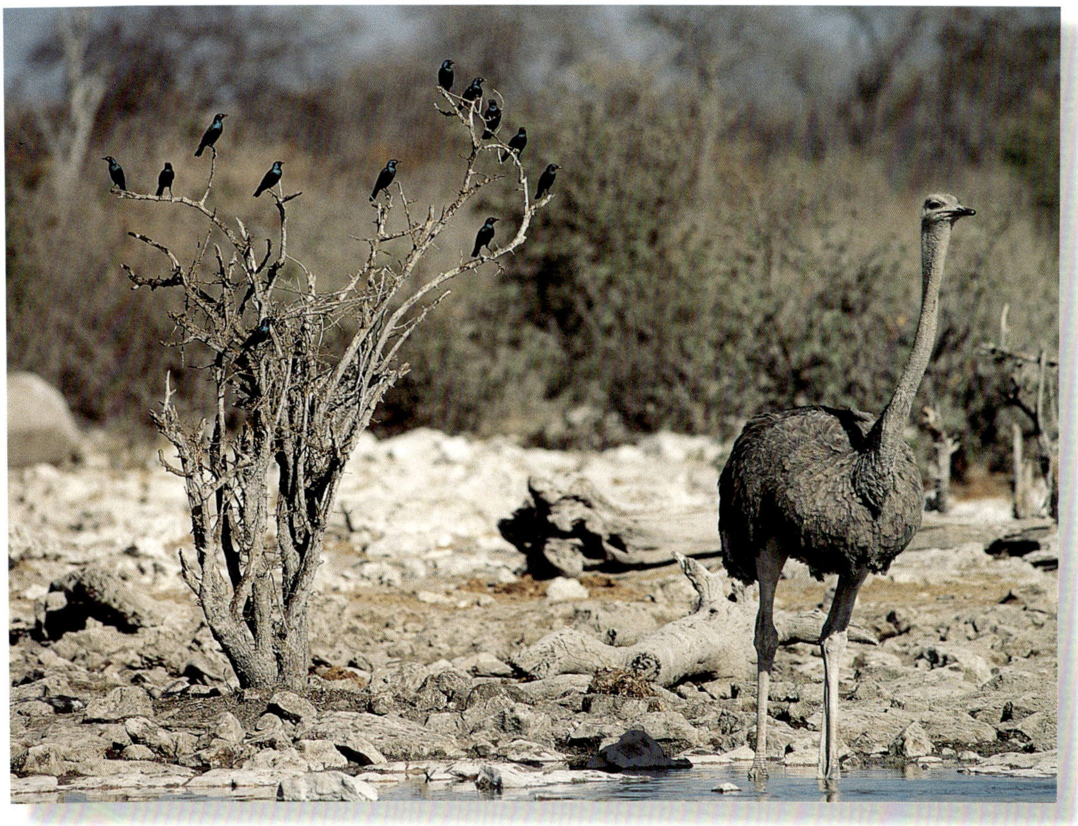

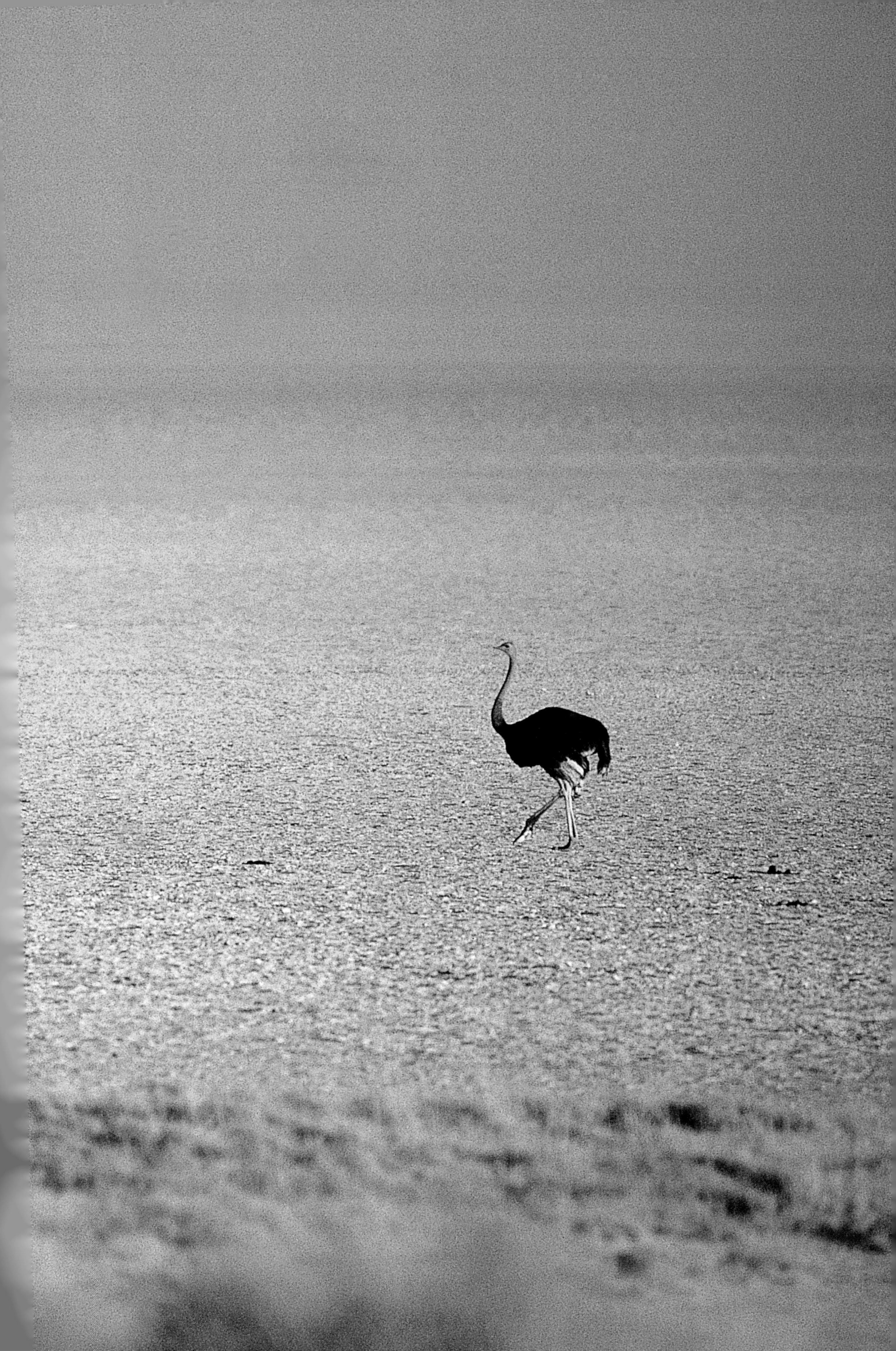

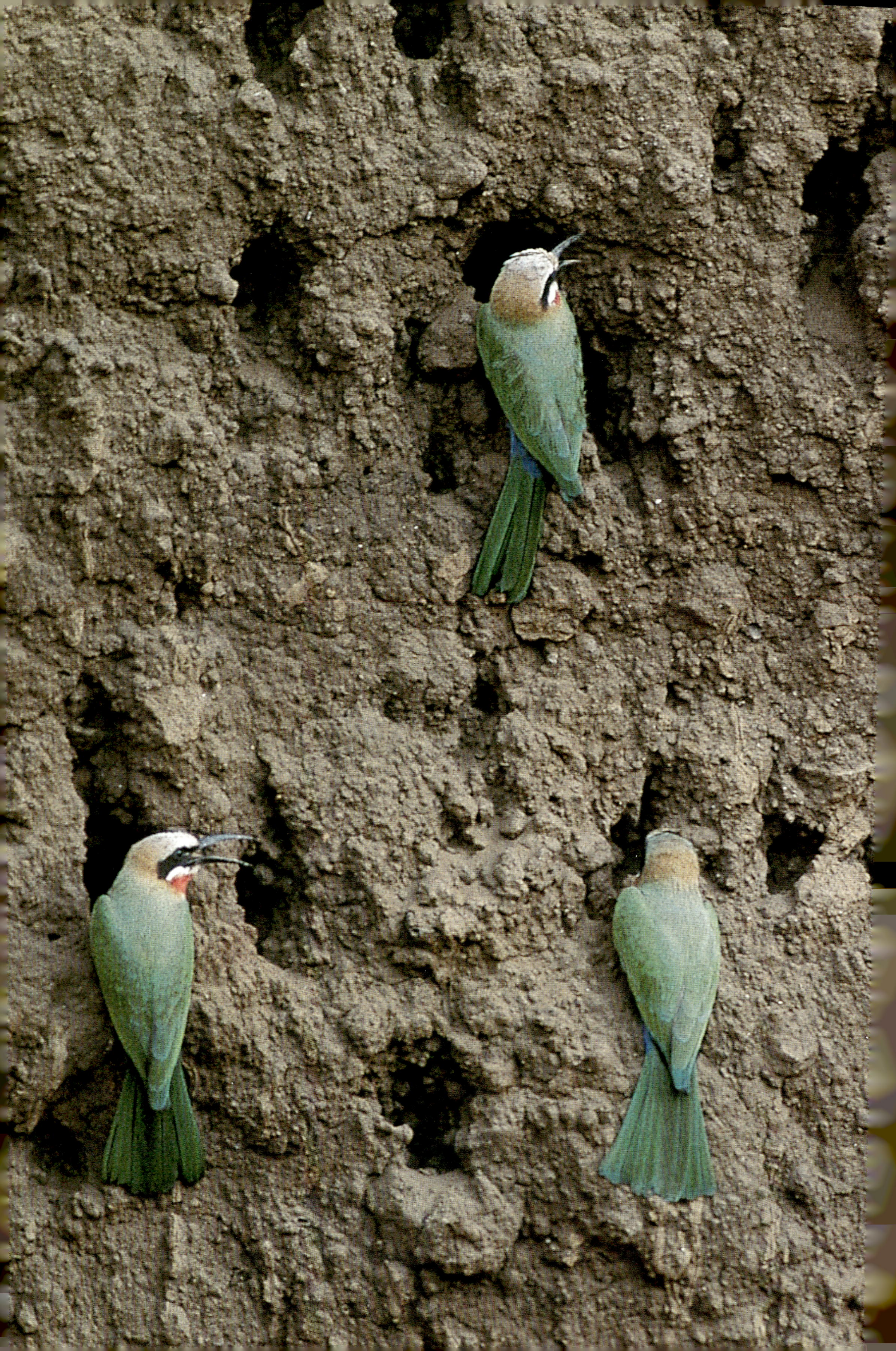

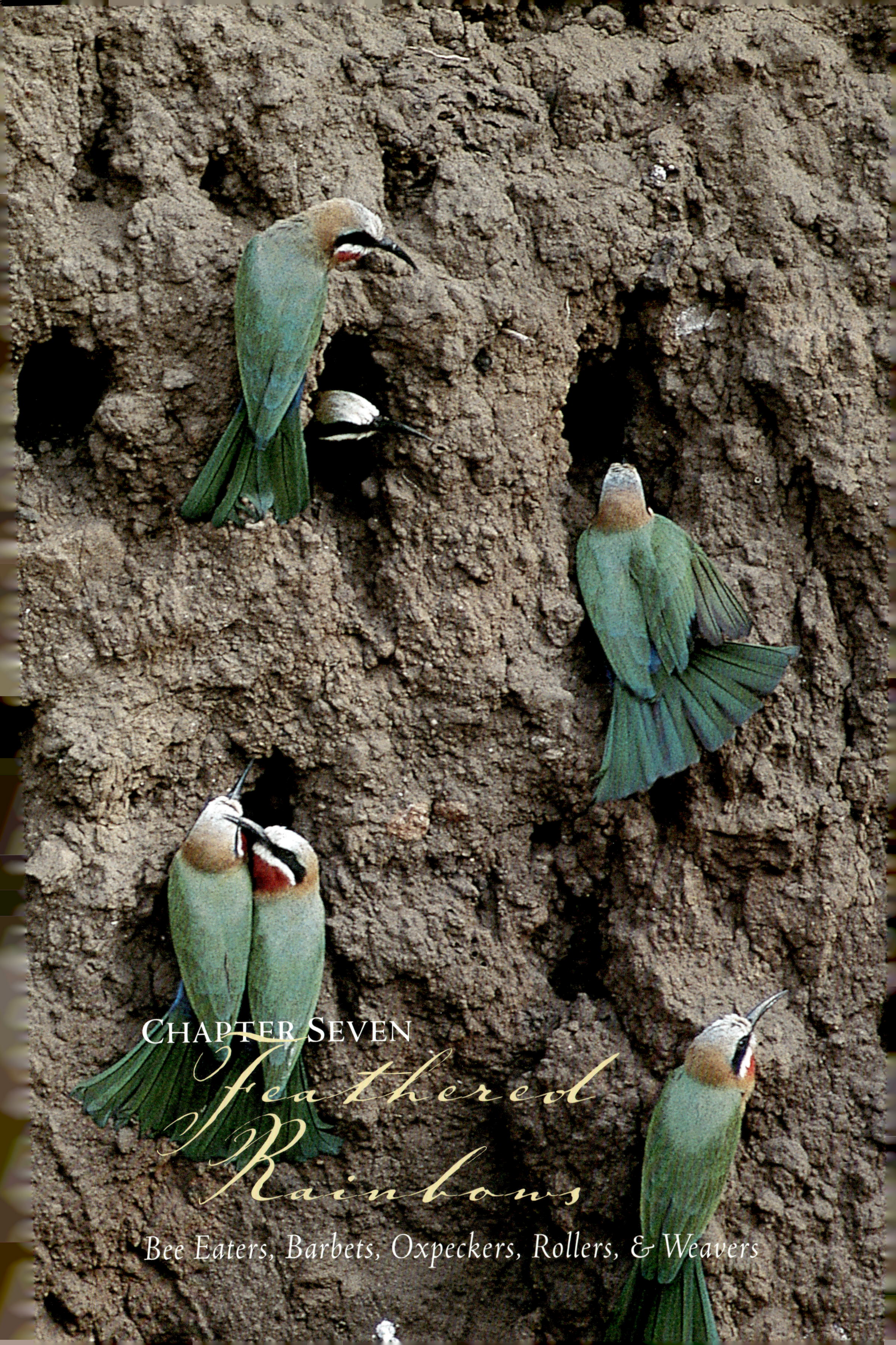

Chapter Seven

# Feathered Rainbows

Bee Eaters, Barbets, Oxpeckers, Rollers, & Weavers

Amid the trees and thickets of the bushveld perch an assortment of rainbow-hued birds, decorating the branches like holiday ornaments. They feast variously on seed and fruit, bug and butterfly. Their nests are architectural marvels crafted from earth, lined with hair or woven like baskets. Their calls are a curious mix – as delightful to the ear as their plumage is to the eye.

The White Fronted Bee-eater likes to sunbathe – its bright green back to the sun, feathers raised. The complementary crimson and white throat shows off a black eye patch and bill. The long curved bill is specialized for catching flying insects. While the several species of Bee-eaters are known for eating Honey bees, the White Fronted has a taste for butterflies – which are caught during graceful aerial maneuvers.

PRECEDING PAGES

*The holes in this donga are the bee-eaters very unusual nests. A donga is a gully created by erosion and these holes can measure up to three feet deep.*

FACING PAGE

*Whitefronted Bee-eaters are especially "fond" of butterflies.*

Bee-eaters live near lakes and rivers that have steep sand banks called *dongas* where they create their specialized nests. Burrows really, the nests are tunnels excavated horizontally into the banks. A donga can be a neighborhood for dozens of Bee-eater nests. Each tunnel ends in an oval chamber where single eggs are laid at one or two day intervals. After hatching, the chicks are fed by both parents, assisted often by helper birds. After fledging, young birds stay with their parents for several weeks.

Bee-eaters are active and normally solitary by day. At night they gather together to roost in trees and thickets, twittering and calling – *quirk, quirk, quirk, qurrrrr.*

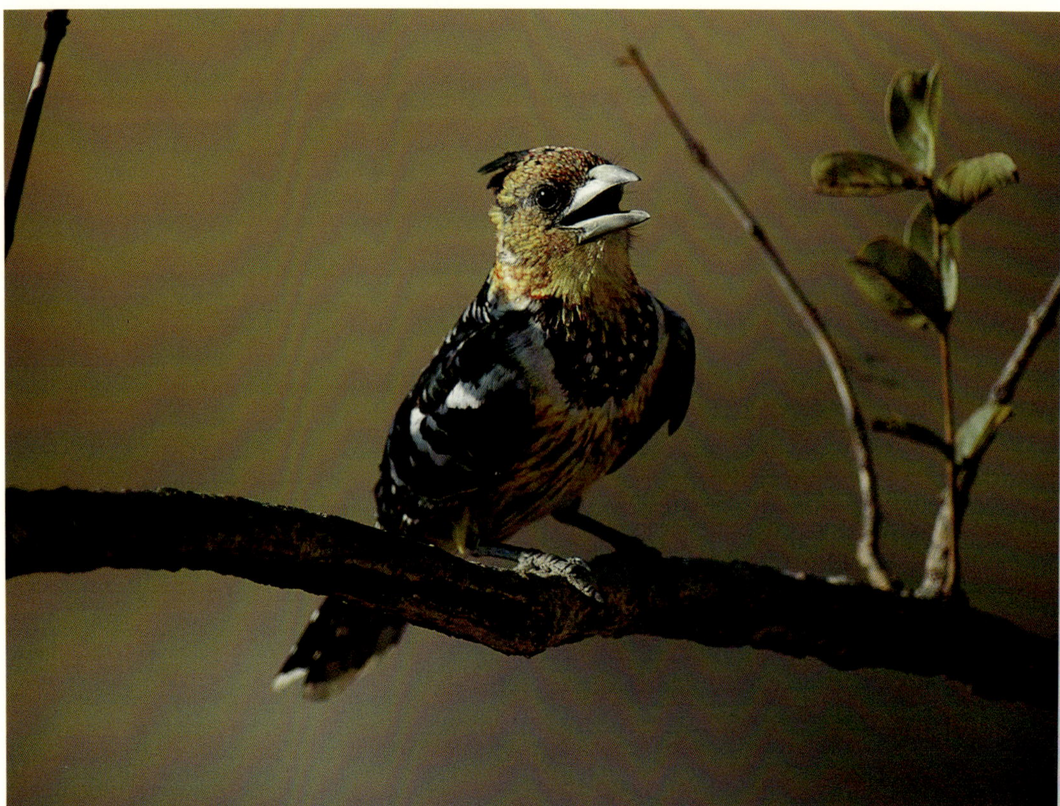

*The Crested Barbet has been dubbed "the fruit salad" bird because of its multi-colored markings.*

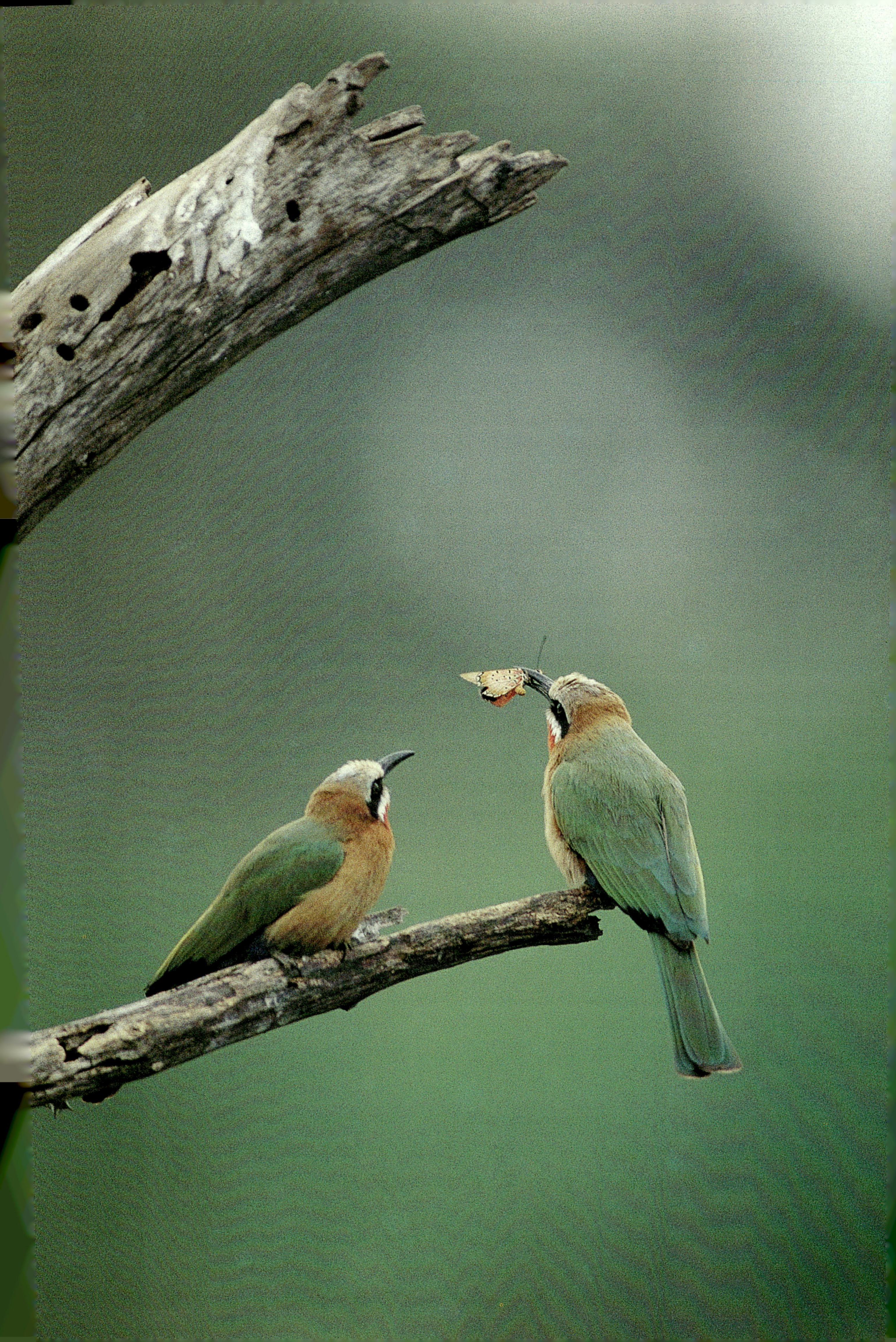

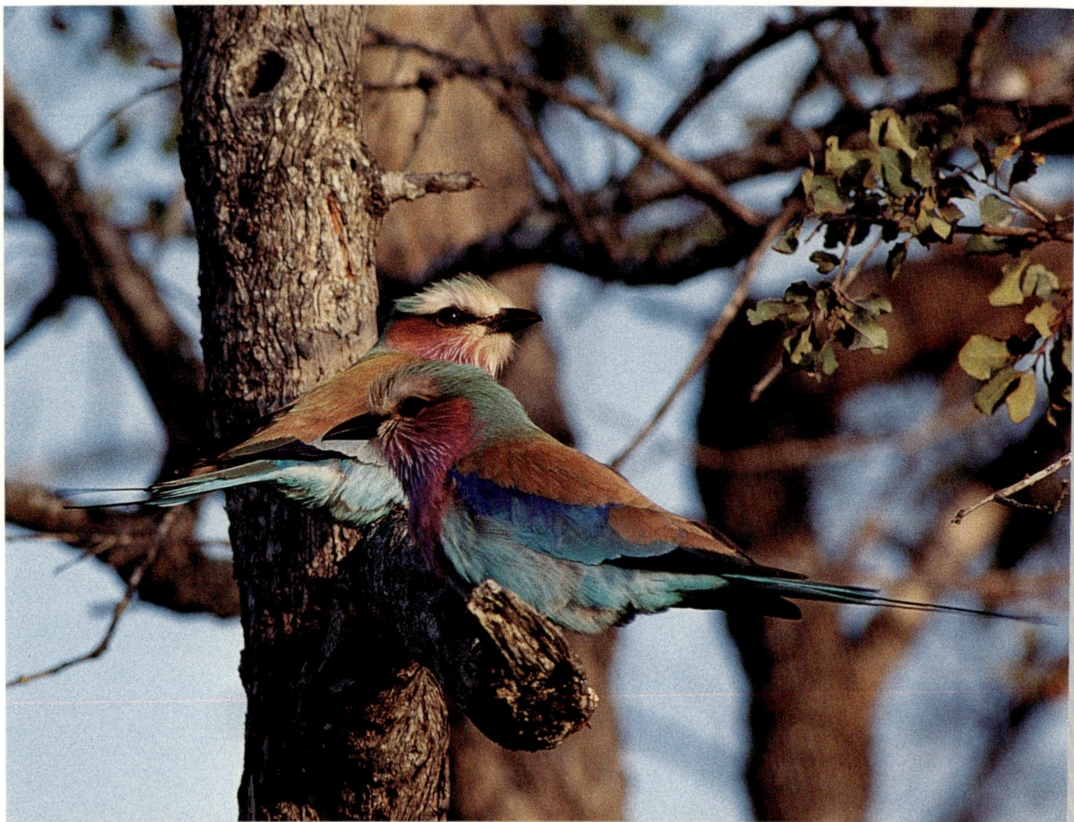

*Lilacbreasted Rollers are a common resident in the area.*

Barbets are robust, colorful birds known for their noisy chatter. Honks, chirps, and hammer-tapping sounds are delivered in rapid repetition. They are often heard calling in duet, a style which is used to announce territory and to reinforce family bonds.

They have large heads and stout conical bills with sharp tips. Like their woodpecker relatives, they climb trees with short strong legs, using their tails for support. Also like woodpeckers, they nest in holes excavated in dead, decaying trees. Barbets are found singly or in pairs. They live in wooded areas, often near water, and they aggressively defend their territory, chasing out other birds.

The Crested Barbet is a medium size bird at about nine inches long. One of the most colorful Barbets, it sports a red and yellow face, a red rump, yellow belly with red streaks, and greenish legs and feet. Its back, tail and wings are black with white patches and its head is topped with a shaggy black crest. When hungry, this multi-hued beauty hunts for termites, worms and snails. Birds eggs are sometimes taken as well as sweet fruits. The male Crested Barbet's distinctive call sounds like a muffled alarm clock – *trrrr, trrrr, trrrr;* to which the female answers – *puka, puka, puka.* 'Barbet' for *shut it off!*, perhaps.

The Starling family includes a member bird most often found perching on the backs of large grazing mammals. The Red Billed Oxpecker finds its meal of parasites such as ticks, and horse flies on the hides of giraffe, rhinoceros, hippopotamus, buffalo, antelope, and other hoofed mammals. This handsome, eight inch, brown and buff bird has red eyes surrounded with yellow and a pointed scarlet bill. The oxpecker is always on the lookout for burrowing ticks while scampering over its host beast, adept at pecking out the pesky parasites. The mammal, in turn, is relieved to be rid of the itchy pests.

Red Billed Oxpeckers are common in game reserves where there are plenty of hides to be gleaned. The hosts also provide more than food for the birds. Oxpecker nests, which are housed in natural tree cavities, are made up of 90 percent animal hair. Inside the soft, warm nest, the female lays eggs any Easter bunny would prize – pinkish-white, speckled with red, blue, purple and lilac.

Although most of the oxpecker's time is spent on its host animal, often roosting there at night, it will fly to roost with a flock of its kind in trees or reed beds. In flight, it will call a shrill scolding *chrrr, chrrr* and a staccato *tzik, tzik, tzik*.

Flying is the particular talent of birds called rollers. Swooping, looping, wheeling, tumbling – the roller displays aerial excellence for defending territory and in courtship flight. Rollers even mate in the air. Loud raucous calls complete the display.

One species, the Lilac Breasted Roller, is large at fourteen inches, and is particularly colorful; an artist's palette of bright blue, lilac, light brown, and green. Although these jewel-toned birds are often seen perched on roadside telephone wires, they usually inhabit open woodland and savanna. At dusk, this roller sits atop bushes or trees, scanning the ground for its evening meal. Locusts, beetles, scorpions and caterpillars are favorites. Amphibians, and small reptiles, sometimes fit the bill. When prey is spotted, the bird spins into flight, rolling to the ground to pounce neatly on its victim.

Lilac Breasted Rollers are found singly or in monogamous pairs. They nest in natural holes in trees, where the female lays a small clutch of eggs. Offering no clue as to the feathered rainbows they some day will be, the Lilac Breasted Roller's eggs are plain and white.

Elaborate nests are the forte of weaver birds. Woven of sticks or grasses, the nests come in many unique shapes and sizes according to individual species. Most weaver birds build nests which hang suspended on the ends of branches in trees. A nesting colony often fills the treetops with nests. Some choose bushes or reed beds for their homes.

The Lesser Masked Weaver constructs a tightly woven ball-shaped nest of grass, reeds and palm leaf strips. On one side, a short vertical tunnel entrance is added. The nest has a unique prickly appearance with short lengths of grass and reeds sticking out all over. The birds nest safely inside the cozy orb. Colonies of ten to twenty nests can be found in trees, reeds or low bushes, usually near water. At about six inches, the Lesser Masked Weaver is one of the smaller species, but its lovely plumage makes it quite noticeable. A breeding male has a green back with black streaks, yellow under parts and blue legs. The head contrasts with a black mask and yellow eyes. Females lack the mask and are paler. Both call with a raspy swizzling sound. They prefer to live near water in savanna, bushveld and acacia woodland and they forage in the canopy of trees for nectar, seeds and insects, especially caterpillars. Because they probe flowers, their faces are often covered in yellow pollen, which only adds to their colorful plumage.

It is said that the best gifts come in small packages. Southern Africa's trees and thickets are full of small gifts – feathery gems to watch, hear and enjoy– revealing, as well, that it doesn't have to rain to see rainbows.

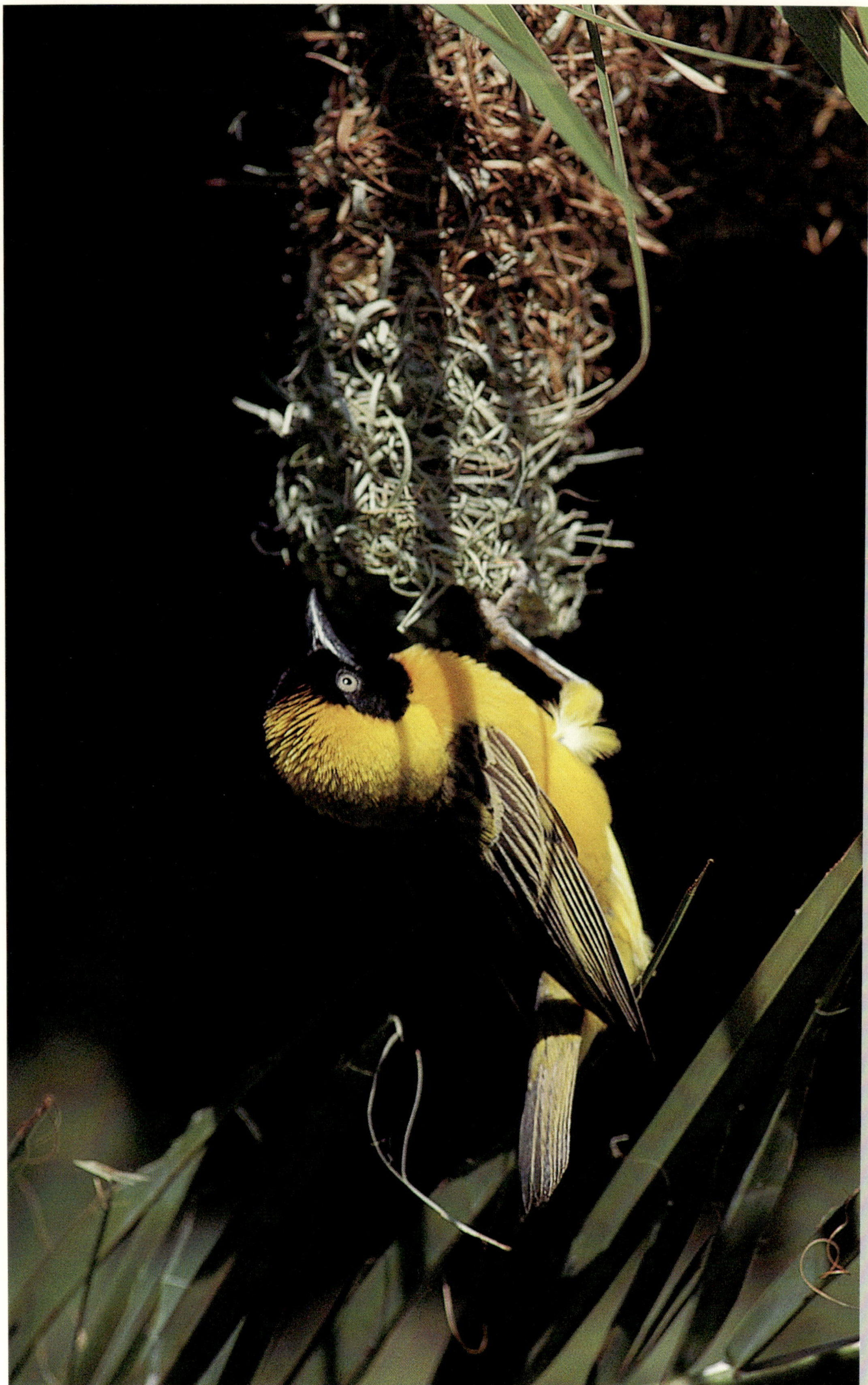

*A Lesser Masked Weaver must enter its nest through a tunnel.*

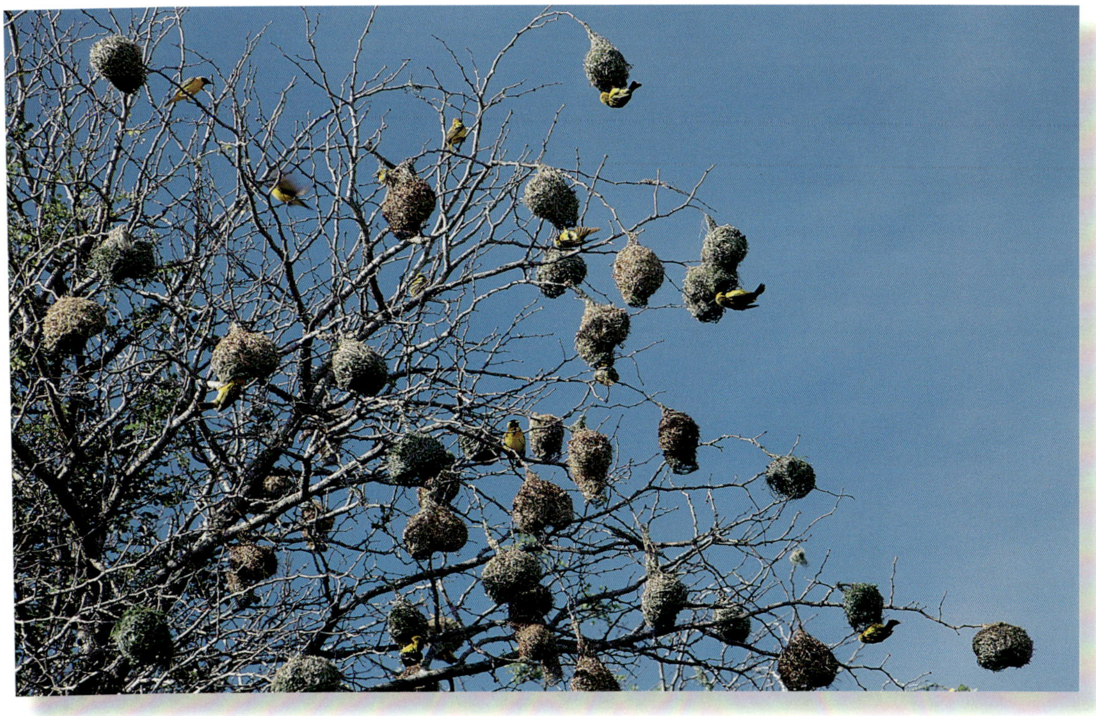

*Weavers build their nests on the edges of thin branches which support the nest but will not support predators.*

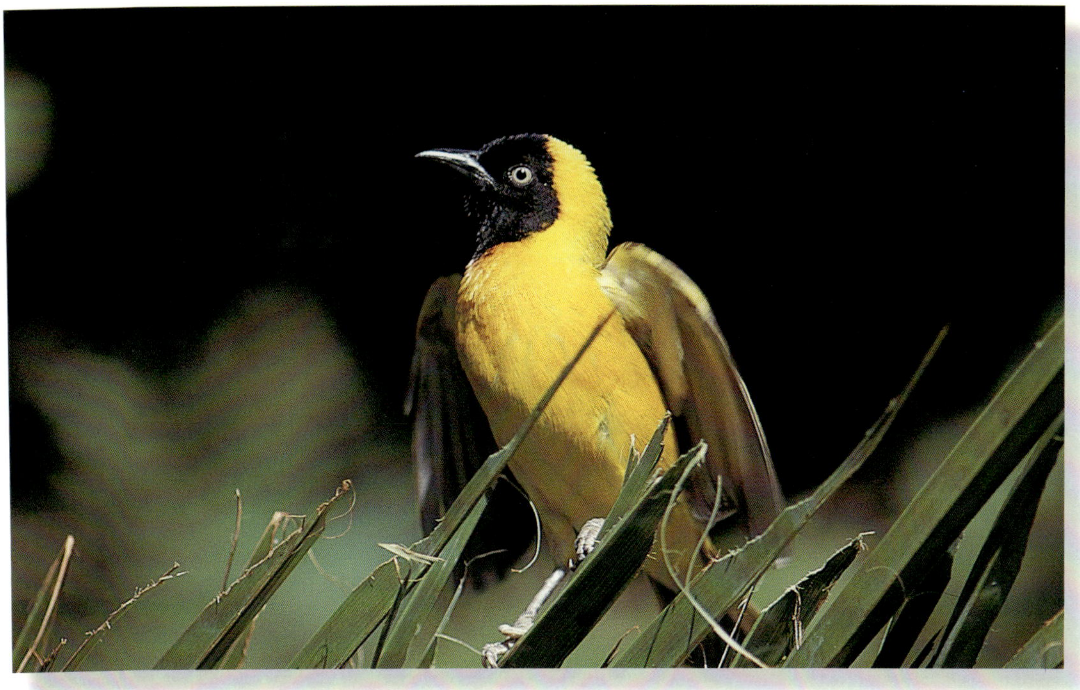

*A Lesser Masked Weaver measures about six inches long and is normally found near water.*

FOLLOWING PAGES

*A giraffe and an oxpecker often live in a symbiotic relationship. One forages for parasites, while the other benefits from getting rid of them.*

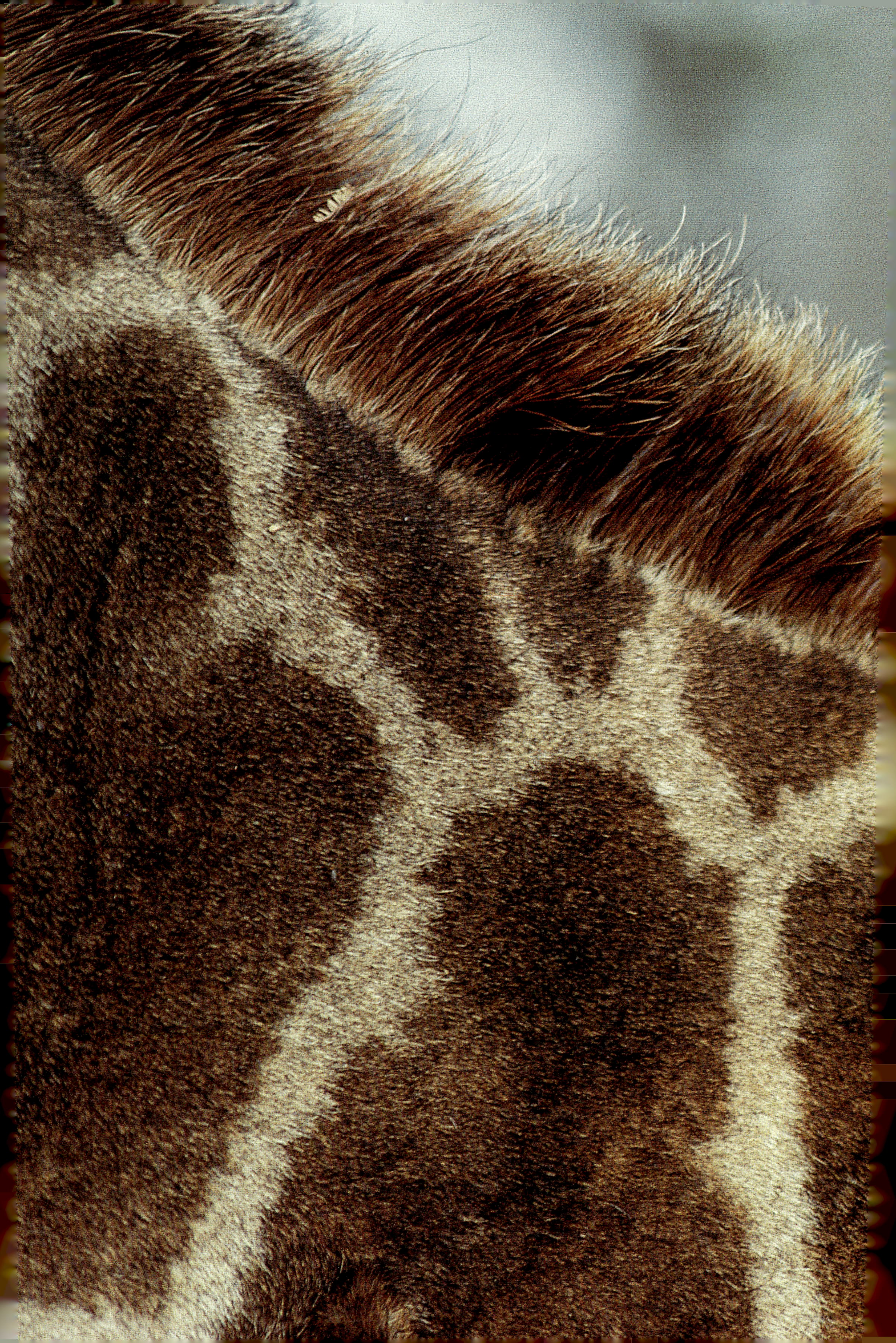

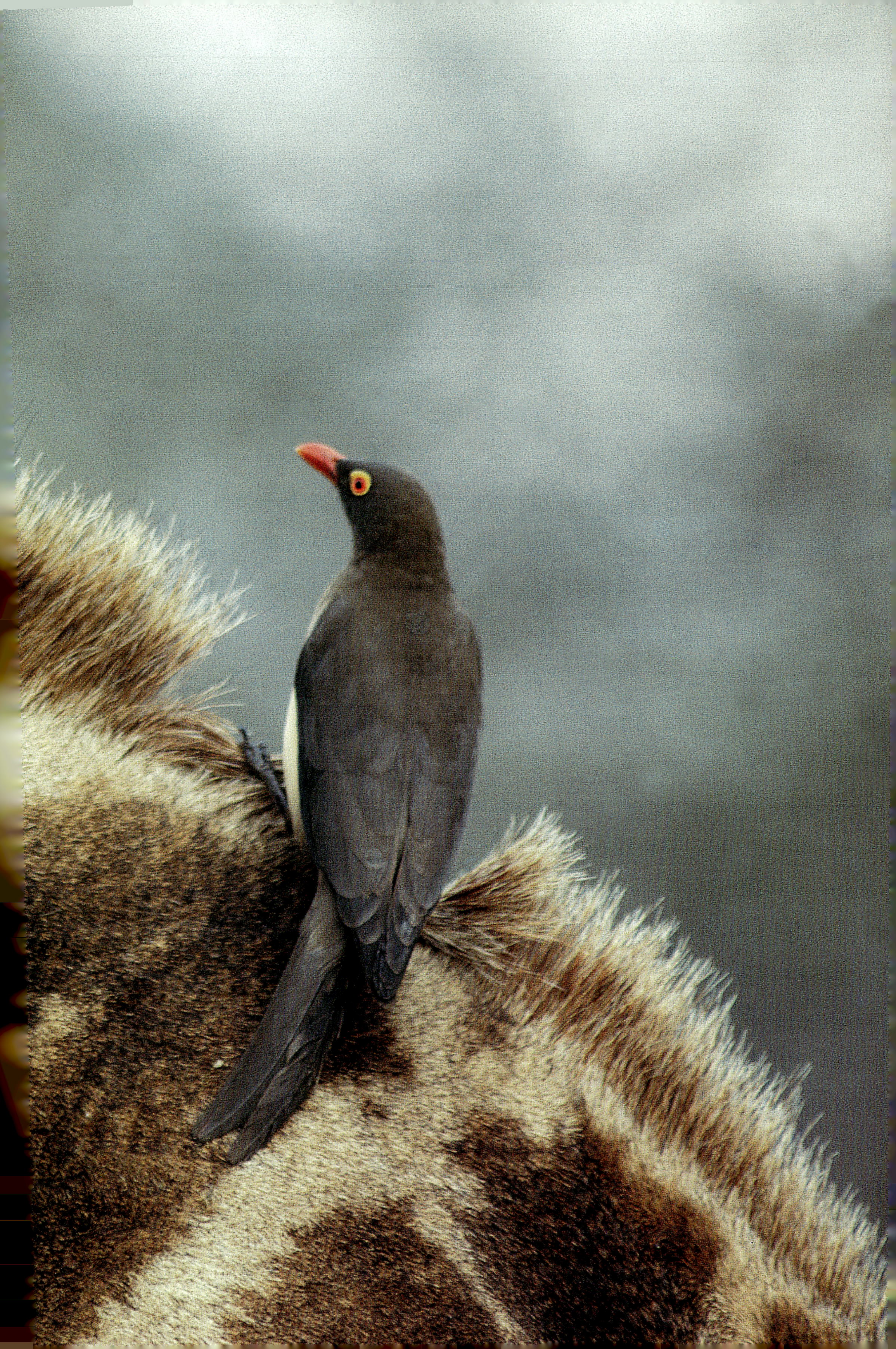

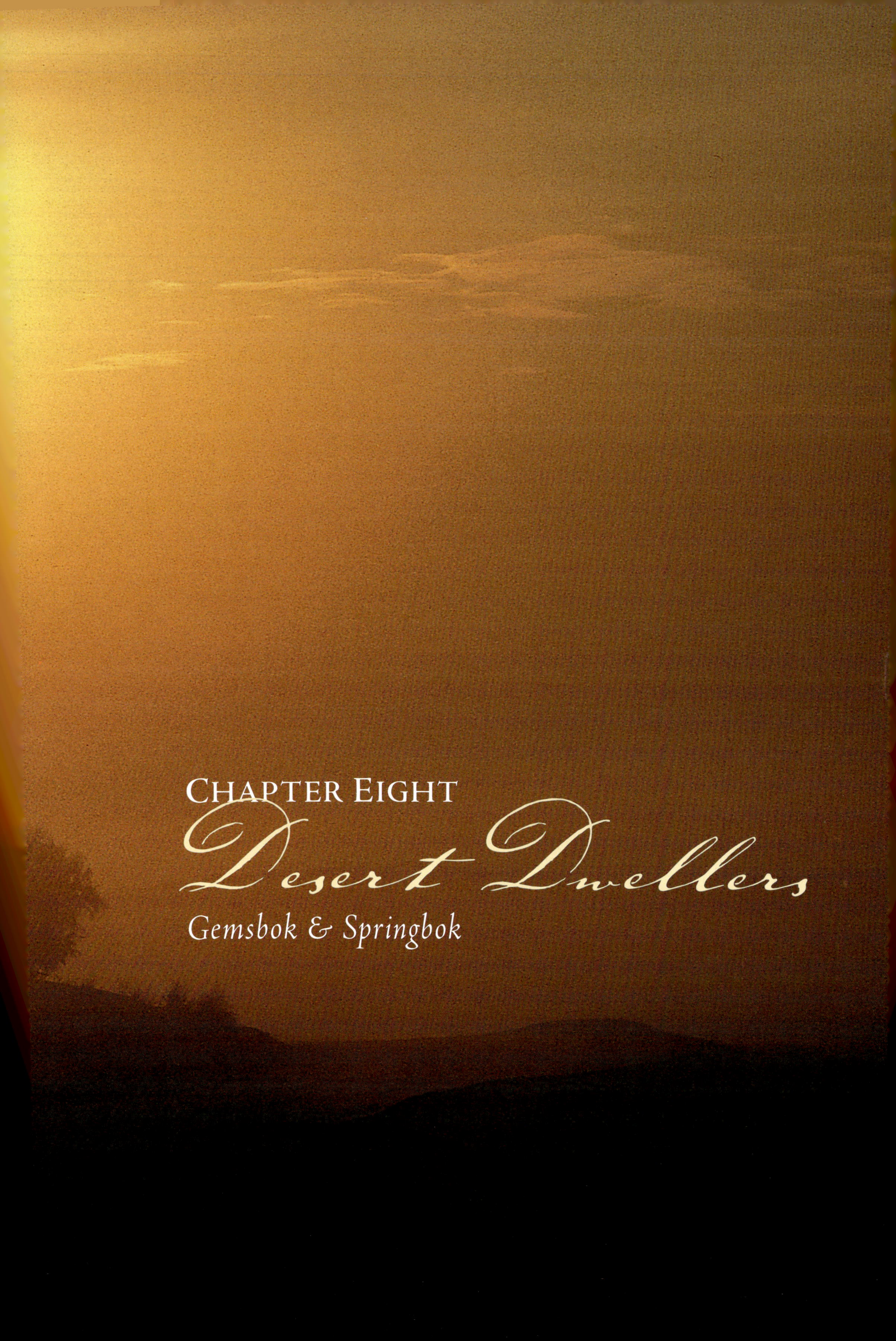

# Chapter Eight
# *Desert Dwellers*
## *Gemsbok & Springbok*

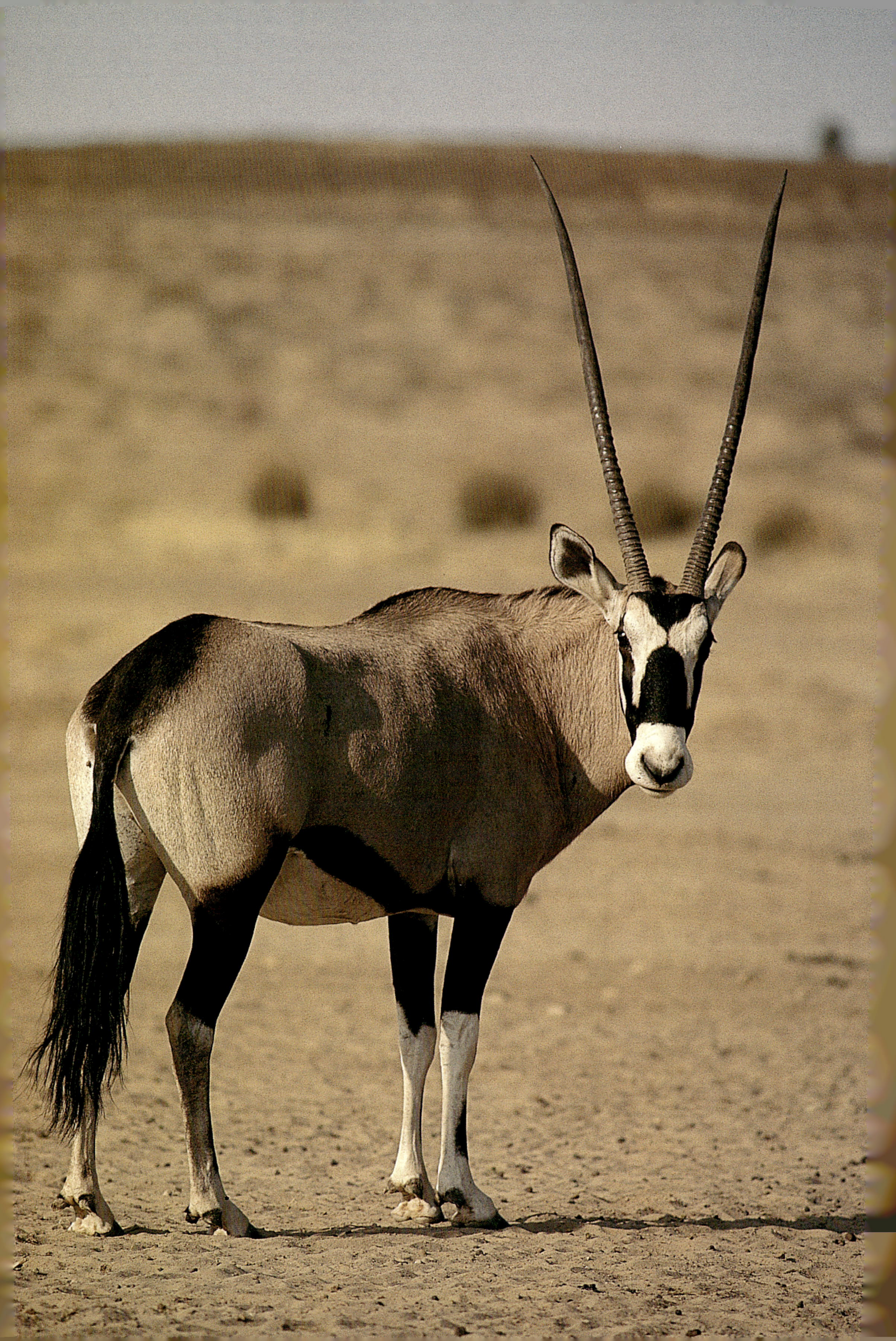

The desert lands are home to creatures that must endure both fiery temperatures and drought conditions. Two such desert dwellers are the gemsbok, an antelope, and the springbok, a gazelle. These bovids are very different looking but have many things in common; traits that allow for survival in a very forbidding environment.

PRECEDING PAGES

*Dust and sand are something we all must endure while in Africa. However, a sunrise with sand hanging in the air can act as a creative photographic filter.*

FACING PAGE

*The gemsbok is one of the most beautiful desert animals in the world.*

Both animals can go without water for long periods of time. Instead of drinking, they obtain moisture from the food they eat; sweet grasses, succulents, and fruit such as wild melons and cucumbers. They also use their hooves to dig up roots and bulbs in the driest season. Plants need have only ten percent moisture to sustain these herbivores. They avoid the heat of the day, preferring to be active from dusk to dawn, feeding and moving when temperatures are lower.

Each animal has evolved mechanisms for metabolizing water efficiently. The gemsbok can function with a raised body temperature so that it does not sweat in the desert heat. When it runs, panting cools blood vessels that send blood to the brain, keeping its temperature down. The gemsbok also instinctively follows rain, trekking over sandy desert, finding patches of newly sprouted plants in the wake of recent precipitation. They search out natural springs and salt licks as well. The springbok is capable of drinking water with a high mineral content – water that most other animals cannot stomach.

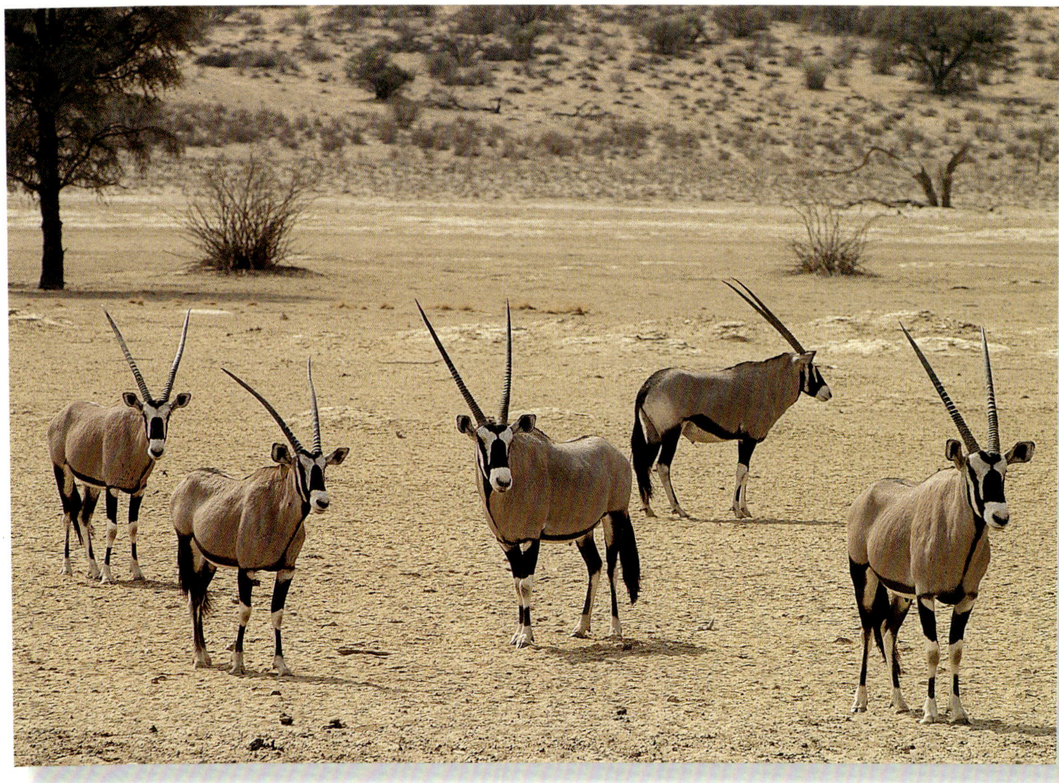

*When a gemsbok stands in profile, its horns merge and it looks like a one-horned animal. Perhaps this is where the legend of the unicorn originated.*

Both gather in small herds during the dry season. When food is in short supply, they range further

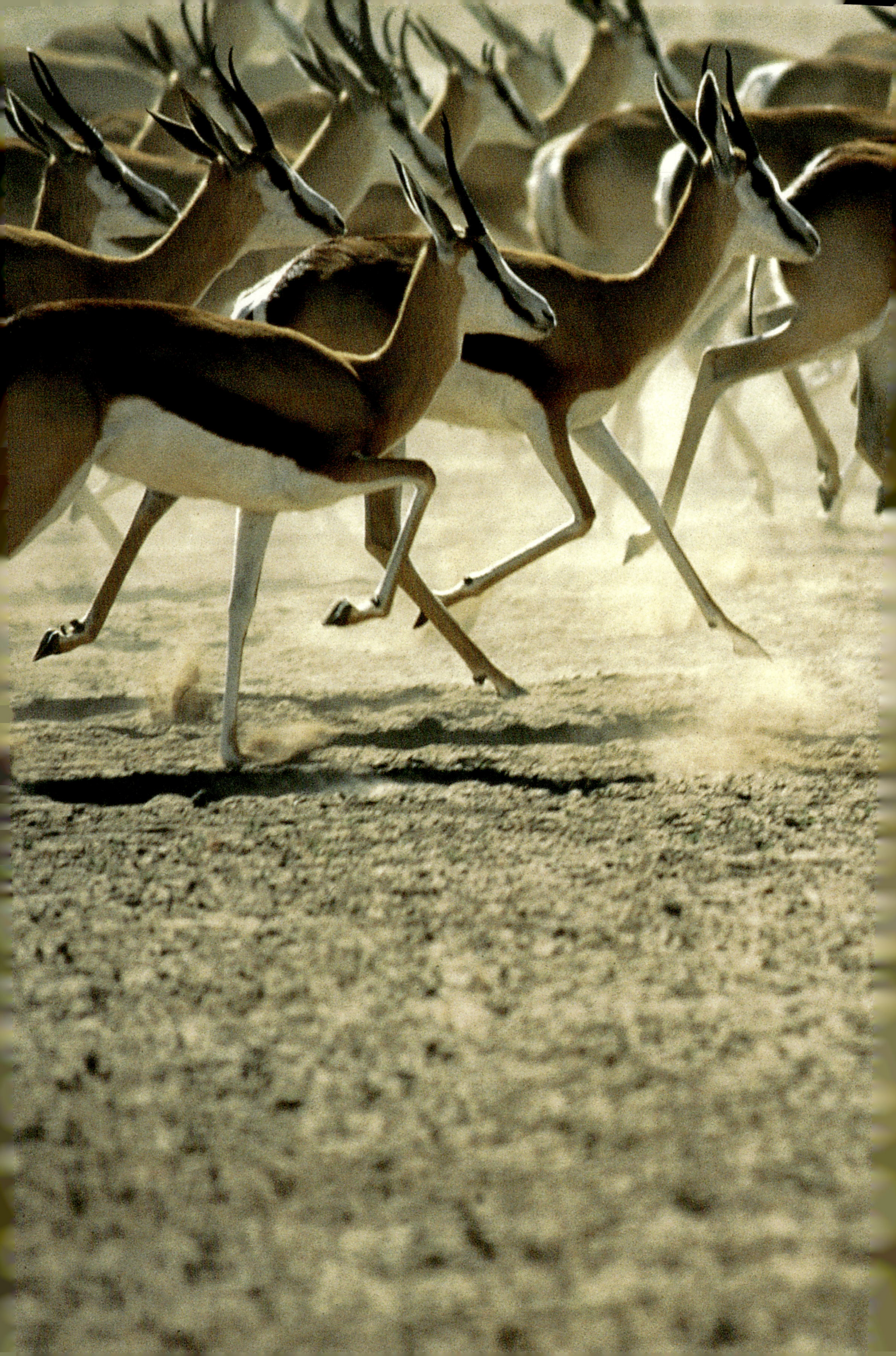

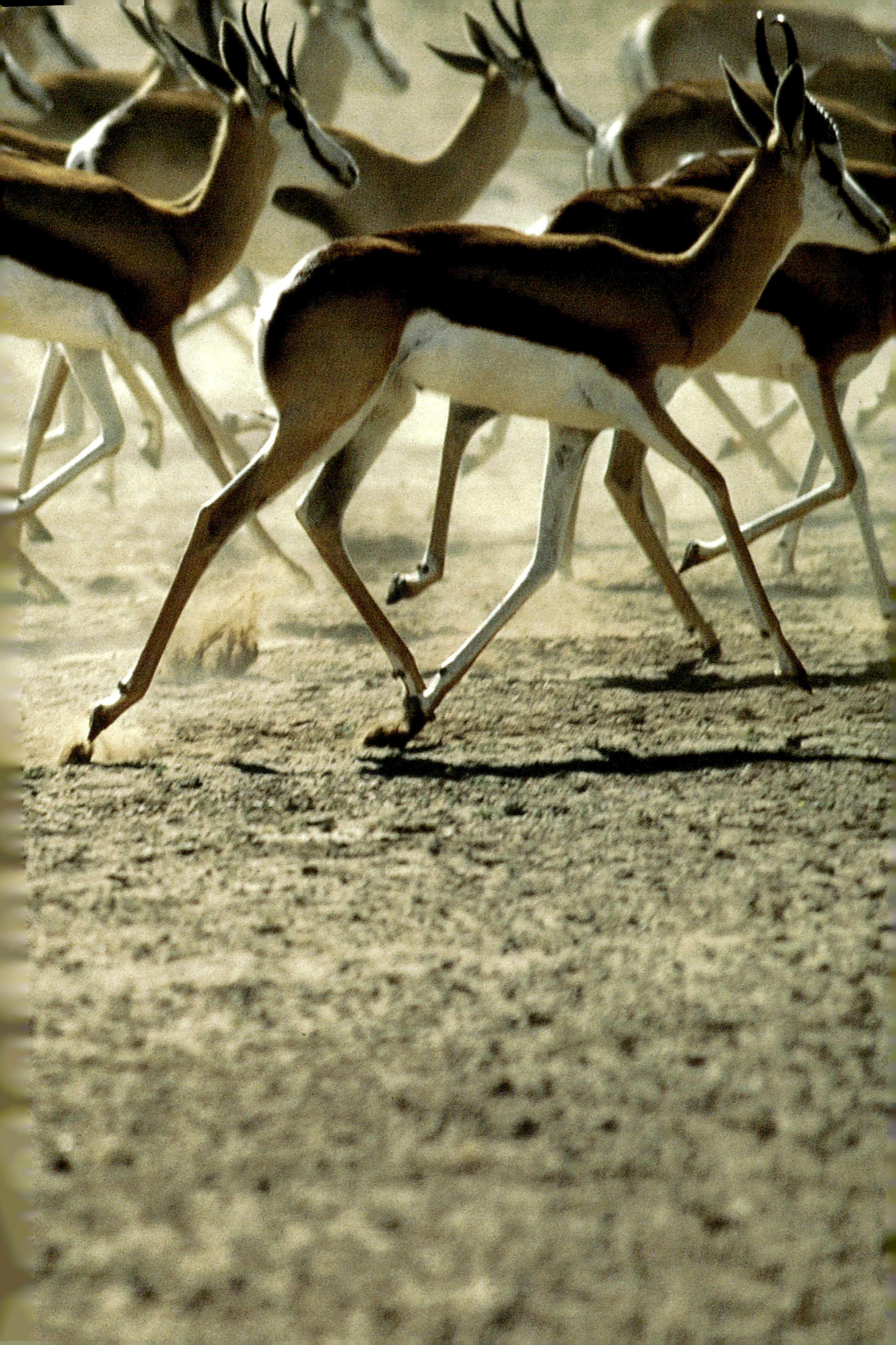

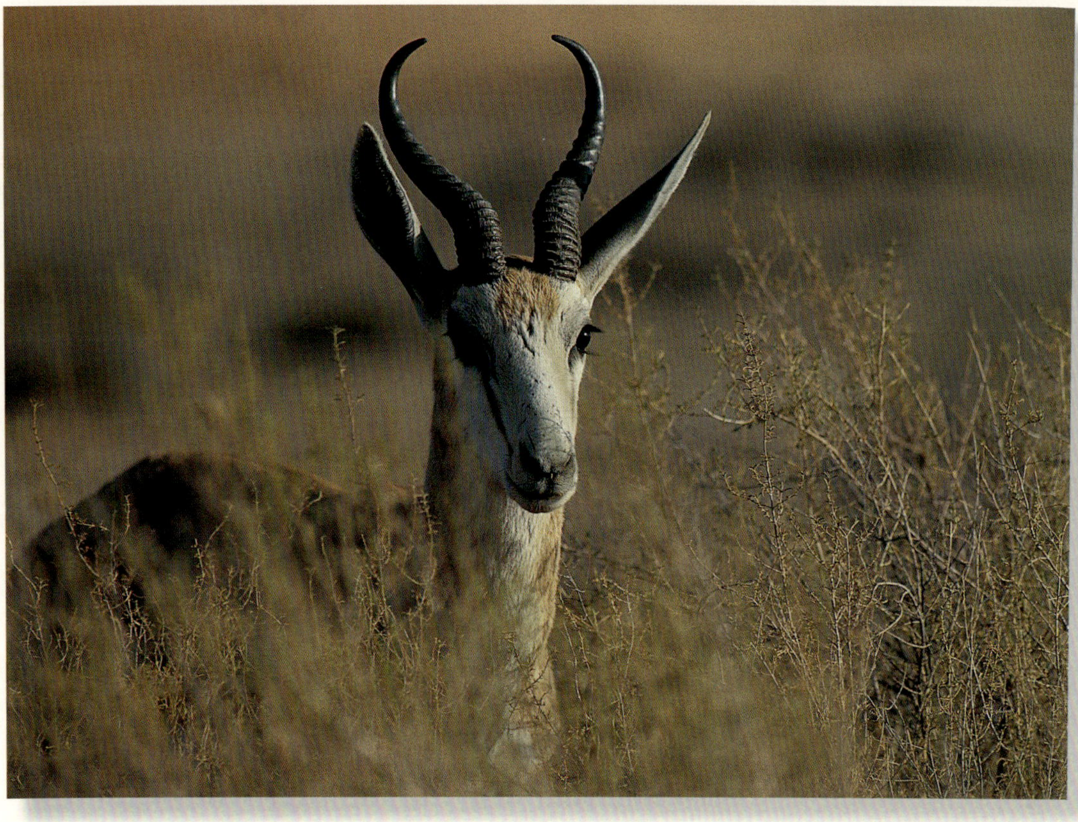

*The springbok can subsist without drinking as long as food plants contain at least 10% water.*

apart so that there will be enough forage to last until the rains come and fresh vegetation is replenished. Then, after the rains, both species gather in large herds to share the bounty. The rains leave pools as well, and the herds will take advantage of a cool fresh drink while they last.

PRECEDING PAGES

*The setting is a dry river bed in the Kalahari Gemsbok Park, where the Auob River flows only about once every one hundred years and after extremely heavy rains. Jan and I were able to see this springbok herd trotting toward us. We wondered what caused them to move so quickly and after they passed us we moved into the direction from which they came. Then we discovered why! There were three cheetahs on the opposite side of the riverbed. The springbok had seen them or possibly picked up their scent and were quick to get out of the area.*

Herds are subdivided into four basic units; single males, bachelor herds, territorial males with their harems, and nursery herds comprised of females with young. Both gemsbok and springbok have similar breeding behavior. Territorial males mate with their females, excluding other males. Young are born in spring at the start of the rainy season and both species hide their newborns initially. Springbok conceal their lambs for only a few days. Gemsbok calves are hidden for up to six weeks, during which time their horns begin to grow. Since they are not seen until after that time, there is a mistaken belief that they are born with horns.

Inasmuch as there are similarities between these bovids, there are also differences.

The gemsbok is a large antelope, averaging four-hundred pounds. With its long straight horns and striking black and white stripe-patterned face, it is a most handsome species. The upper hide is fawn grey with a dark horizontal stripe below that continues down the legs to the knees. The under parts and lower legs are white. On the neck, a short ridge of dark-brown mane tapers to a stripe that extends along the spine, spreads to a saddle at the rump and then narrows along the tail, which ends in a flowing horse-like brush.

The springbok is a petite eighty to ninety pound gazelle. Its well groomed coat is bright cinnamon-

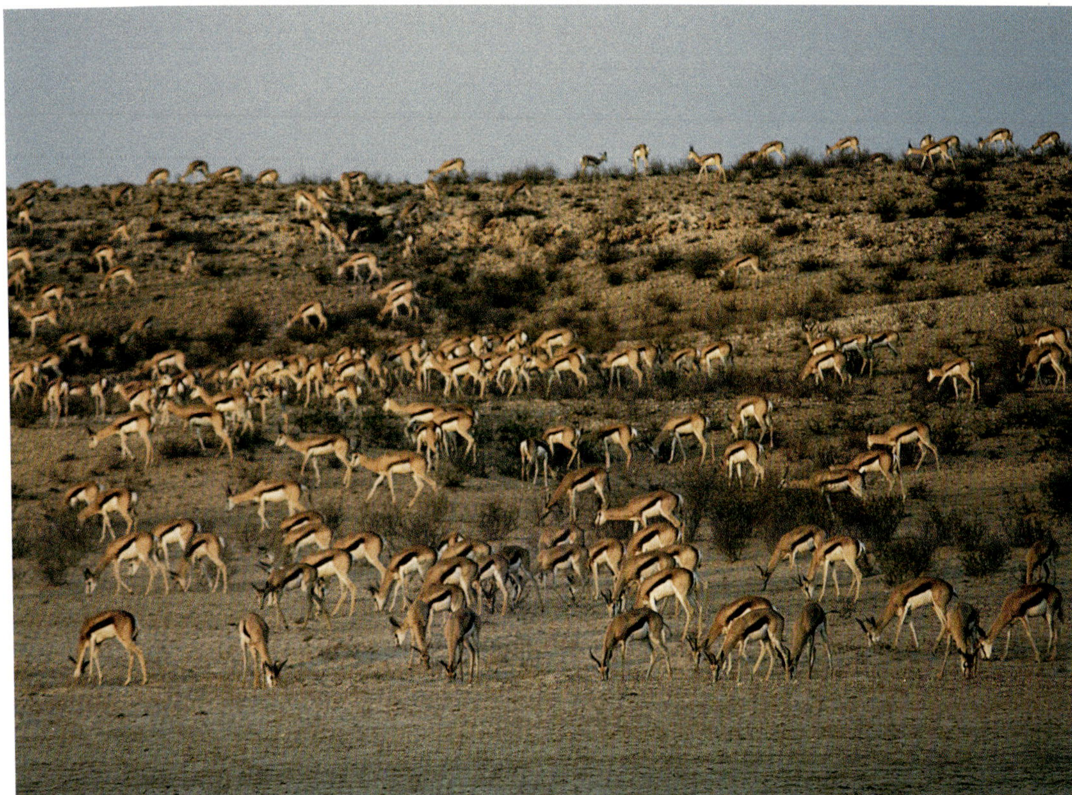

*Dawn's soft light envelops a large herd of grazing springbok.*

brown with a reddish-brown horizontal band extending between foreleg and hip. The under parts are white. Its face is white with a brown line running from eye to nose and atop the head are two medium length horns, curving gently inward. There are sometimes occurances of black or white individuals. The springbok's neck, ears and legs are long and it has a short tail which flicks and quivers continuously while the animal grazes, especially when it is nervous.

A nervous springbok one moment may be a startled springbok the next. It is then that the herd is alerted to flee – demonstrating a behavior unique to the springbok. Called pronking or stotting – from an old Scottish word meaning 'going by leaps' – the gazelles run in a springing manner. With head low, back arched, and long legs stiff, they vault six feet off the ground, land briefly with all four feet at once, and leap again and again. Combined with fast trotting, they can travel up to fifty-five miles per hour. While pronking, their backs display a bristly flag of short white hair, giving them the appearance of some strange type of desert flying fish. Alarm and flight is accompanied by grunting bellows and whistling snorts.

Fleeing from predators is a frequent necessity for both gemsbok and springbok. Lions and Spotted hyena hunt gemsbok. Leopards take their young. Black-backed jackals, Spotted hyena, wild dogs, lion, and cheetah prey on springbok. Their horns, while capable of intimidating rivals, have little effect defending against hungry carnivores. They must depend on running and dodging in efforts to escape.

Nature has provided for these hardy beasts to survive an arid land, intense temperatures and sparse provisions – with adaptive physiology and behavior. But nature provides also for the survival of the predators. The ultimate fate for many gemsbok and springbok is under fang and claw – as prey. It is nature's way – survival of the fittest beneath the indifferent desert sun.

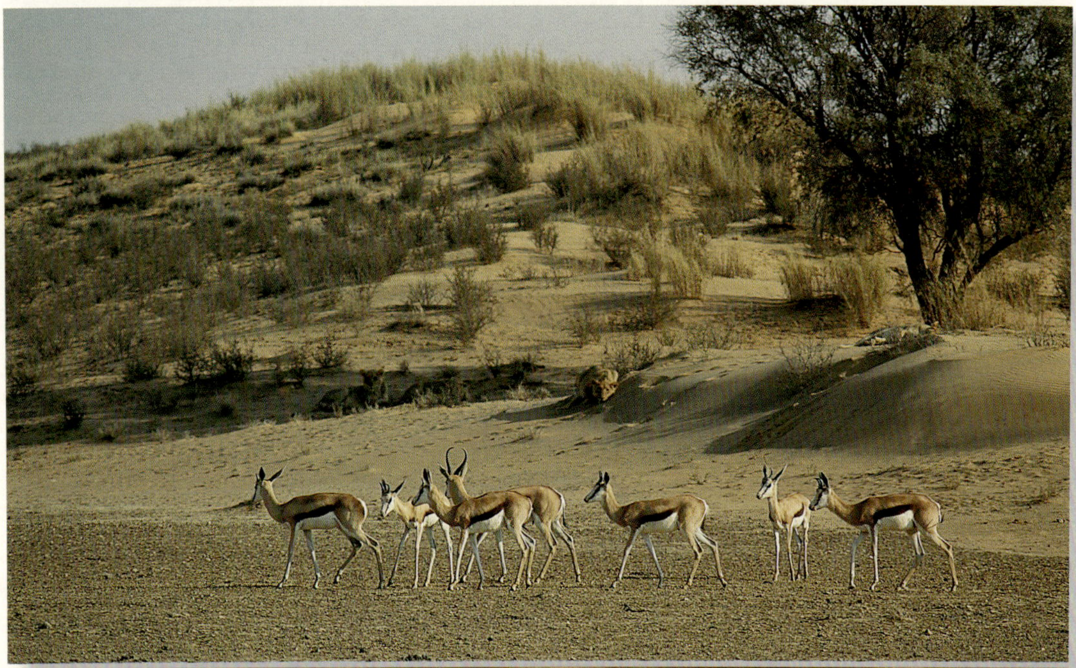

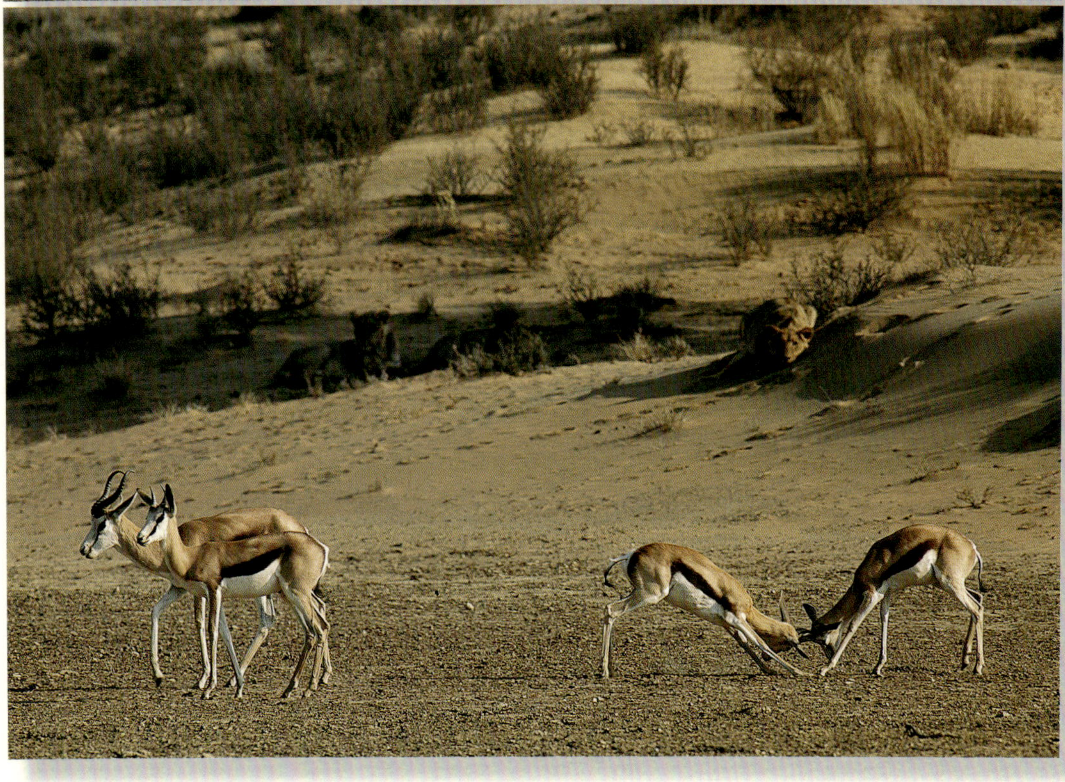

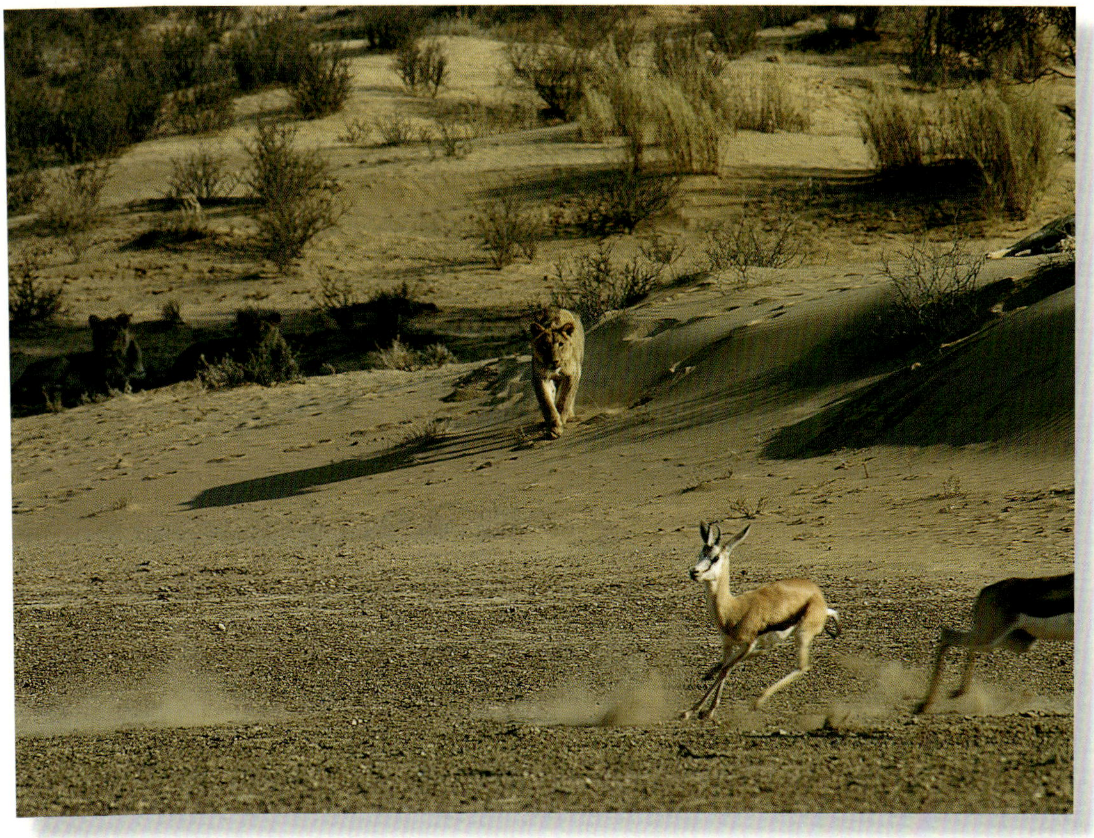

FACING PAGE & ABOVE

*In the Kalahari desert Jan and I approached a water hole called Leeuwdril, which loosely translates to "the lion's hangout." A small pride of lions was lying on a dune close to the water hole sleeping off a big meal. One lioness became alert as a small herd of springbok approached. Perhaps she could not tolerate the audacity of springbok playing in front of her. The lioness made a half-hearted attack, but the springbok got away.*

FOLLOWING PAGES

*Young male springbok often indulge themselves in playful sparring.*

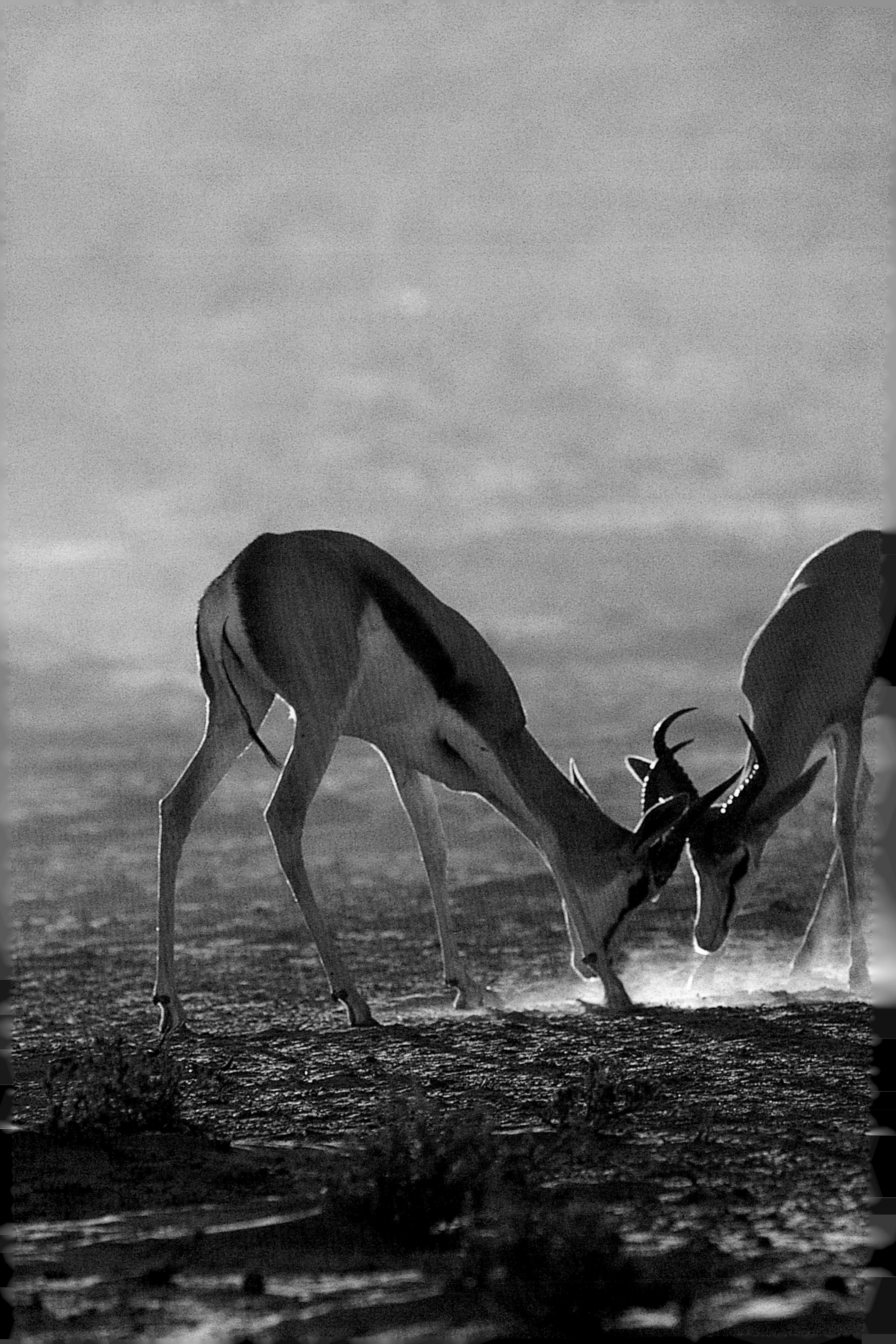

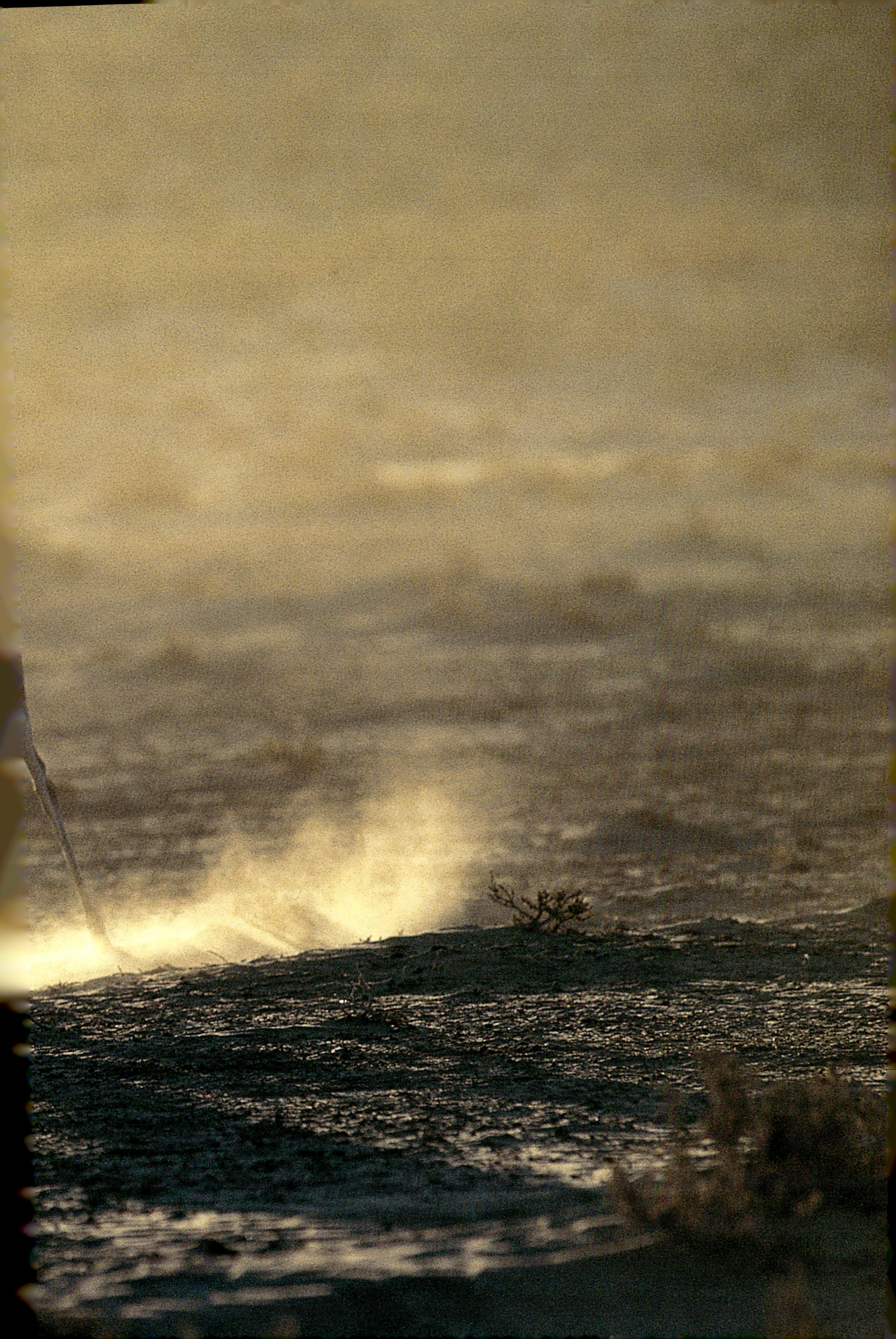

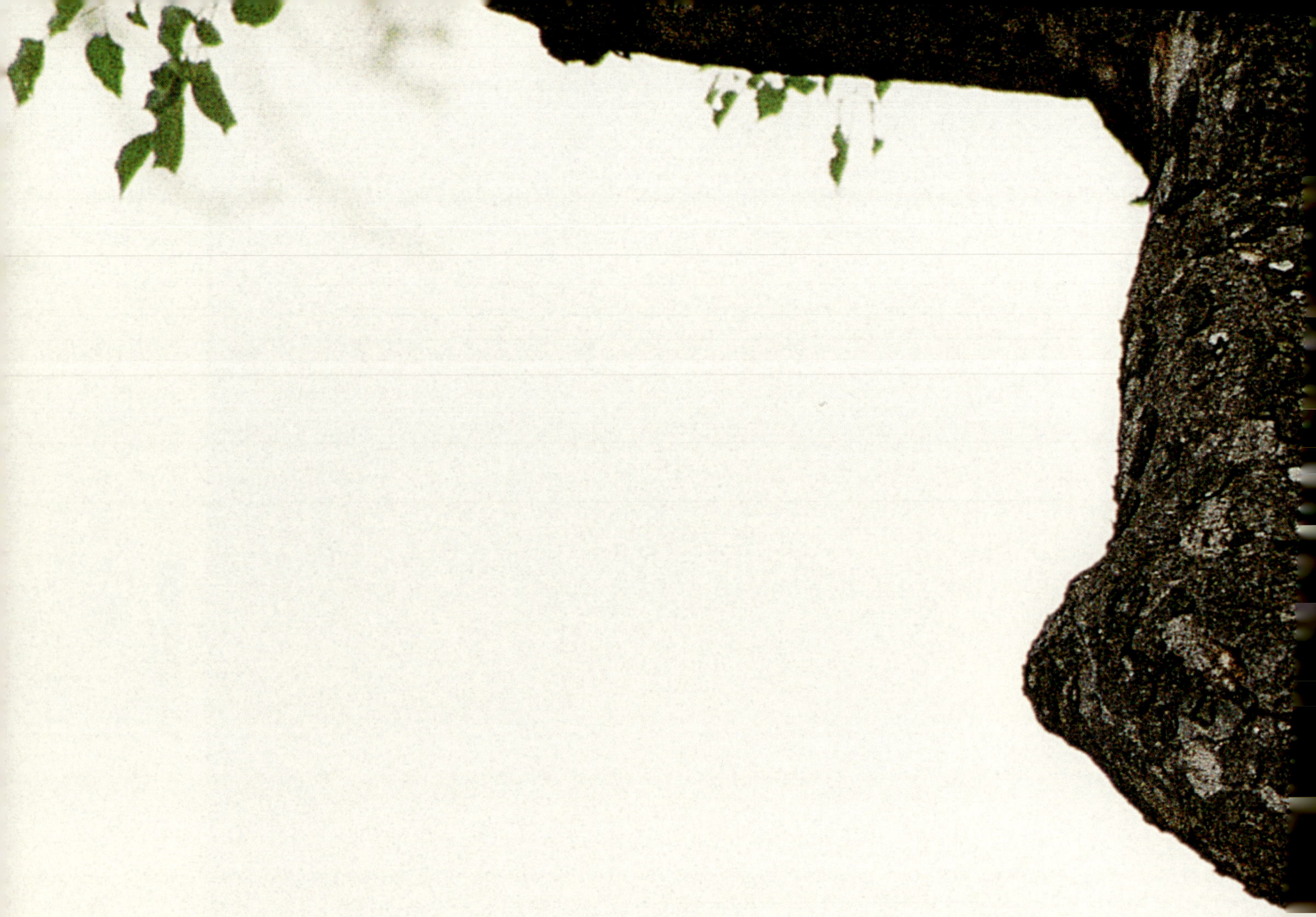

CHAPTER NINE

# Hunter Felines

## African Wildcat, Cheetahs & Leopards

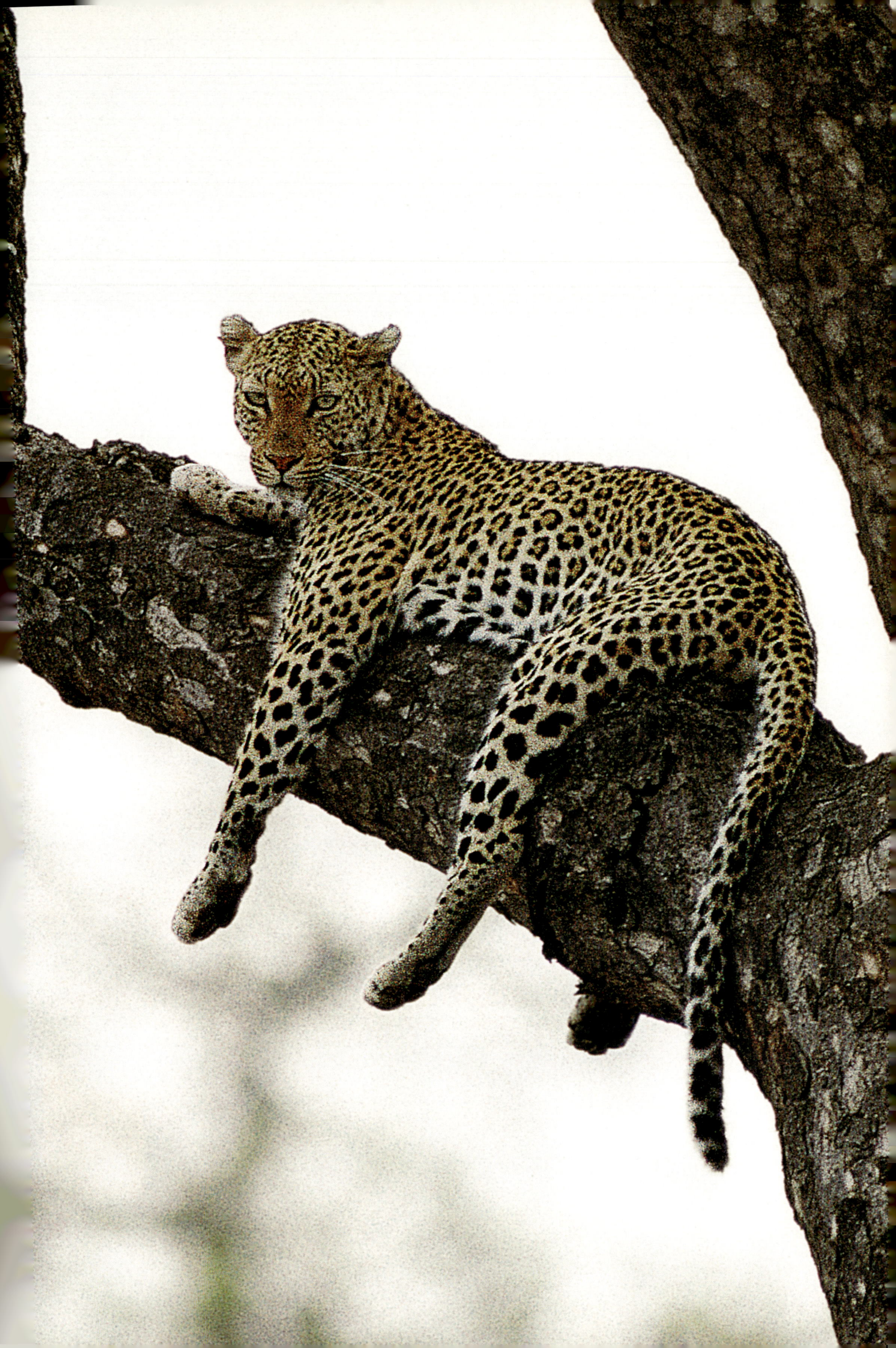

Crouched silently, concealed in the high grass and scrub of the bushveld, the hunter patiently waits, watching for prey in a moment of inattention or careless straying from the herd. Then, when the likely victim is chosen, the hunter slowly, secretly, creeps closer on soft padded paws, supple muscles tensing. Hesitating for a moment to wriggle the hindquarters and plant the rear feet for traction, the hunter springs from hiding to ambush its target. The hunter is feline; one of many wild African cats that prowl the bush and plains looking for hapless prey. The cat relies on stealth, surprise and speed – for the intended quarry is also watching, also fleet-footed, and also has a few tricks to elude a hungry predator.

PRECEDING PAGES

*It was a rainy afternoon in the Satara area of Kruger National Park. Our spirits did not remain dampened after spotting this leopard.*

FACING PAGE

*My chief scout and I spotted this African Wildcat early one cloudy morning. Since they often breed with domestic cats, we were eager to discover if this was a true African Wildcat. We were delighted to see the rufous ears on this one and the long legs, which are diagnostic traits of a true African Wildcat.*

All cats, except the lion, are largely solitary. They hunt alone and live alone except when mating or, for females, when raising a litter of young. Nearly every running, flying or swimming animal need be wary of cats. Large or small - every feline has the heart of a hunter and a taste for meat.

The African wildcat is a small cat – the ancestor to all domestic cats. Four-thousand years ago the Egyptians took African wildcats as pets. Over time, mutation has produced thirty-three domestic breeds. Although similar to a household cat with striped tabby markings and ring-tail, the African wildcat still exists in its original form – distinct from the domestic in two ways. Its ears are reddish brown on the back and its legs are long, which allows for an upright seated position, a posture which domestic cats cannot assume.

Like all cats, the African wildcat prefers a habitat with cover. Grass, underbrush, holes and rocky crevices provide hiding places from which it can wait and watch. Sometimes a tree provides a lofty perch from which to pounce. Sight is the wildcat's best sense. It hunts nocturnally with excellent night vision, looking for minute movements of rodents, birds, and reptiles on which it feeds.

Sound plays a part in hunting and social behavior. Territorial caterwauling, hissing, growling and screaming are sounds as common in the wilds of Africa as they are on the neighbor's fence. Meowing and purring are the same for the wildcat as for the contented lap cat.

African wildcats share the body language and various facial expressions that are distinctive means of communication for all felines. Sense of smell is important for interpreting scent marked territory and finding mates.

Mating and birthing occurs mainly in the rainy season. Litters consist of two to five kittens born in hollow trees, dense vegetation or borrowed dens. At three months kittens are taught to hunt. Females practice communal care until the kittens gain independence at five months. Then the kittens move on to establish individual territories and to hunt alone.

The cheetah is a big cat famous for its speed. The fastest mammal on earth, it can accelerate to forty-five miles per hour in two seconds. It can travel at seventy miles per hour only for short distances; but that is enough to outrun its quickest prey. Like a greyhound, the cheetah's body is built for swift pursuit. At seventy-seven to one-hundred forty-three pounds it is slender and light with a small head, short neck and

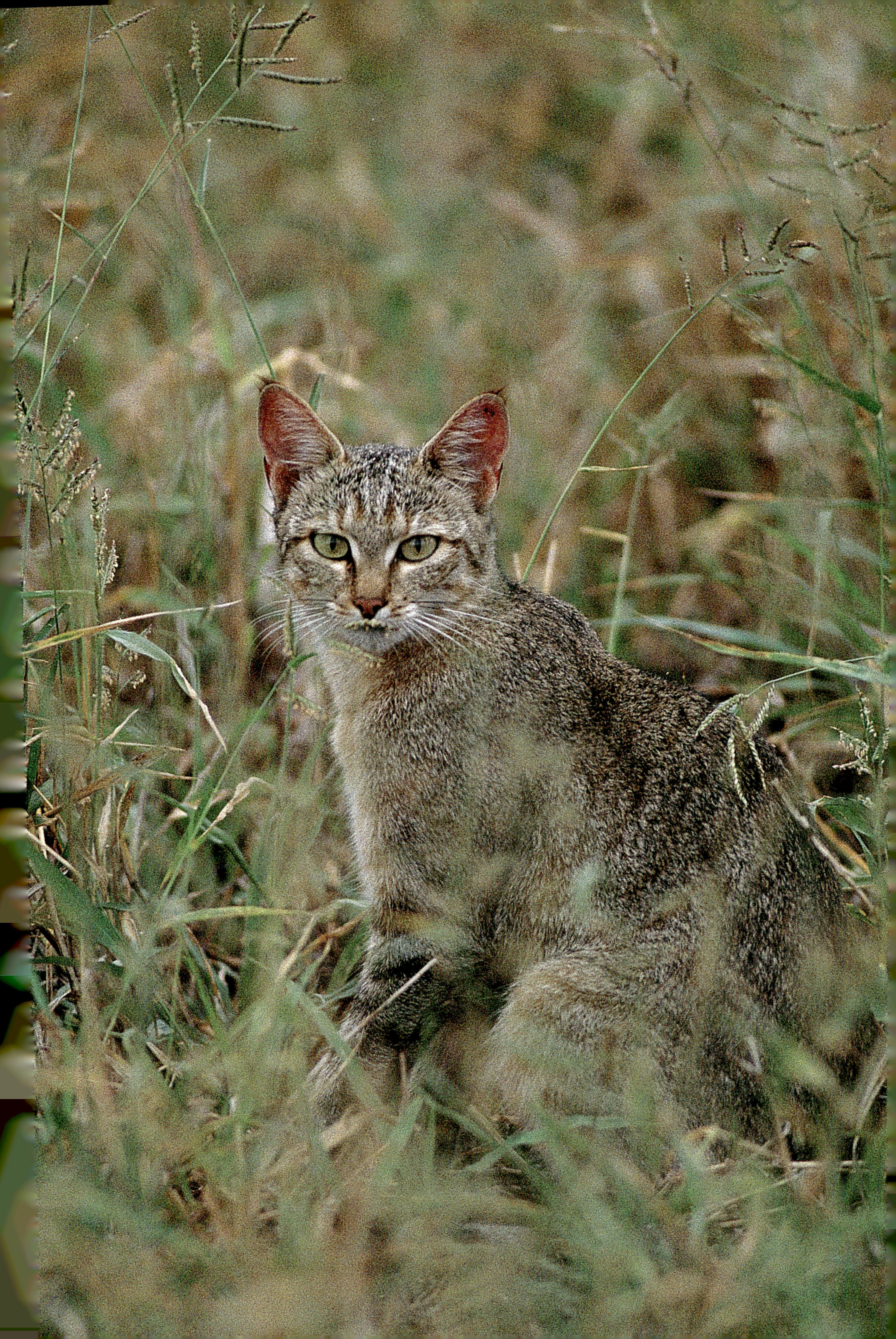

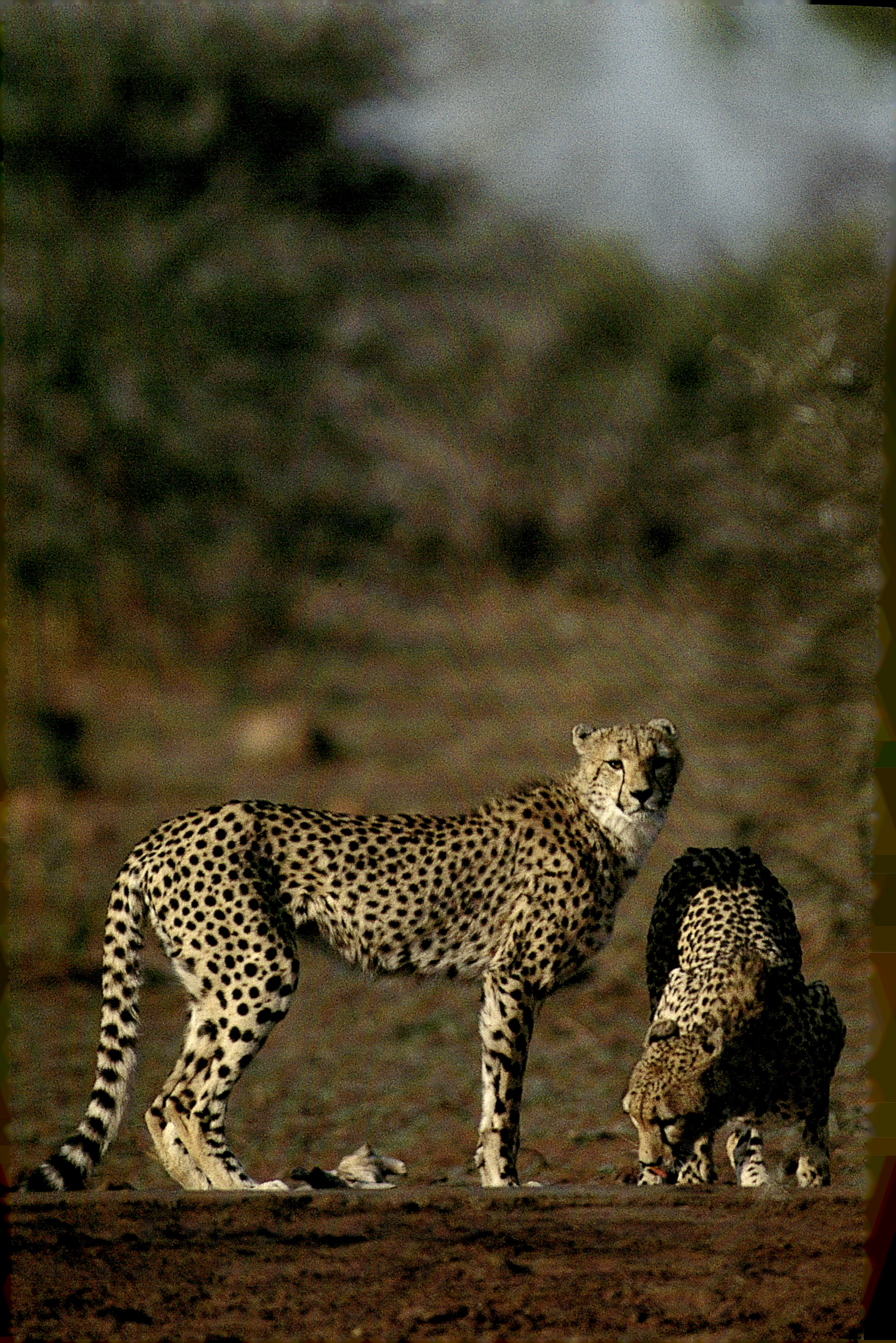

muzzle. The spine is like a spring that flexes from an arch to a bow – launching the animal into twenty foot strides on its long legs. Like all cats it runs on its toes, but unlike the others, its unretractable claws grip the ground and pull it forward ever faster. Nimble-footed gazelles, impalas and other small antelopes challenge the cheetah with evasive maneuvers. At high speed the cheetah's thirty inch tail acts as a rudder for balance when reacting to the sporadic turns of its prey. Even so, only one half of the pursuits results in a kill. When a kill is made, the extreme effort of the hunt leaves the cheetah breathless, and it must wait up to thirty minutes before it can eat. The cheetah prefers to hunt herd animals, but will also take birds, rodents, lizards and frogs. Although they need not drink often, when water is scarce they eat tsama melons for the water content.

Facing page
*Most male cheetahs that form coalitions are brothers. These two are taking time for a cool drink of water.*

The cheetah's appearance allows for camouflage in grass and woodland as they hunt by day. Tawny fur is patterned with a confusion of black spots – for which it is named. The Hindu word *'chita'* means *spotted one.* Black emphasizes the nose, eyes and ears. Prey animals watch for the flicker of the black banded ears twitching in the high grasses. Distinctive 'tear-stain' markings extending from the cheetahs eyes, down the face to the mouth, are effective for reducing the sun's glare for better vision. A Zulu story tells of why the cheetah's face is stained with tears:

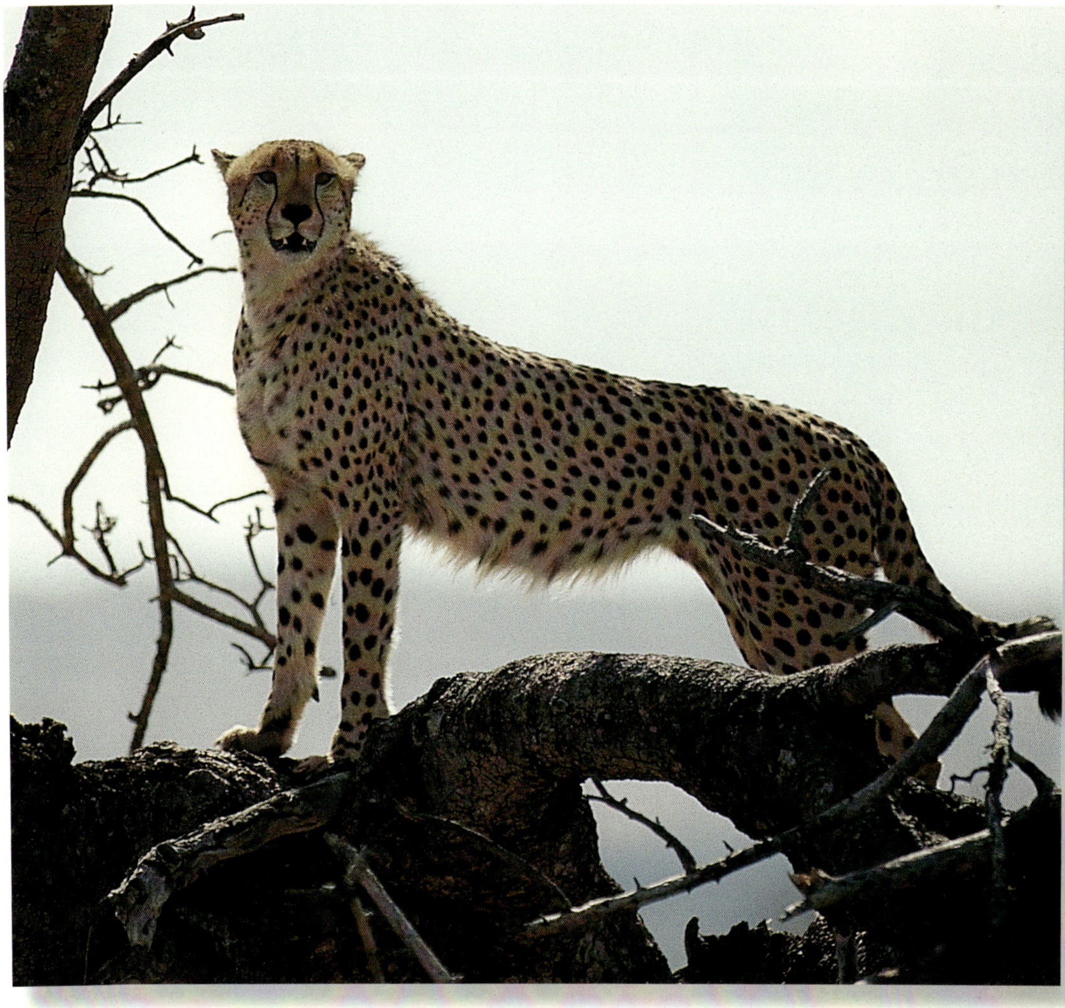

*By standing on a thick low tree limb, a cheetah is able to survey an area for prey.*

## *A Zulu Fable*

*One day, many seasons past, there was a lazy hunter resting on a hill, not eager to expend any energy on hunting. Below, on the plain, he watched as a female cheetah stalked a herd of gazelles. Her stealth fascinated him and he continued to watch her single out, chase and kill her quarry. Then as the cheetah dragged her kill into the bush, he noticed three cubs hiding and waiting for her – and their dinner.*

*The lazy hunter was jealous, wishing he had someone to provide food for him. He waited for the next time the mother left the cubs so he could steal them. His plan was to train them to hunt for him – and he took them away. When the mother cheetah returned and found her cubs gone, she began to cry. She cried so long that the tears stained her face.*

*Another hunter discovered the distraught cheetah, and when he realized what had happened, he went to the village and found the lazy hunter with the cubs. The lazy hunter was ashamed because he knew that it was against tradition to hunt with anything but his own abilities.*

*The cubs were returned and all was well, but since then, every cheetah has retained the tear-stains of that mother's sorrow. When hunters see the stains, they are reminded to hunt with honor.*

*Continued maternal care prepares young cheetahs for complicated adulthood and hunting behavior.*

Although just a fable, the story also alludes to the caring nature of the cheetah. Most gentle of all the big cats, they rarely fuss among themselves. Unaggressive, they prefer to flee than to fight. Cheetahs are unable to roar like other big cats. Instead, they have a language of chirps and yelps to greet and summon one another. Threat calls consist of coughs, snarls, growls and hisses – usually emitted before flight.

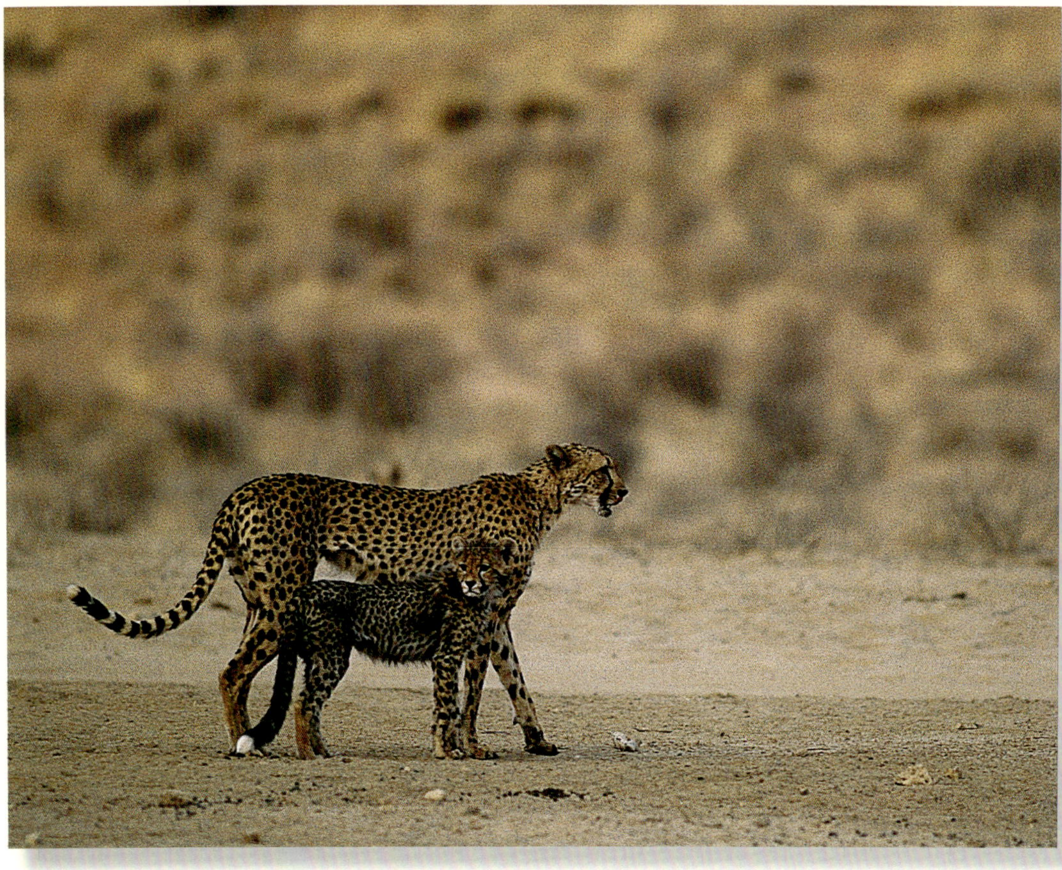

The female is a very good mother. When litters are born, they are well concealed by the mother and moved to new hiding places every few days. A litter averages one to four cubs, but can be up to eight. Each cub weighs about one pound when born and has a blue-grey or silver-grey mane on its head, neck and back, which recedes at three months old, when weaning takes place. Cubs are very playful; chasing, stalking, pouncing and playing other games that mimic hunting behaviors. At six months old, the cubs learn to hunt. When they stray too far, the mother calls them with chirping sounds. At fifteen months, they hunt well but stay with their mother until two years old. Even with mother's protection, only one-third survive.

Grown female cubs stay in their natal territory, living solitary lives, while males leave to find their own range or join a coalition of males consisting of two to four members. The male coalitions are usually joined for life. Coalitions strengthen the possibility for holding territory, for mating success, and for gainful hunting.

Unfortunately other predators are more aggressive than cheetahs and often steal their kills, attack adults, and kill cubs. The life span for a cheetah in the wild is only four or five years. The cheetah is extremely endangered. Habitat loss, non-aggressive behavior, and genetic uniformity make it a fragile species. Once kept by the thousands as pets and hunting cats, they did not breed well in captivity. It is thought that cheetahs living now may have descended from one female. Disease in the wild population could be devastating. Today, captive breeding programs are hoping to bring back a healthy population of cheetahs.

*Seeing three cheetah cubs playing is invariably a big thrill.*

Like the cheetah's, a spotted hide is effective camouflage for another big cat as it prowls through the

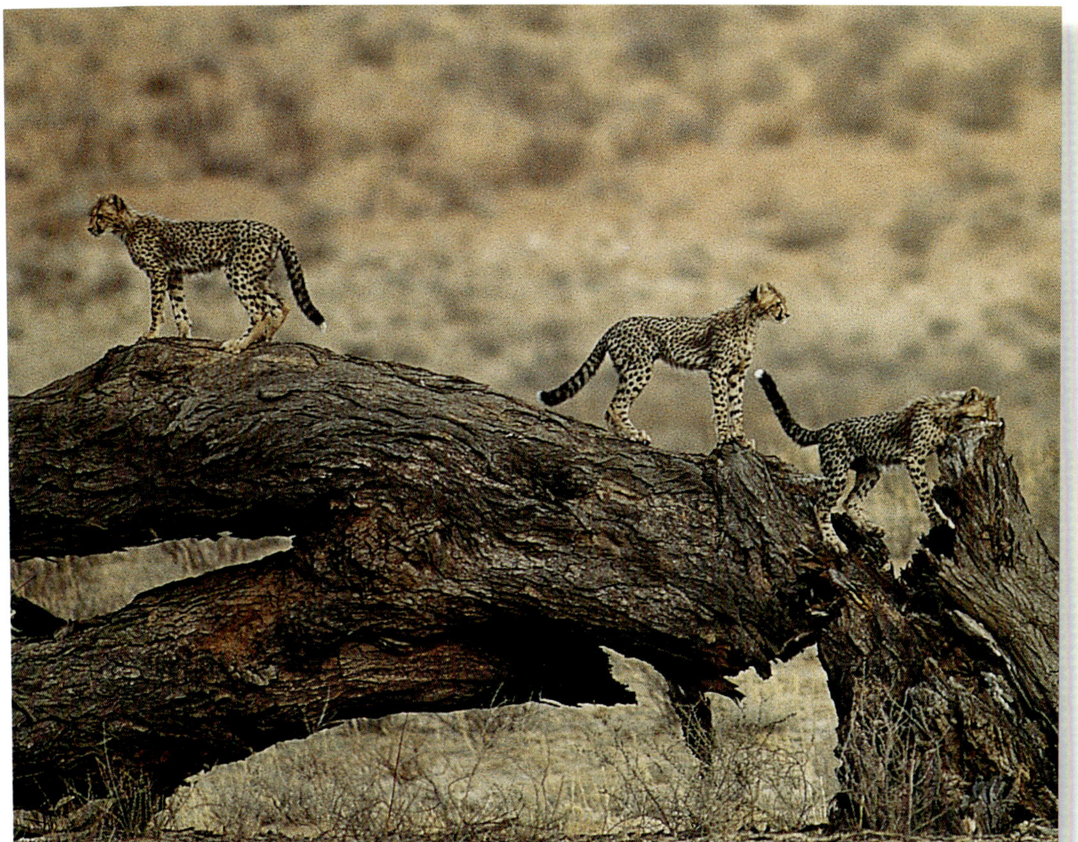

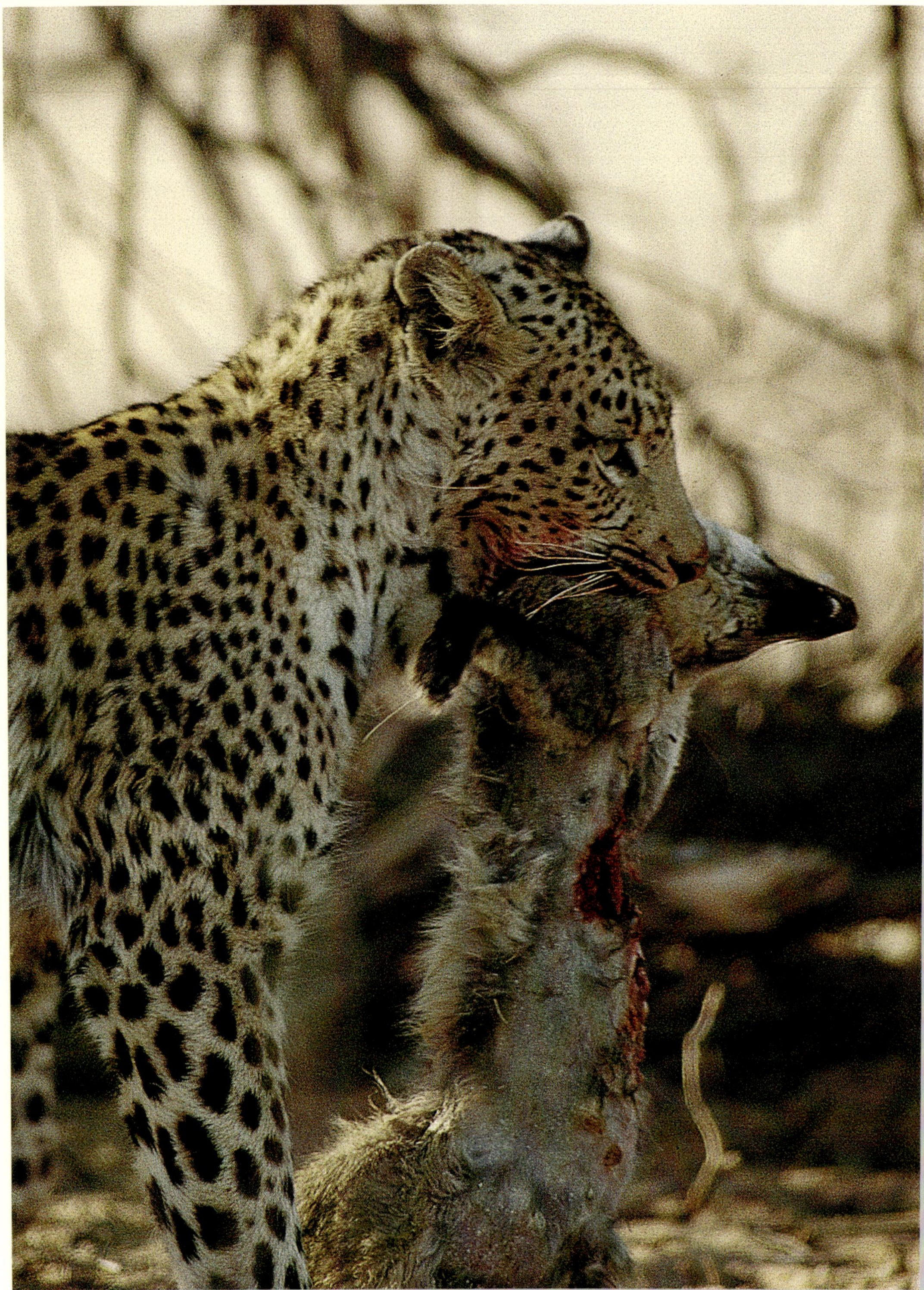

ABOVE & FACING PAGE

*After capturing a Bat-eared fox, a female leopard settles down to eat her prey. When she is filled she will sit up, clean herself a bit, and carry the remains of her prey to a stash in a tree where other predators cannot get to it.*

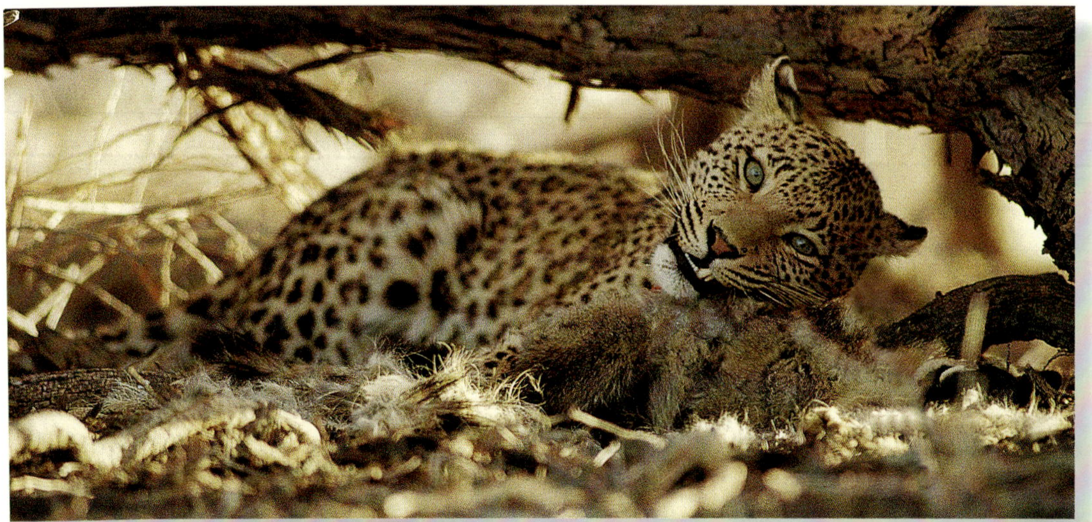

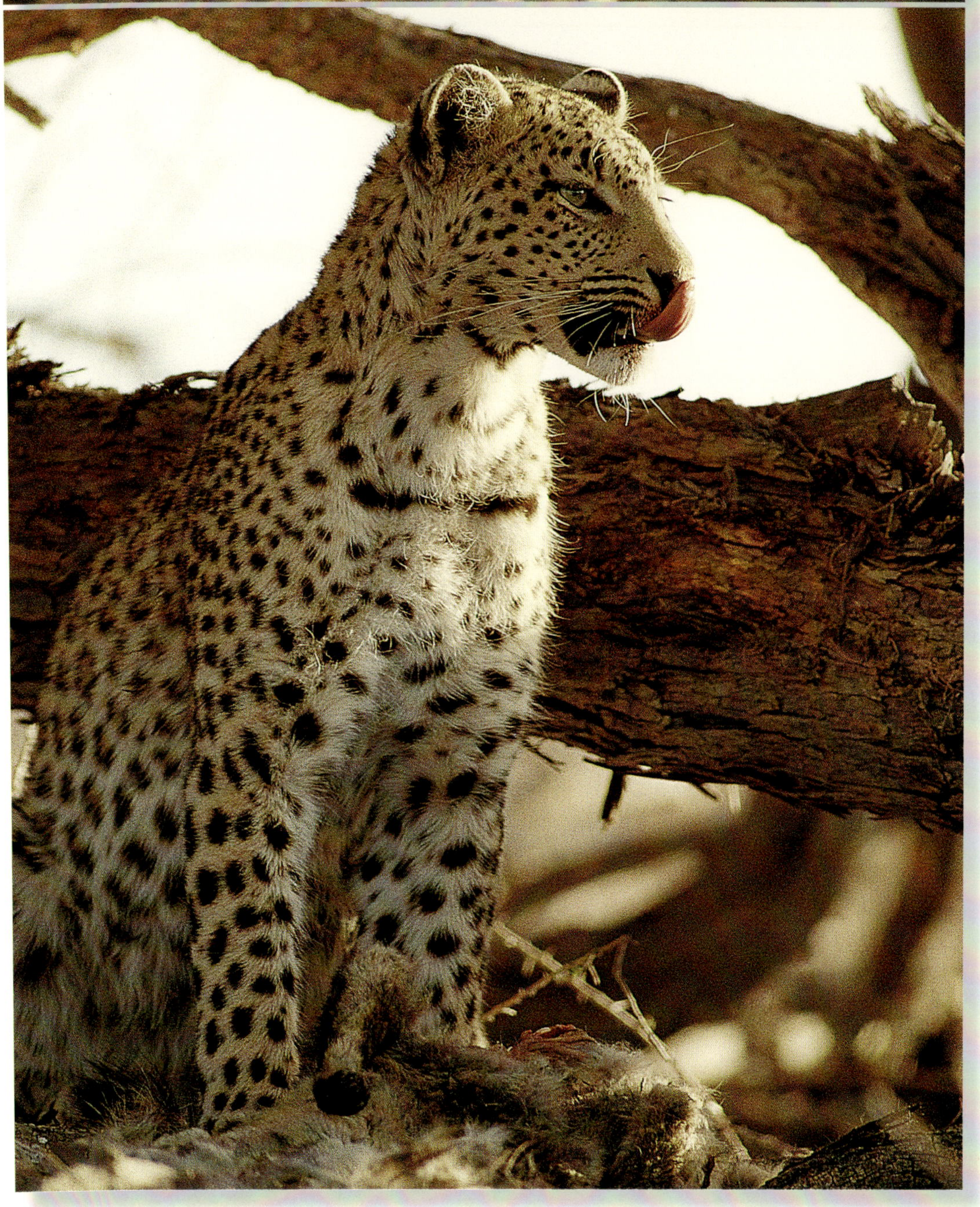

bush. Combined with the patience and ability to silently approach its prey unawares, the dappled feline waits until it is within fifteen feet before a surprise attack. The leopard is the master of stealth.

The largest spotted cat in Africa, the leopard is named for the mistaken belief that it is a cross between a lion (*leo-*) and a panther (*-pard*). The panther is actually a black leopard and not a separate species. The leopard is a long, low slung cat with short muscular legs. Spots are arranged in a rosette pattern across the tan to reddish-brown torso and upper limbs – no two patterns are alike. The head is wide and the jaws are powerful.

FACING PAGE
*Long grass helps this leopard to camouflage himself. He is the quintessential stalker.*

This hunter feline is found in regions from rain-forest, to mountain, to desert, to bushveld. It will inhabit land adjacent to water courses – for it swims readily.

By day, the leopard seeks dense cover, caves or aardvark burrows in which to rest. It will sometimes sunbathe on rocky outcroppings with a good view, or snooze 'spread-eagle' on broad tree limbs. As evening approaches the leopard stirs. Upon waking from a day's nap, it may be lucky enough to simply drop from its perch onto some unsuspecting animal - an appetizer perhaps. But, most likely, it will begin to prowl, slipping into dense brush to hunt. As it walks, the leopard emits raspy sounds like sawing wood – until prey is spotted - and then, silence is necessary.

The leopard hunts impala, klipspringer, antelope, gazelle, and wildebeest. It has a taste for pigs, rodents and monkeys. Sometimes a leopard will take canids like the Bat-eared fox. When a kill is made, it may carry the animal up into a tree to eat. It can climb carrying a one-hundred fifty pound impala. Moisture is obtained from the meat and so the leopard rarely needs to drink. But where water is available, it will take advantage of the resource. The leopard is active until a few hours after dawn. Then, as the sun climbs into the sky, it will fade into the brush to rest.

Leopards are seen together only to mate. Occurring at any time of year, the union is a noisy affair. The pair growls and snarls during mating – a caterwaul even louder than that of lions. After a gestation period of about one-hundred six days, the female bears one to three cubs. Each weighs fourteen to twenty-one ounces. Thickets, caves, or hollow trees provide hiding places for the helpless cubs for six weeks. At three months they are weaned and for one year mother and cubs stay together. They follow her everywhere. Padding silently in the dark, through the tall grass, the cubs keep track of their mother, watching the white tufted tip of her upright tail – like a beacon in the night. Following patiently, they learn to master stealth.

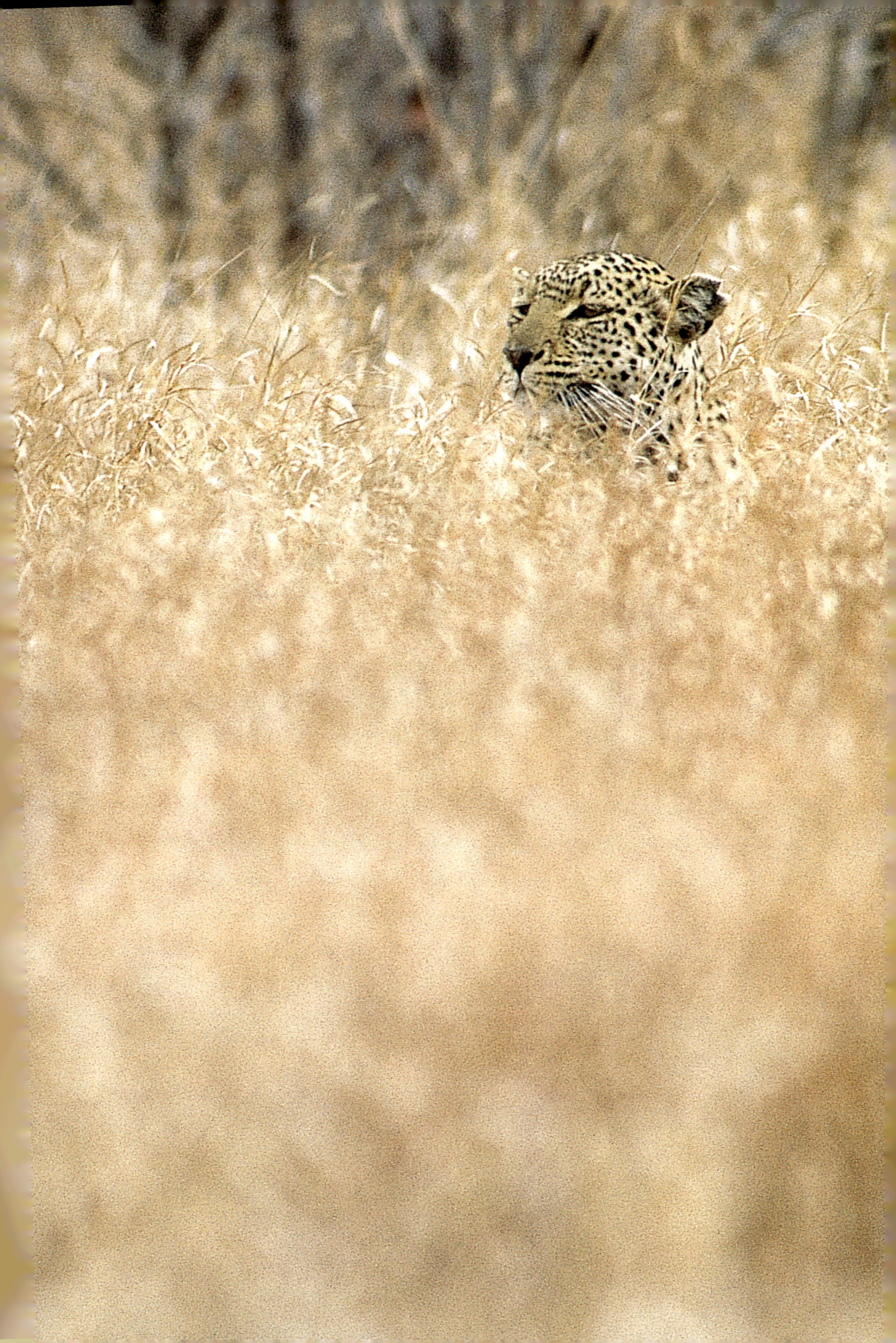

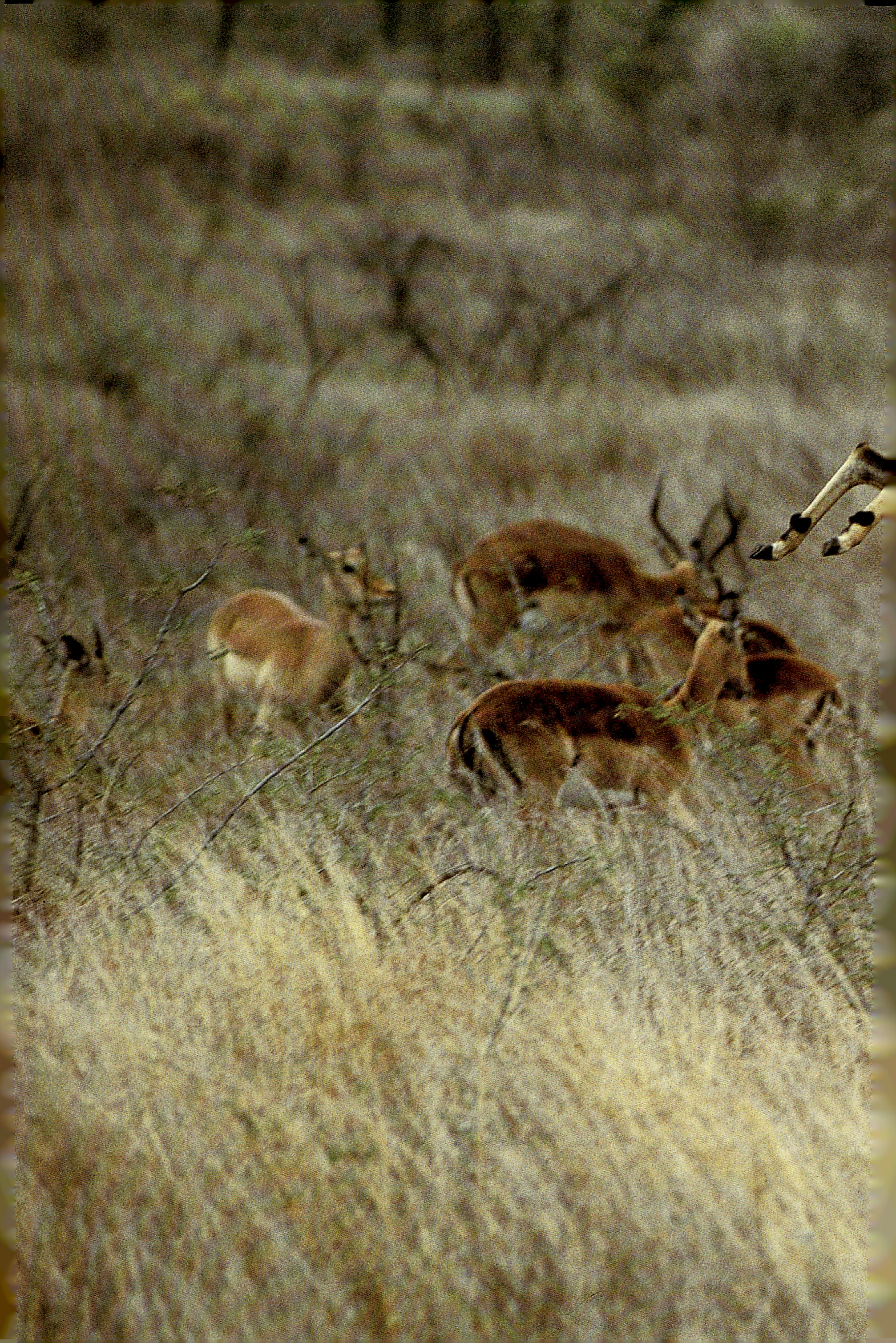

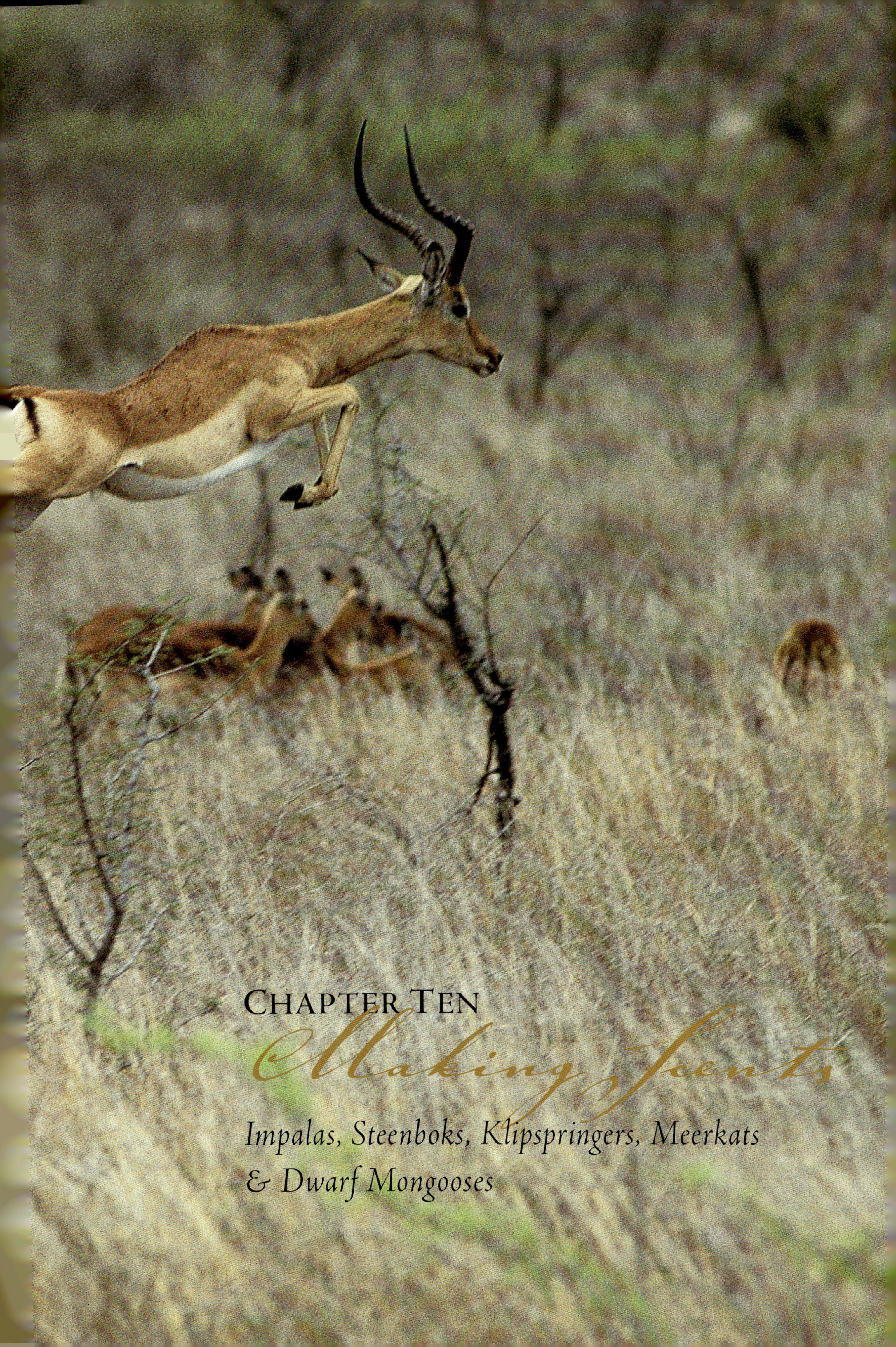

# Chapter Ten

## Making Scents

*Impalas, Steenboks, Klipspringers, Meerkats & Dwarf Mongooses*

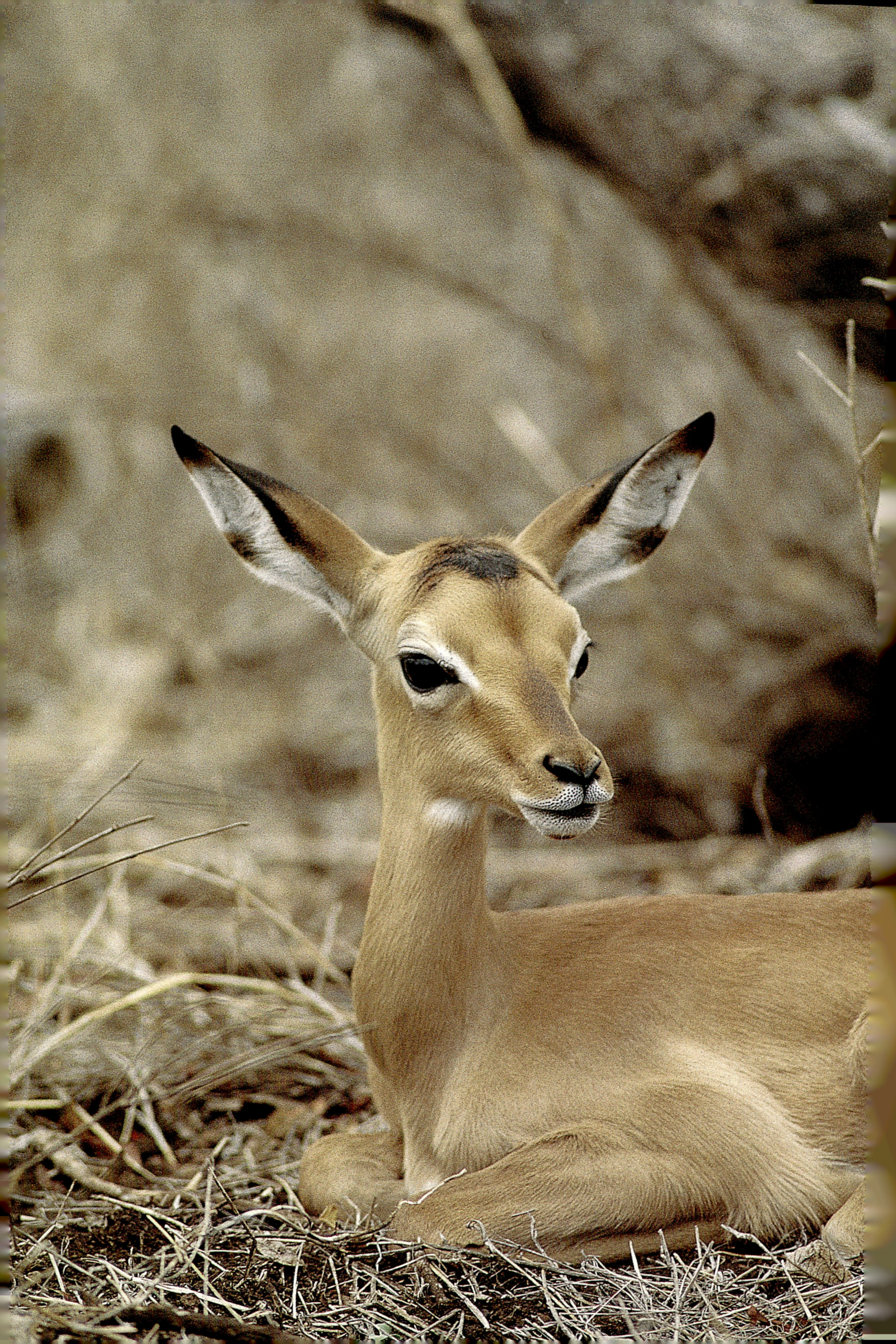

All across the bushveld, invisible trails and signs guide many creatures of the wild. Trails mark the way to follow the herd. Signs reveal – Welcome Home! – Time to Mate! – or, Keep Out - Private Property! The information is spread by scents excreted from specialized glands and deposited on grass and twigs, or near burrows. The aroma of dung middens serves to advertise who's who. Each scent is species specific, ignored by animals other than that for which it is intended. This is the secret language of the beasts, a silent but powerful communication.

PRECEDING PAGES

*A leaping impala reminds me of the great ballet stars. The impala can soar to a height of about ten feet.*

FACING PAGE

*When I photographed this lamb, I wondered if it would ever grow into those ears.*

The impala is equipped with scent glands on its fetlocks, visible as black tufts above the hooves. As it walks, scent oils are laid on leaves of grass. Herds of impala can keep track of each other as they graze, and if dispersed due to a scare, they can easily sniff their way back to the herd. Young impalas have undeveloped glands so as not to be detected by predators.

The gregarious impala is the dominant antelope of Southern Africa. It is one-of-a-kind, having no close relatives. Weighing about one-hundred pounds, the impala has a graceful build, with a long neck and long limbs. Only the male wears handsome ridged, 'S' shaped horns. The coat is smooth and shiny – a two-toned tan to reddish brown. A black stripe runs down the thigh and short bushy tail. The face and ears are marked with becoming stripes and patches of black and white.

Impala herds live on the edges of open woodlands bordering grassland – often near water. They graze for tender grasses and herbs during the rainy season and browse leaves, fruits and seeds at other times. They are active by day. At night they rest lying down in open areas.

At all times impalas must be wary of predators. Spotted hyenas, Wild dogs, and cheetahs hunt them. Martial eagles will take lambs. When attacked, the escape strategy of an impala herd is to explode in all directions, confusing the predator. Amid the confusion, individuals will leap up to ten feet high, landing over thirty-five feet away. These spectacular leaps are the impala's forte.

Another habit distinguishing the impala is its roaring display. Territorial males emit a series of explosive snorts, followed by guttural grunts. These sounds, directed toward subordinate males, can be heard more than a mile away. Roaring increases during rut.

At breeding time, only the territorial male mates with his harem. He uses roaring, body language and scent marking to ward off rivals. He has scent glands on his forehead which he rubs on plants, marking his property; signs that warn – Keep Away! Nevertheless, fights sometimes break out.

When mating occurs, females time their conceptions by the full moon. It is advantageous for all births to occur simultaneously so that predators, being easily sated with the sudden abundance of prey, will take fewer young. Even so, half of newborns are taken.

Of those that survive, the females will remain in their natal group. Male young have another challenge to face at a tender age. At four to five months old, their horns begin to appear – a tell-tale sign of their gender. By eight months, their father sends them away to join a bachelor herd. There they remain until they seek to find territory of their own or challenge a landlord for his harem.

Not all antelopes gather in herds; some prefer to live solitary lives. The steenbok is such an animal. Considered a dwarf antelope, the steenbok is only about twenty inches tall and weighs a light twenty-five pounds. It has large rounded ears, sharp upright horns and long limbs. Its coat is brick-colored, for which it is named. In the Afrikaans language, *steen* means 'brick.' Scent glands are highly visible in front of the eyes. It also has glands on its throat and feet.

The steenbok lives in acacia savanna where there is cover available when needed. It is active day and night – hiding at midday to rest. Woody plants, seed pods, berries and tender grass are some of the steenbok's favorite foods. These plants supply much of the antelope's moisture requirements making it less dependent of water.

Male and female come together only at mating time, which can occur any time of year. After less than six months gestation, a single, very precocious calf is born. It nurses within five minutes of birth and begins grazing in two weeks. At three months the calf is weaned, but stays with the mother a while longer. Females can mate as early as six or seven months and can produce a calf every eight months. Calves are well hidden from predators and learn that cover is their best defense against their main predators, Martial eagles. Cheetahs, Wild dogs, leopards and lions hunt adults. If discovered by a predator, the steenbok first runs, bounding and dodging in a zig-zag pattern – then looks for the safety of a dense thicket. Life can be short for this small, timid antelope.

Another dwarf antelope finds safety in rocky terrain. The klipspringer is the only antelope that lives on steep rocky outcrops called 'koppies.' The ability to navigate such terrain has given the klipspringer its name, which, in Afrikaans, means 'rock-jumper.' The klipspringer's body is built for quick climbing – a short body with muscular hind-quarters. It is only two feet tall and weighs only twenty-five pounds. A small head features large rounded ears with a pattern of black and white stripes, and small pointed horns. The coat is made of special hollow hairs that insulate the animal from both heat and cold, and, give it a rough grizzled appearance.

Klipspringers form monogamous pairs. Both have large pre-orbital scent glands that they use to mark twigs every day or so; and, they visit dung middens to remind any intruding klipspringers of their territorial boundaries. Klipspringers traverse the rocky hills with surefooted grace, bounding up and down steep slopes and jumping from rock to rock. They perch precariously on hooves with flat tips, all four feet together; a stance that identifies the little klipspringer from afar. The male stands watch as the female grazes below on flatter ground. She stays within fifteen yards for safety. Foods for klipspringers consist mainly of leaves and shoots, fruit and berries, flowers, seed pods, succulents, grass and herbs. They graze in early morning and late afternoon in hot weather. When its cool, they graze all day.

The closely bonded pair voice a whistling sound which they call in duet. With it they can advertise their presence or reinforce their bond; and, they duet to signal the presence of predators. The same carnivores that stalk steenbok hunt klipspringers. When they have a young calf, it is hidden for several months, as they are small enough to be captured by eagles. When pursued, vulnerable young are hidden safely in rock crevices while adults scramble to safety within their craggy fortress.

In another high place on the vast savanna, a different sentry stands its watch – scanning both earth

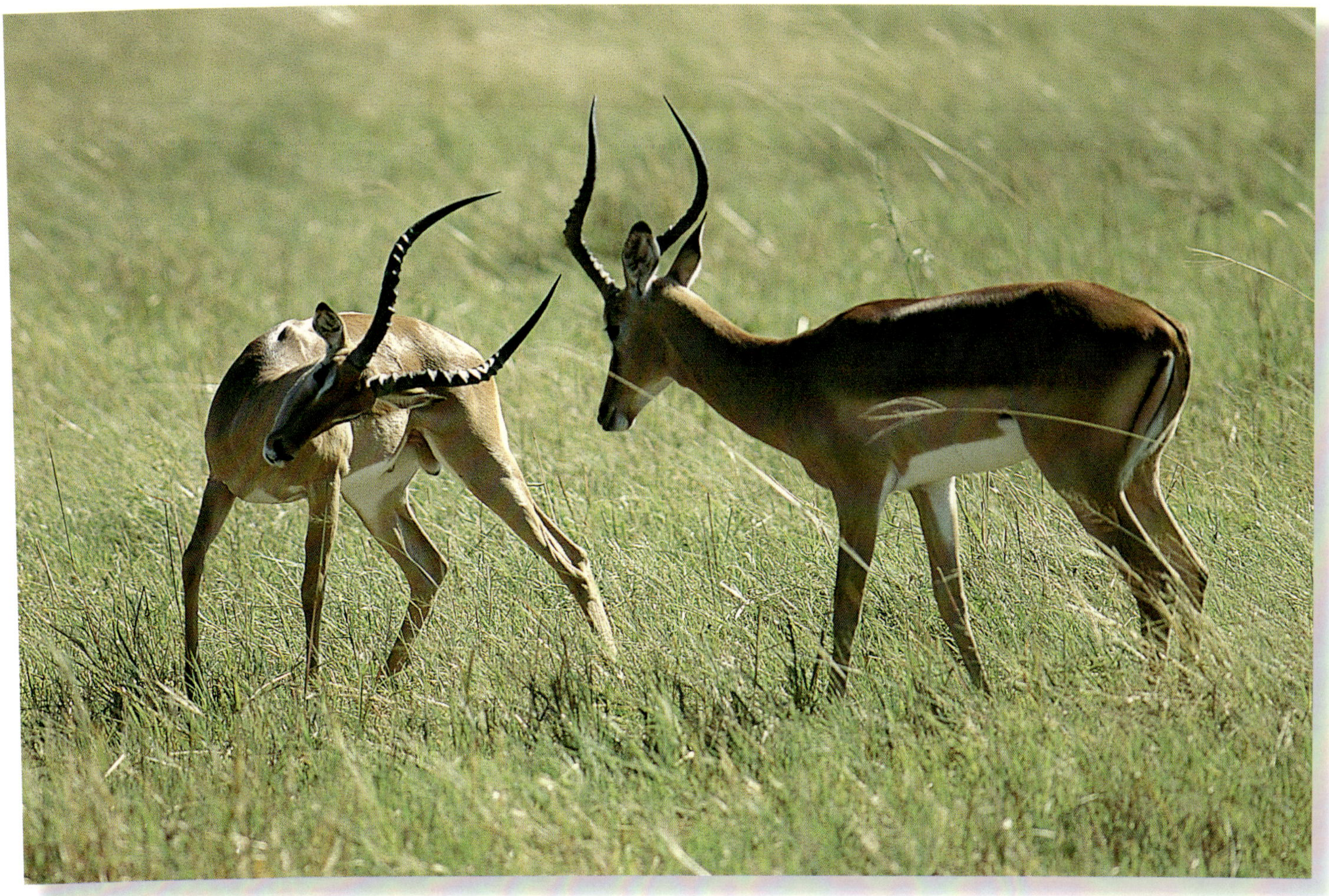

*These rams were about to engage in a fight for the right to mate with the females of a territory.*

and sky for danger. The height of a low bush mound will suffice for a foot tall suricate to sit upright and study the surrounding landscape. Good hearing and excellent sight can detect any predator by land or by air. Below the mound, his pack will be safe to forage, reassured periodically by the sentry's peeping sounds, as if to say, "All's well."

The suricate is a mongoose, a small carnivore taxonomically placed between the weasels and the cats. Also called a meerkat, it is a thin, one pound, long-legged mongoose. Its coat is grizzled grey or tan with darker bands on its back. The underside is sparsely furred and yellowish or buff colored. The head has a sharp snout, low, dark roundish ears and eyes accentuated with dark eye patches. Its long tail is tipped in black.

The suricate is a social mongoose living in packs of ten to thirty members. They inhabit dry open savanna. They dig a labyrinth of tunnels and chambers for burrows in the hard ground.

A day in the life of a suricate pack begins at sunrise, when one by one, they leave their dens and visit the communal latrine. Then they gather at the burrow entrances to groom and socialize; and, to sunbathe – sitting up tall to let the rays warm their bellies. When they're ready, they leave to forage. In pairs or small groups they hunt for insects, scorpions, termites, snakes, lizards, small birds and their eggs, and mice. Their excellent sense of smell helps them find buried beetle larvae and other underground inver-

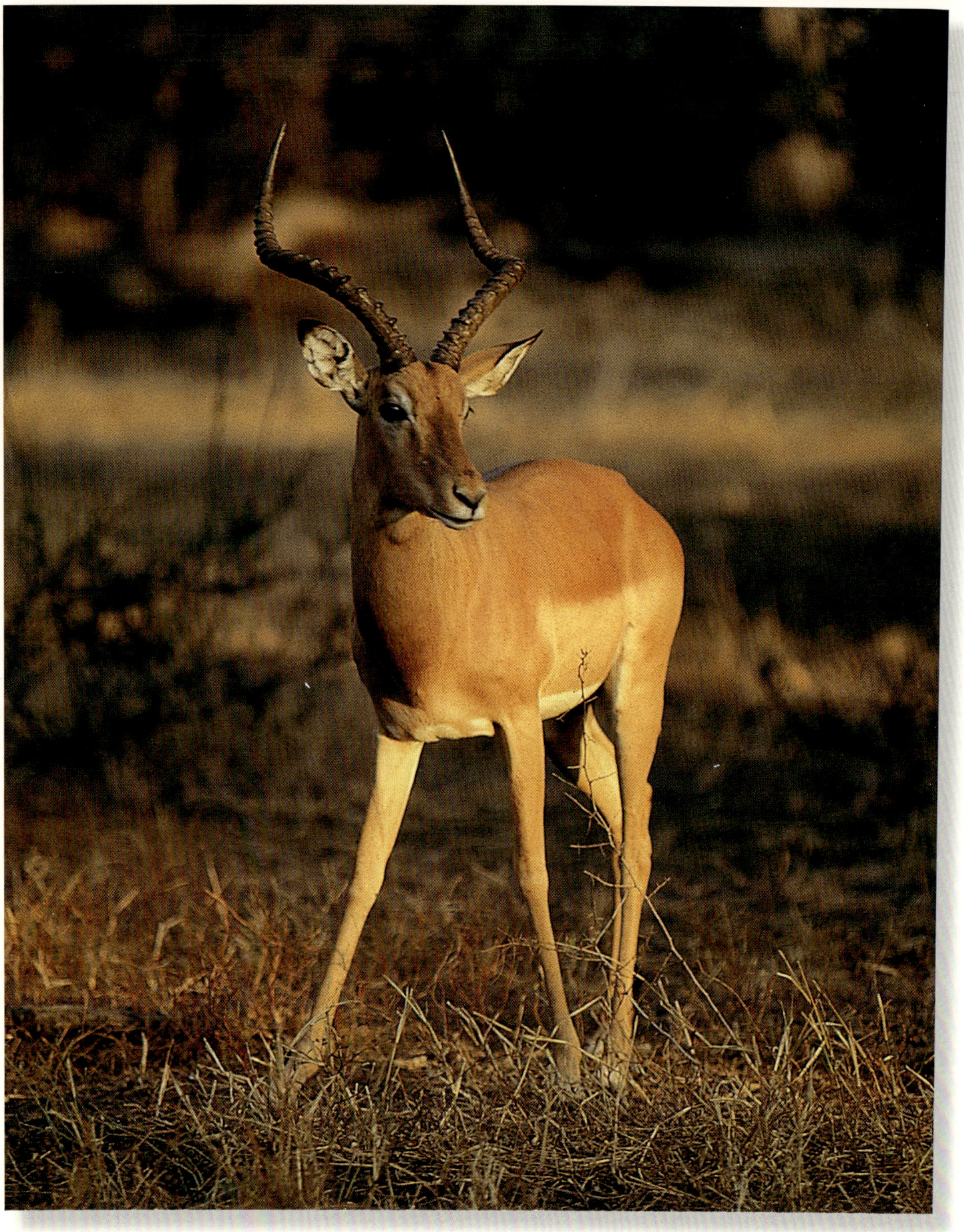

tebrates, which they vigorously dig up with long clawed forefeet. Their water is provided in the form of tsama melons, roots and tubers. While foraging, they murmur contentedly with a low trilling sound and soft bark – *wurruck, wurruck, wurruck*. A sudden, sharp bark from the sentry is a predator alarm which sends them scurrying for shelter – screaming all the way. The main predators of suricates are hawks and eagles.

ABOVE & FACING PAGE
*Late afternoon sunlight always seems to impart a warm glow to any subject.*

When danger is past, they will return to foraging or break for rest at midday. Afternoon foraging is followed by grooming and socializing until dark. They then retire to their burrows to sleep – huddled together, sitting up with their heads hanging curved over their chests.

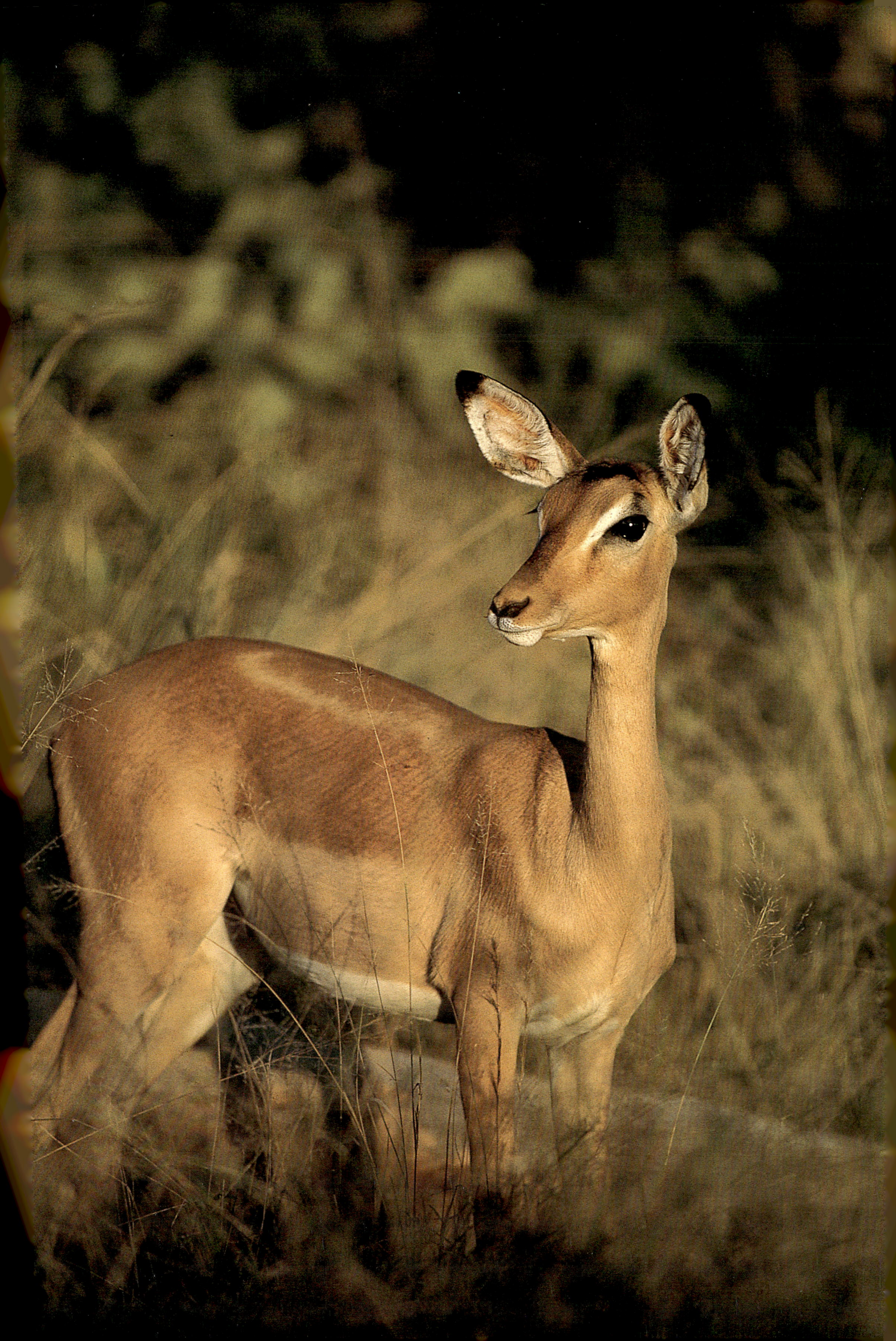

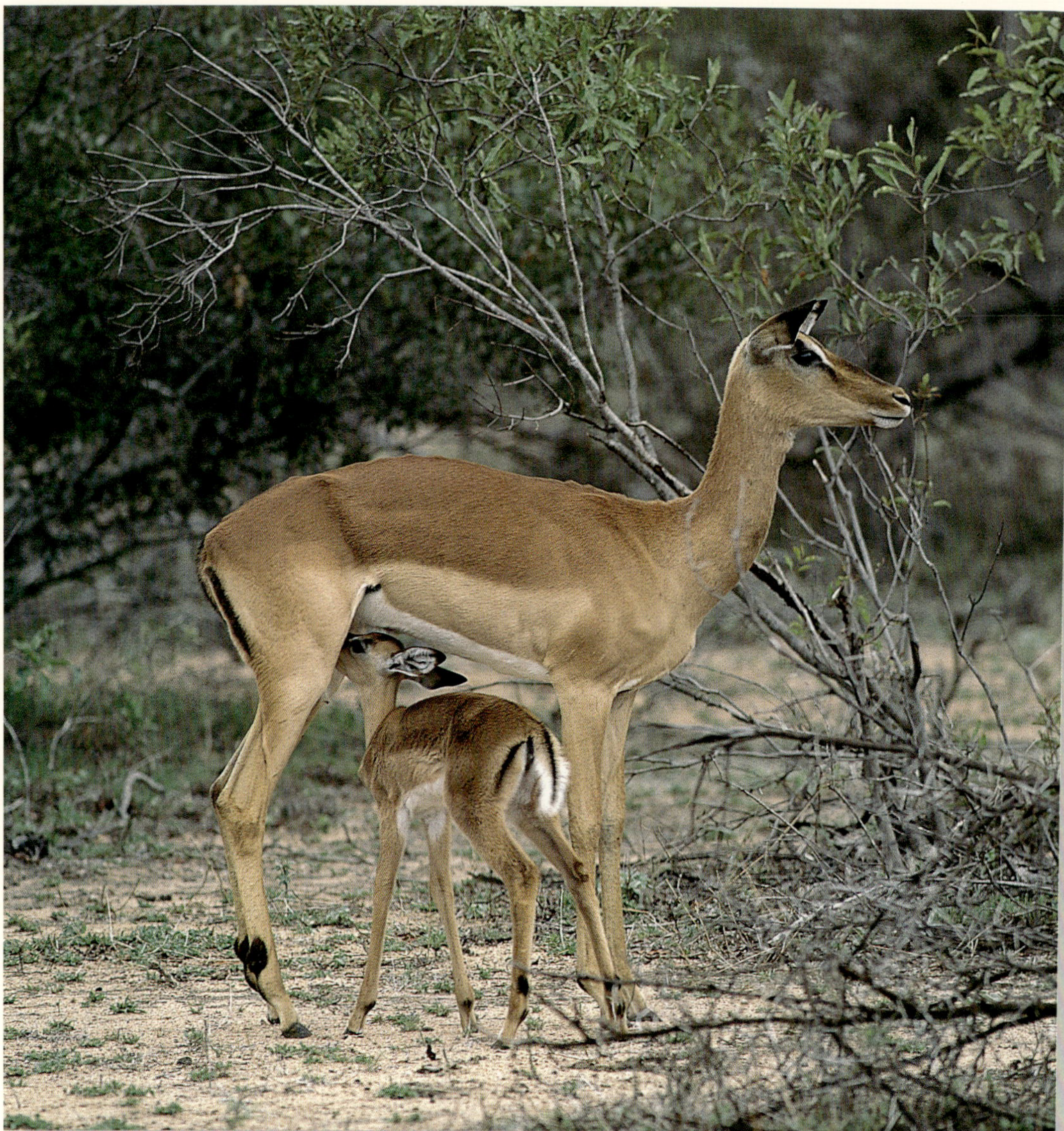

*Most impala lambs in the Kruger Park are born just after the onset of the first rains, usually in late November or early December. Sweet grasses will appear and food will be plentiful.*

Suricates are territorial, Each pack has a dominant pair that uses scent marking to establish territory. The strong scent is detectable for up to three weeks. The pair also groom each other with saliva which serves to strengthen their bond.

FACING PAGE
*When Steenboks are seen they are almost always near plenty of cover.*

The dominant pair breeds in spring. Two to five young are born eleven weeks later. They are born blind, but two weeks later their eyes open. They use the characteristic sitting posture when they suckle. After weaning at three to four weeks, they begin to forage with the adults – carefully watched by parents and baby sitters. Baby sitters care for the young while the pack forages; bring beetle larvae and insects to them; groom and play with them as well. Continuous chirping, purring sounds that young suricates emit, help adults keep track of them. Adult suricates have been known to lay across juveniles to protect them from predator attacks.

Another member of the mongoose family holds the distinction of being the smallest African car-

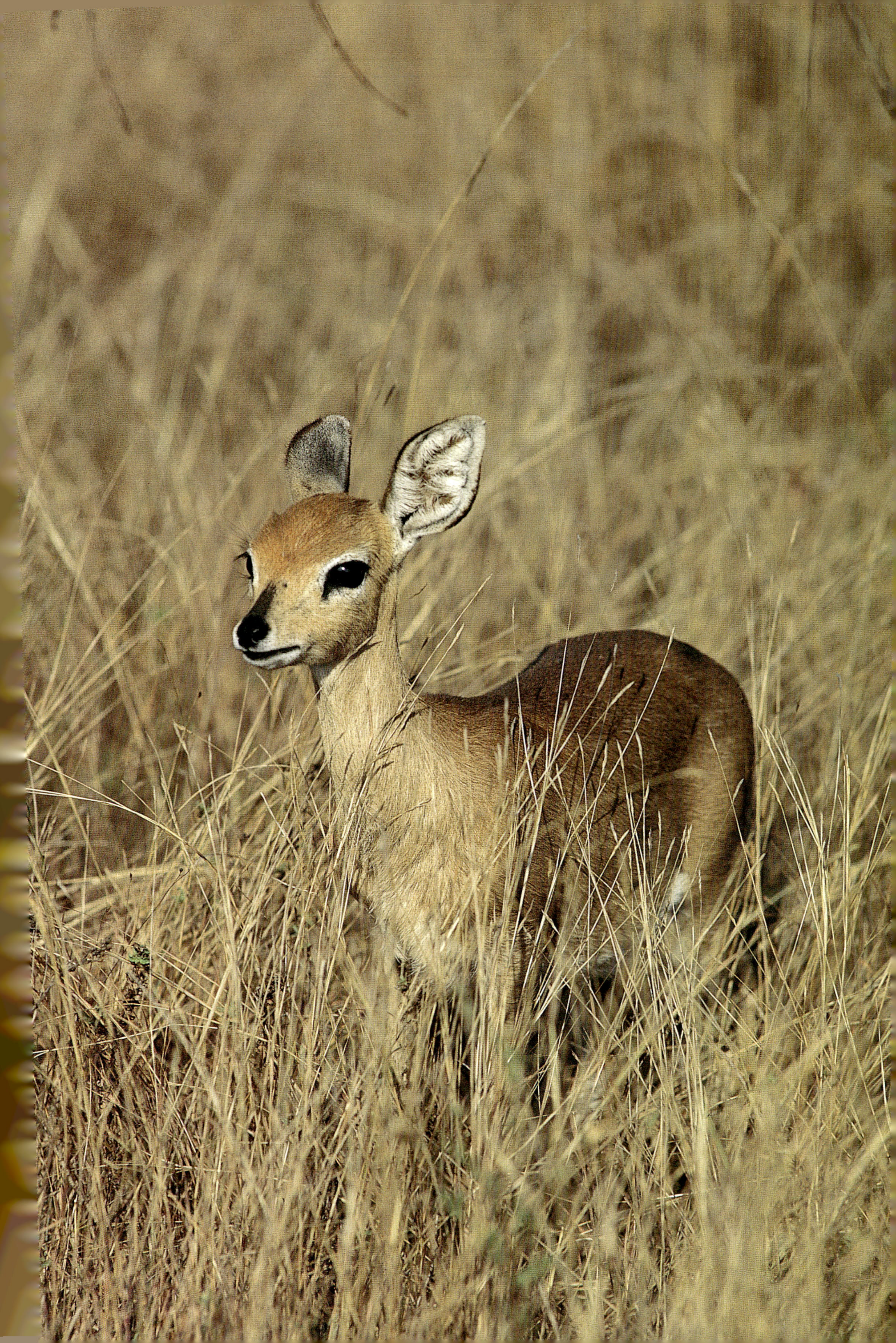

nivore. The Dwarf mongoose weighs in at eleven ounces and stands only nine and a half inches tall. It is the size and color of a red squirrel. Its features, behavior and pack structure are much like those of the suricate, but there are differences.

Dwarf mongooses favor thickets and woodlands over dry savanna. They use termite mounds for burrows. They are curious creatures, and when they discover new things they twitter with excitement. The alarm calls of Dwarf mongoose sentries differ depending on whether the danger is from the sky or the ground. Due to their small size, snakes and other, larger mongooses hunt them as well as birds of prey. Running to cover is their main defense, but they are also able to jump up to three feet high. Growling, spitting and trying to look larger is an effective deterrent when done en masse.

Like the suricate's, the Dwarf mongoose pack has a dominant pair. They are the land holders and they alone mate and produce progeny to the exclusion of all other pack members. The other members often emigrate from pack to pack in hopes of some day being part of a dominant pair. Since Dwarf mongooses can survive up to ten years in the wild, the odds are good that many of those individuals will live to mark invisible scent trails and leave signs on the boundaries of their very own territory.

Signs that say, – No Trespassing.

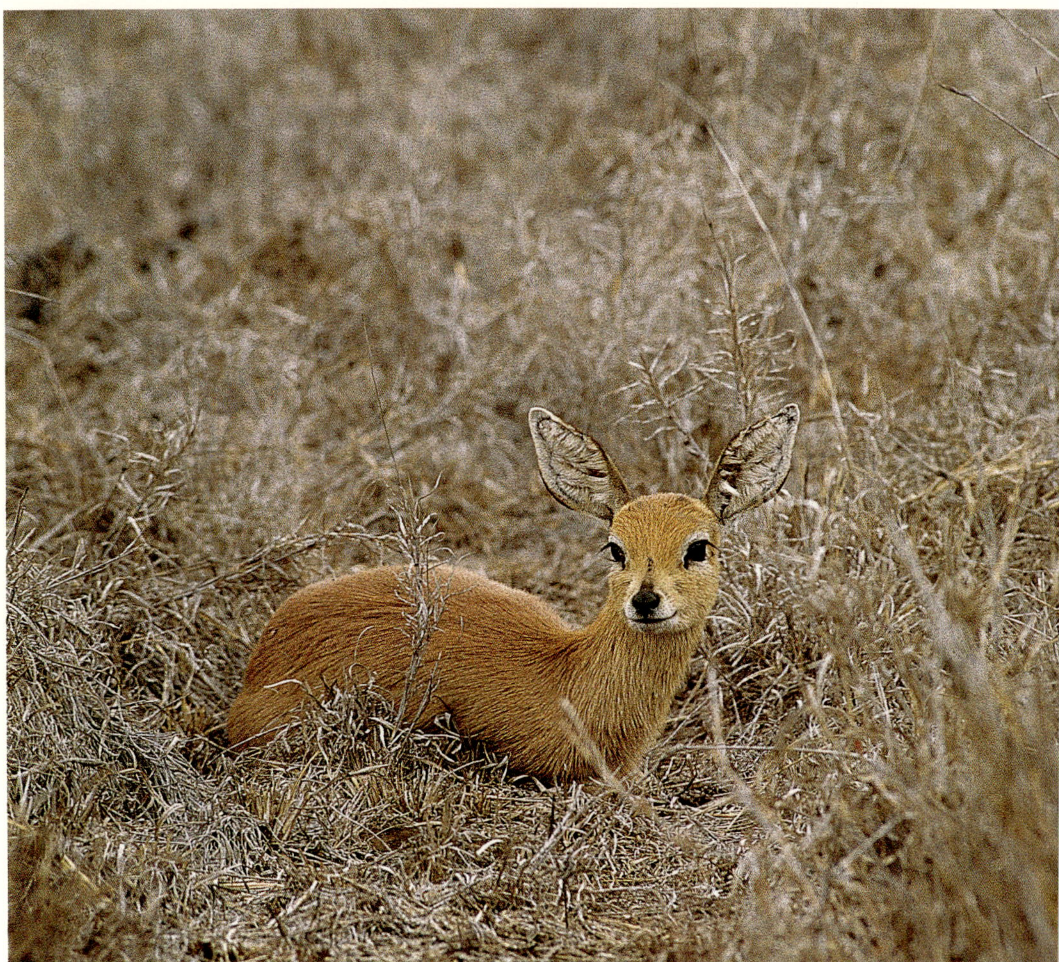

*A steenbok will use regular resting sites as long as there is enough cover to hide from predators. The young, like this one, have woollier coats than do adults.*

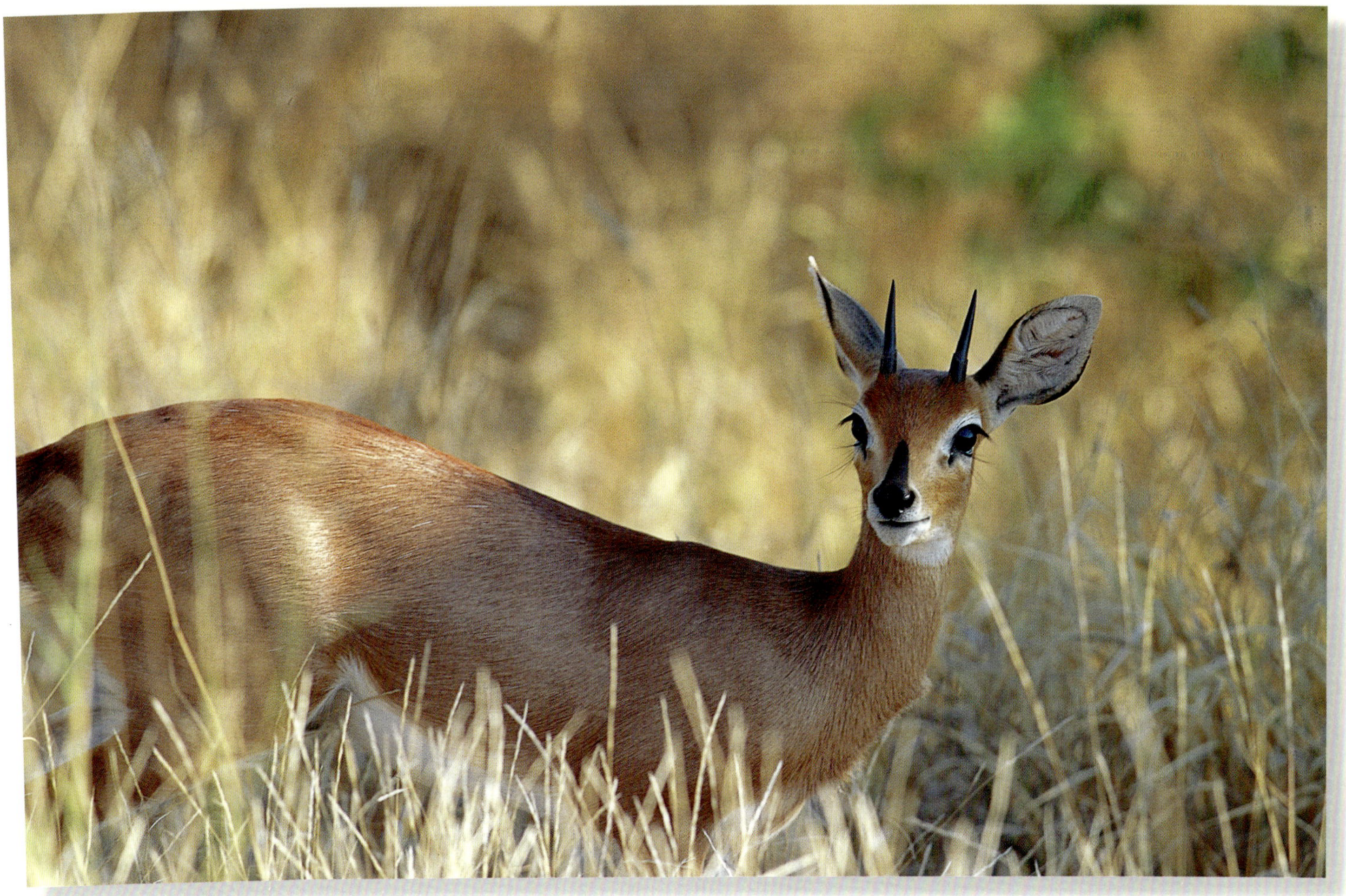

*Steenboks are monogamous but more often than not, they are seen singly.*

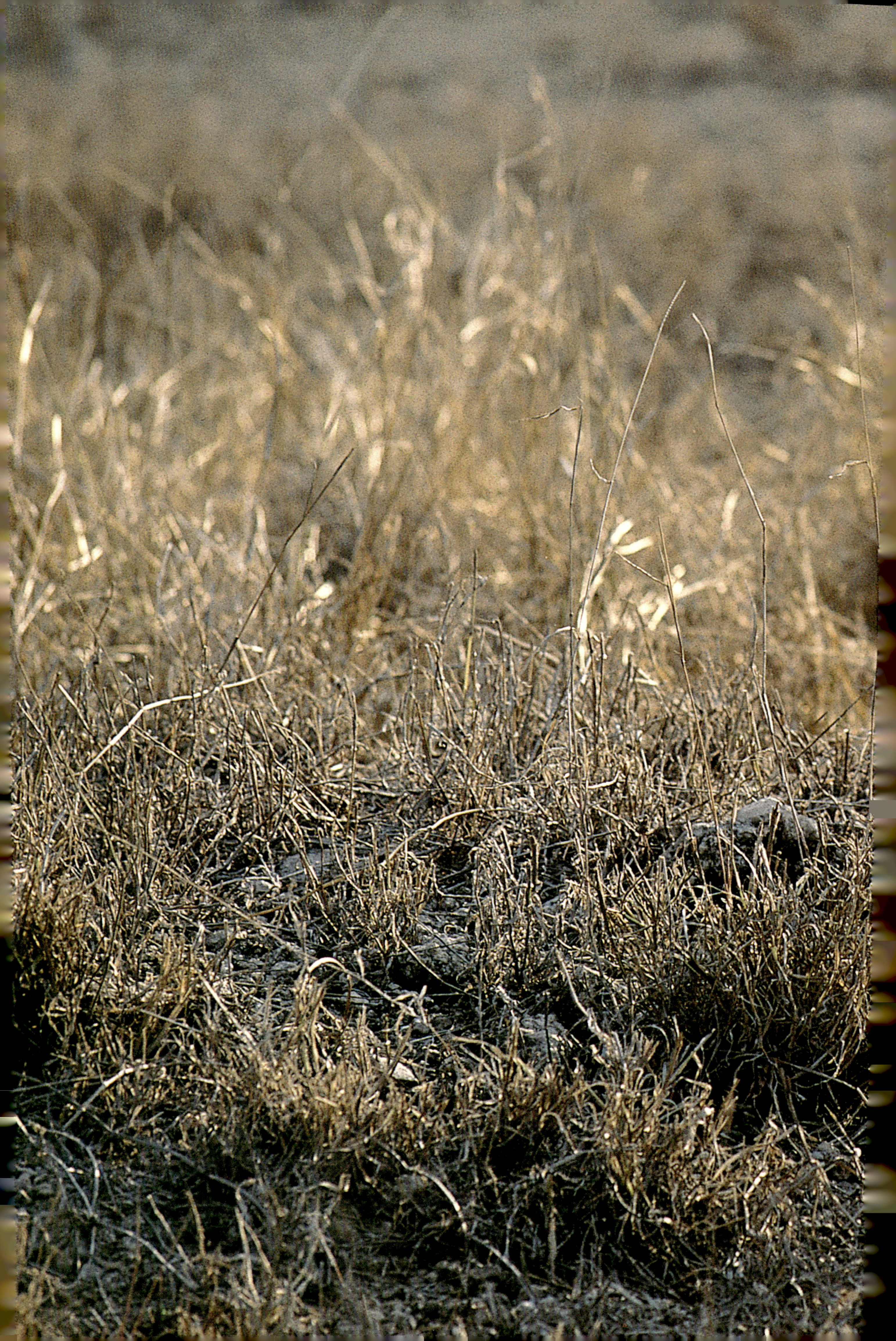

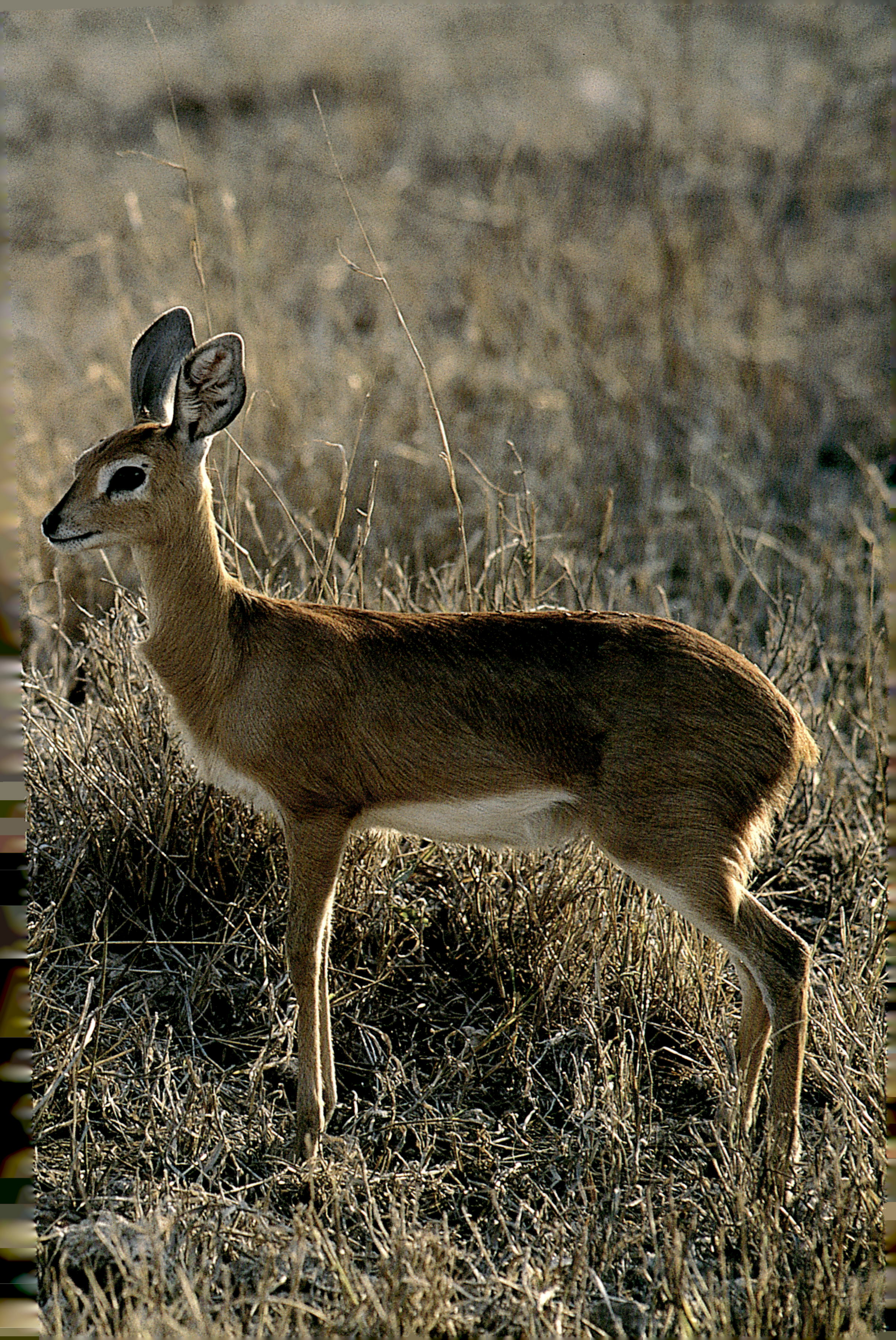

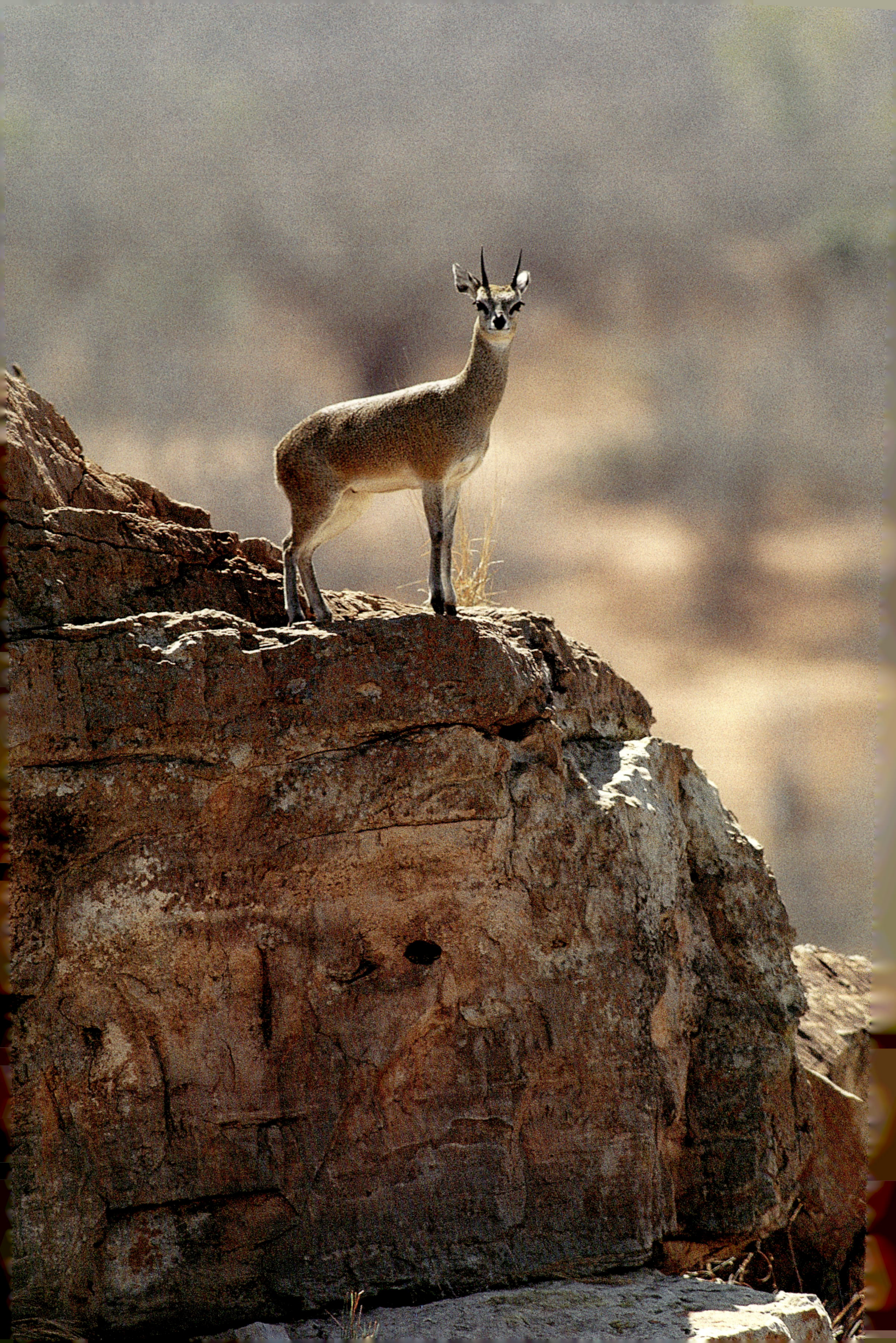

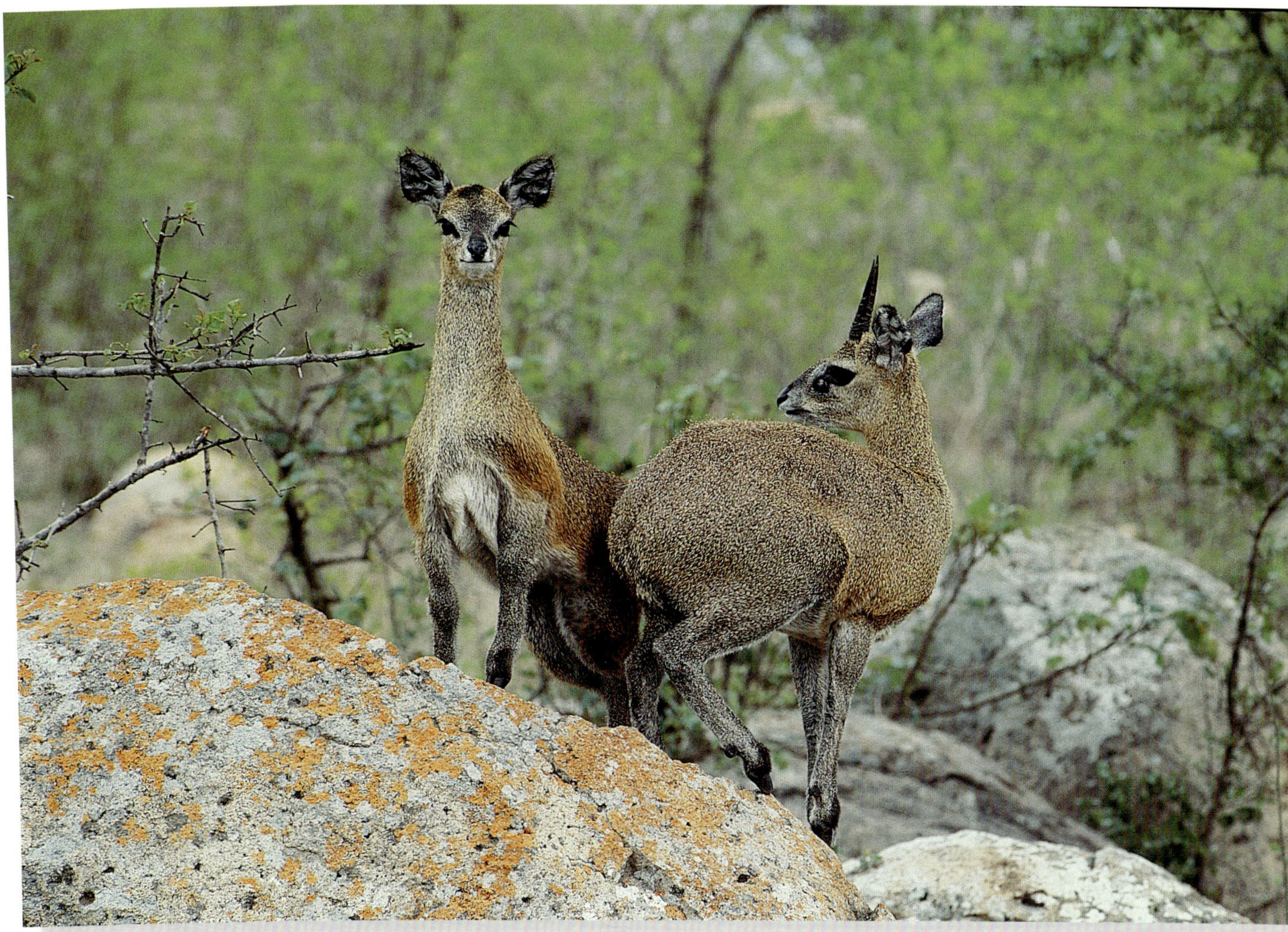

*Klipspringers are territorial and monogamous. They are vigilant and must be prepared to flee any invading predator.*

PRECEDING PAGES

*The harmony of nature's colors between foliage and the steenbok's coat vividly portrays how well this small antelope blends into its surroundings.*

FACING PAGE

*I am amazed at how still klipspringers remain when standing sentry duty. At times they may remain motionless for up to twenty minutes.*

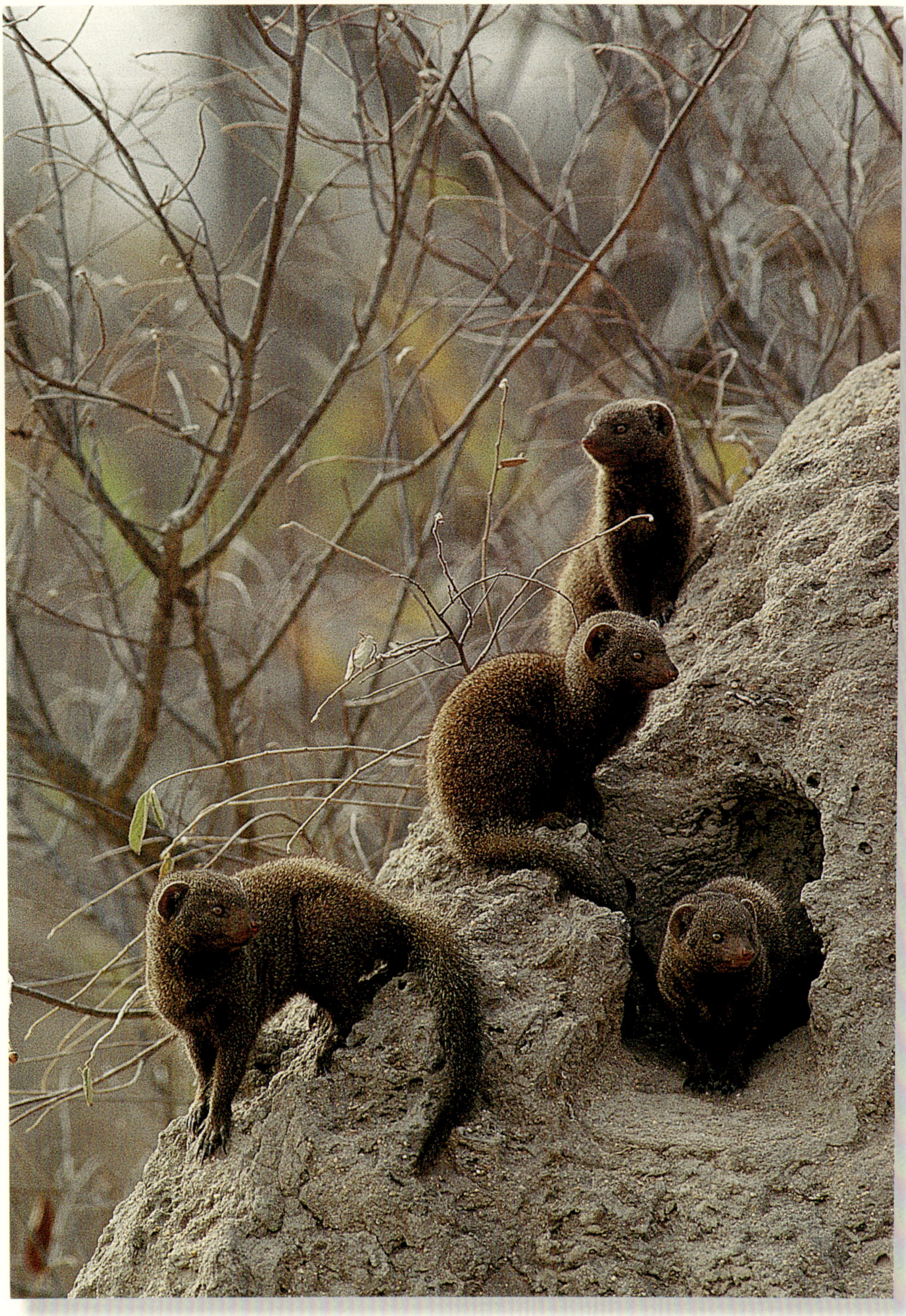

*Dwarf Mongoose packs will use abandoned termite mounds for sleep and refuge.*

FACING PAGE

*While one meerkat digs for tasty tidbits, such as beetle larvae, another meerkat will stand guard, mainly against hawks and eagles.*

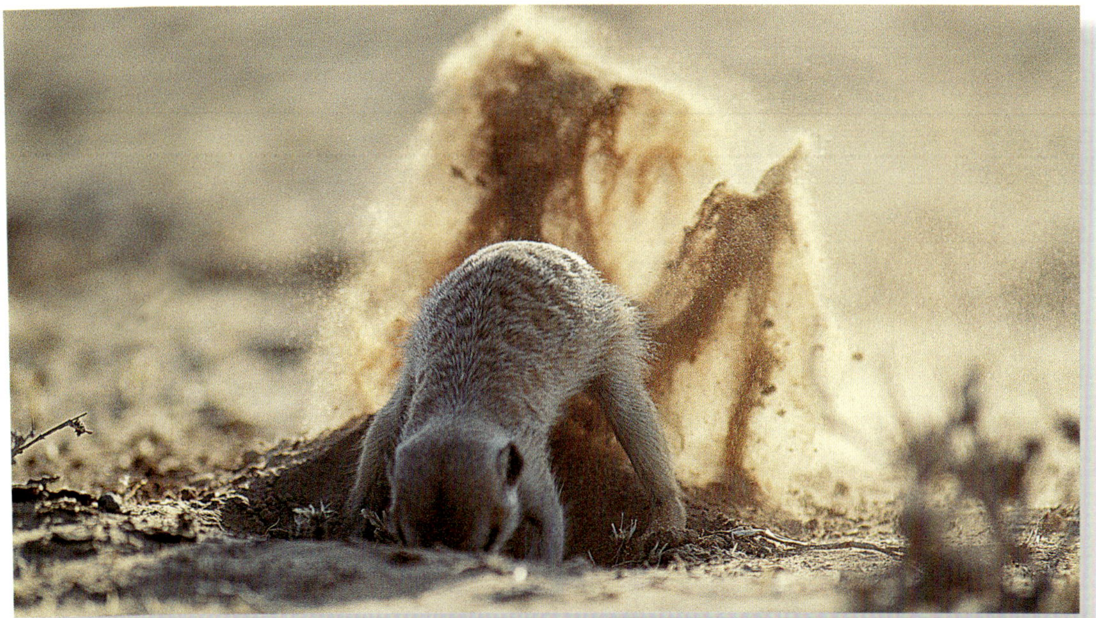

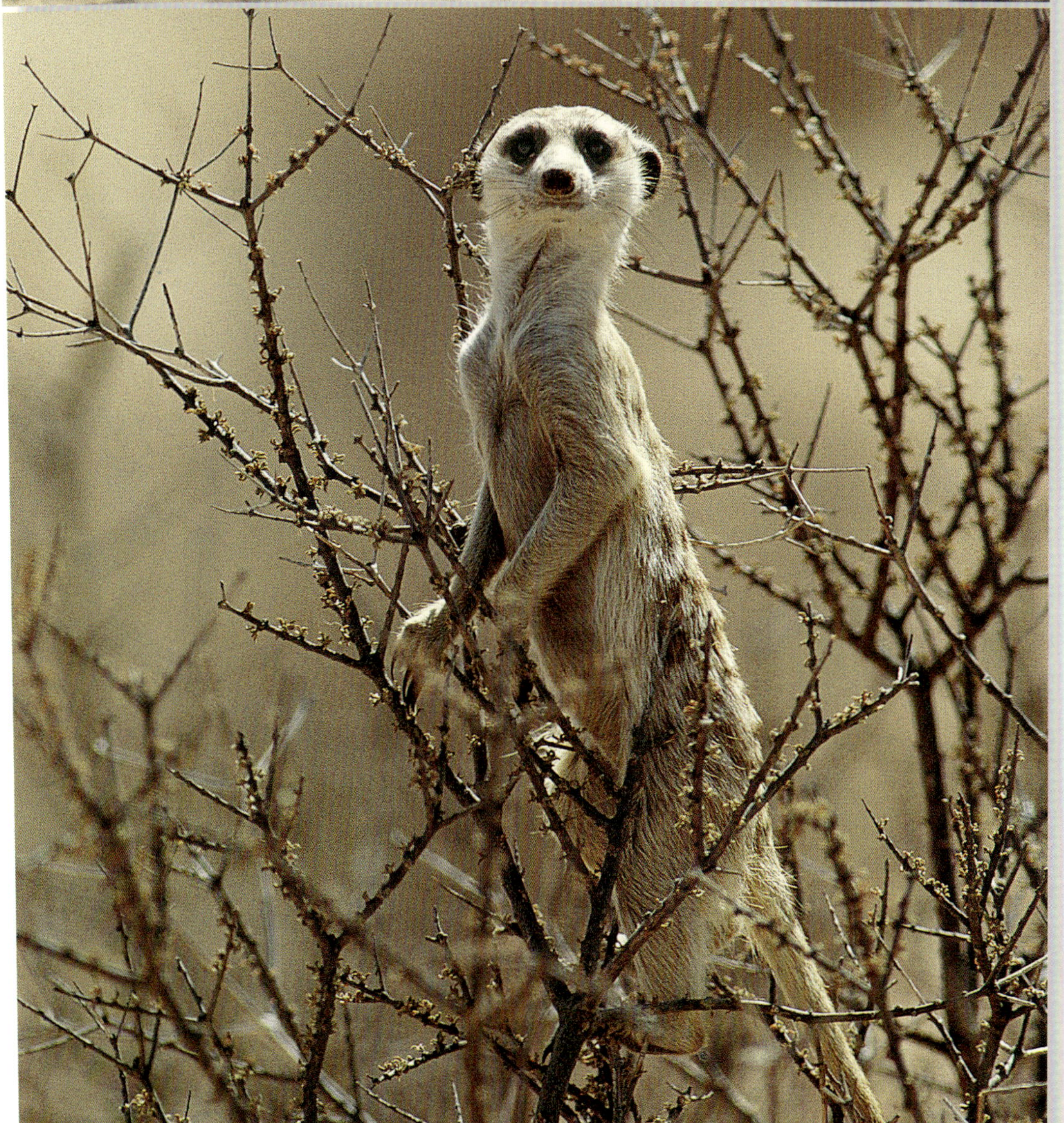

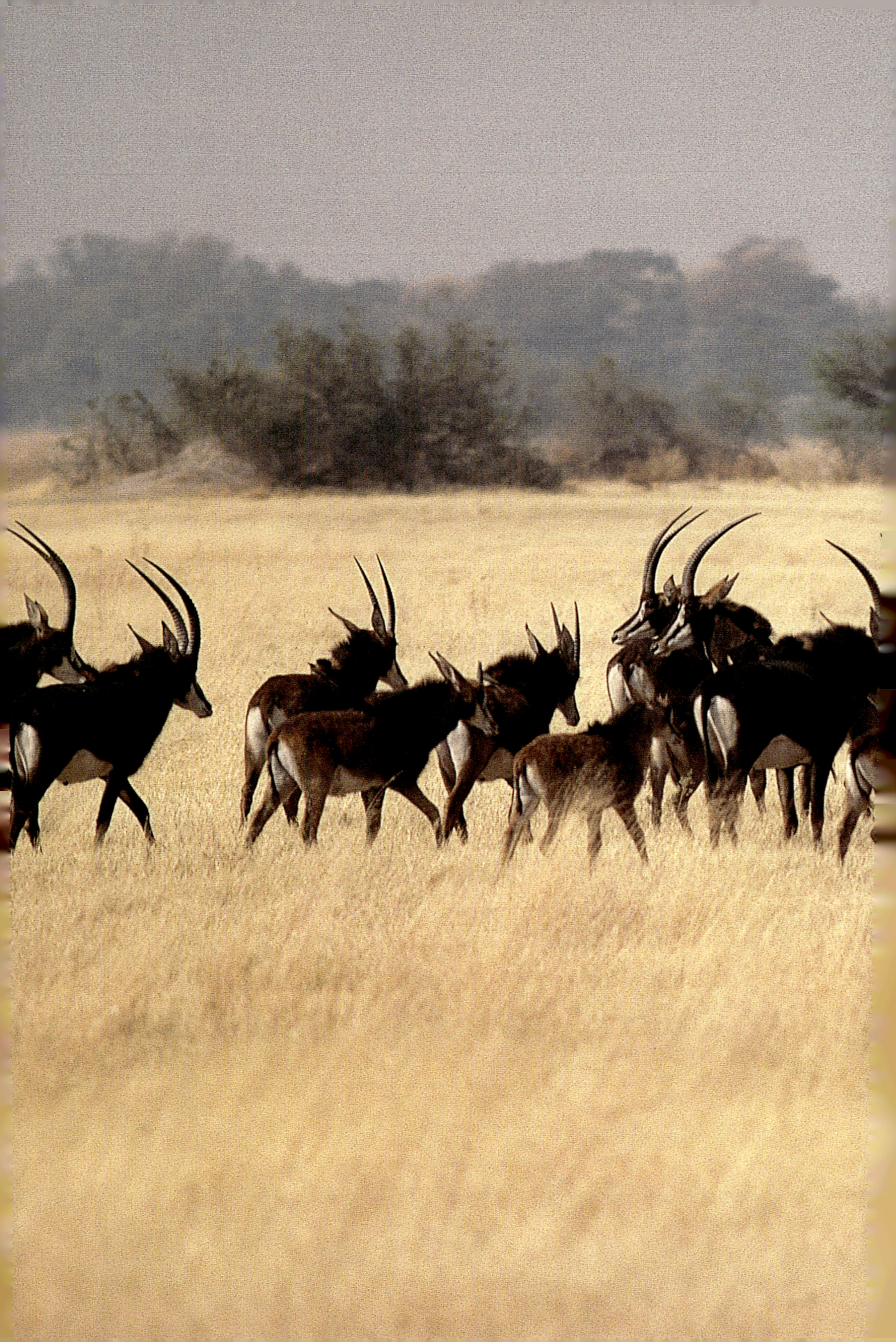

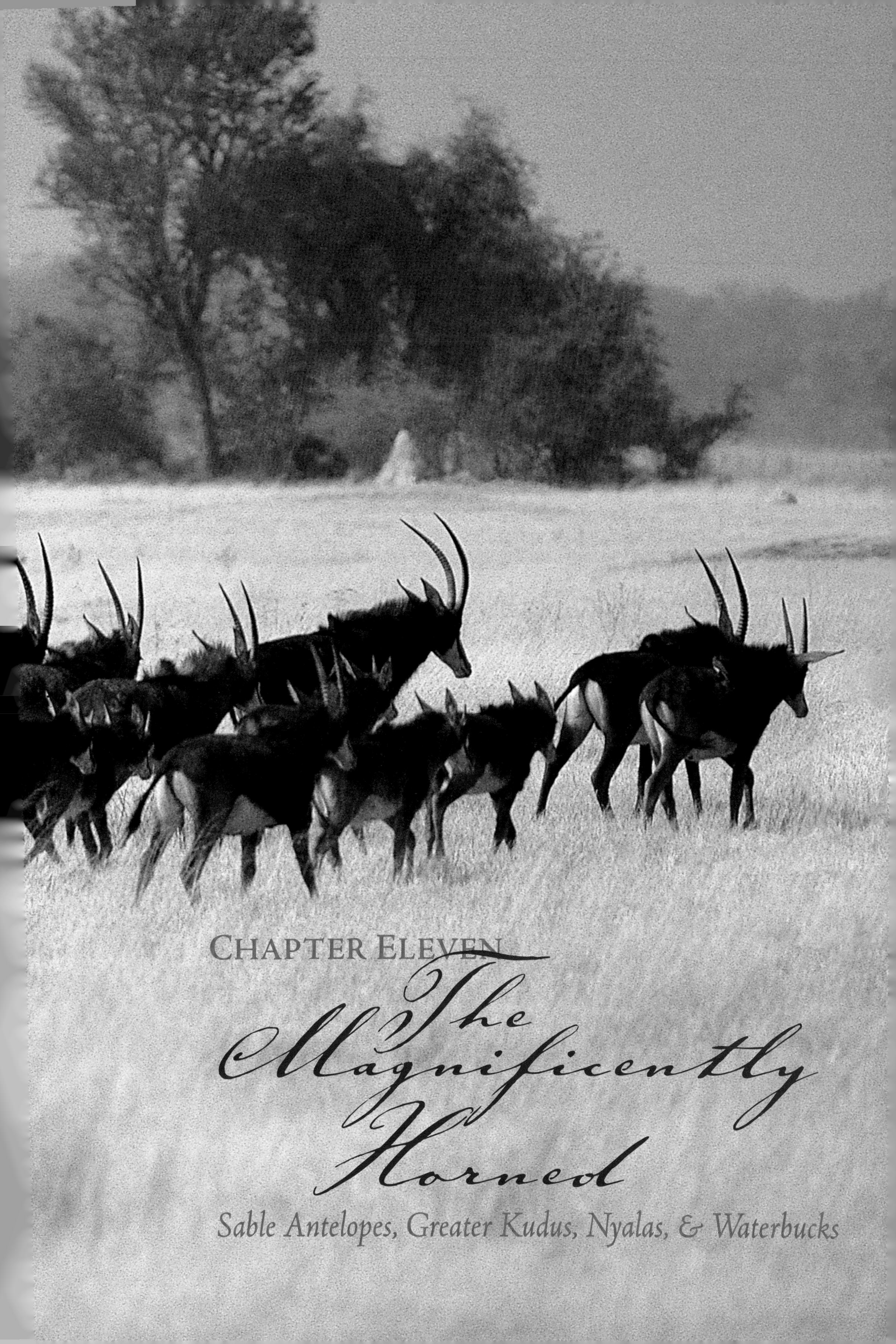

CHAPTER ELEVEN

# The Magnificently Horned

## Sable Antelopes, Greater Kudus, Nyalas, & Waterbucks

Like graceful crowned princes, the magnificently horned antelopes reign over Africa's grasslands and woodlands. Adorning the head of each species is a unique type of horn, aptly described by shape and form; crescent, spiral, corkscrew and scimitar, to name a few. Horns are important in antelope society. They can determine rank – the bull with the longest horns is likely to be lord of the land. He wields his horns in displays of aggression toward the lower ranks. They become weapons in combat against those that challenge his authority. Females of some species recognize the master by sniffing his horns as a prelude to mating. In many species, cows wear horns as well.

PRECEDING PAGES

*The Okavango Delta was home to this Sable antelope herd. It is made up of cows and immature males and females.*

FACING PAGE

*A unique feature of the Sable antelope is the set of scimitar shaped horns that can grow to five foot lengths.*

The majestic scimitar-like horns of the Sable antelope can be up to five feet long. They are massive, ridged and have sharp points. The sable bull is large and powerful, weighing up to five-hundred pounds and standing fifty inches tall. It has a thick neck with an upstanding mane. The bull's coat is black while the cows coat is a chestnut color. Both have white under-parts, white face patches and tails with long fringes.

Sable antelopes live in savanna woodland near grassland and a source of water. An area where cover is available is essential. In early morning and late evening, sables graze on fresh green grasses and browse on foliage and herbs. They frequent natural salt licks and will chew bones for their mineral content.

Sable herds are usually small, consisting of fifteen to twenty-five cows and calves. The territorial bull stands apart from the herd and is often overlooked. But, it is heard, at times, emitting a piercing bird-like whistle. Young males live in bachelor herds until they are large enough and strong enough to claim their own territory or to challenge a reigning bull for his.

In the mating season, the master bull warns off intruders with dominance displays. He raises his neck crest to look larger and swings his scimitar horns. Any challenger is met with loud bellowing and horn to horn combat. To the victor go the cows.

At the end of the rainy season, eight months after mating, each female gives birth to a single calf. For three weeks, the calves are kept concealed. If danger is perceived, the young sables will freeze, hoping to escape detection, rather than attempting to flee. At six to eight months old they are weaned and will join the herd. Sable predators are few, owing to the antelopes' large size and their hiding behavior. The young, however, fall prey to spotted hyena and leopards.

One tribe of antelopes, the Bushbucks, are characterized by spiral shaped horns – usually worn only by males. They are not territorial, tolerating the presence of younger males and other antelopes in their area. Herd rank is determined by size and seniority – the largest and oldest holds the highest rank. All Bushbuck tribe members have spinal crests, stripes and spots decorating their hides. They are built for maneuvering deftly through thick undergrowth rather than for running. They live in thickets and dense woodland, securely concealed from most danger. Hearing is their best sense, followed by sight and smell. The tribe is social. Licking and grooming are important bonding activities for young and adults alike.

The grandest member of the Bushbuck tribe is the Greater Kudu. The bull's spectacular ridged

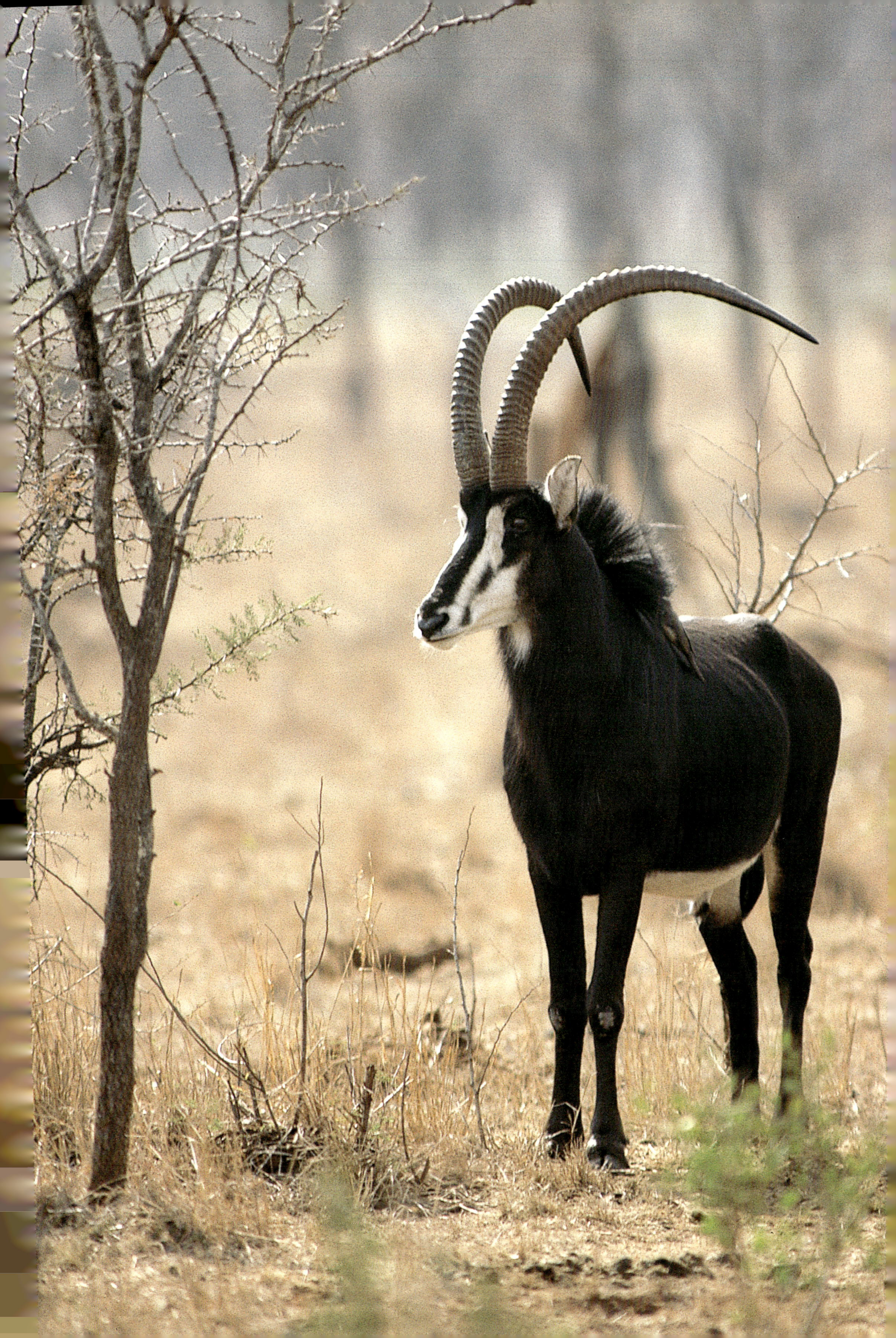

horns spiral up to six feet long in two and a half corkscrew turns. Cows sometimes have smaller horns. The kudu is the second largest of antelopes, behind the eland. The male can weigh over six-hundred pounds and stand forty-eight to sixty inches tall. The female weighs two-hundred pounds less and is six to eight inches shorter. Proportionate to its long narrow body and long legs, the kudu has a small head, albeit with large fringed ears. The face has a white nose chevron and white cheek spots. The coat is smooth, red-brown to blue-grey, with five to fourteen white transverse stripes. The tail is of medium length, white on the underside and black at the tip. Both sexes have a spinal crest, but only the male wears a beard and belly fringe.

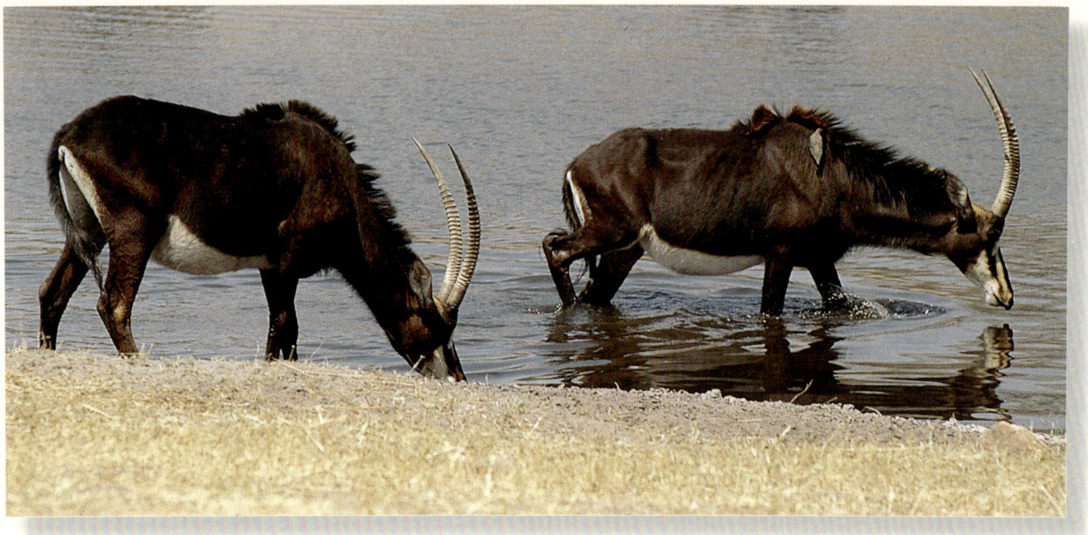

*Sable antelopes only leave a good pasture for water or to find shade during the hottest part of the day.*

Kudus prefer woodland with a mix of scrub and trees such as acacia. Within their densely wooded habitat, the shy kudus gather in small herds, usually with no more than four females and calves. Sometimes senior males are included. Bachelor herds of from two to ten members are loosely organized, forming and dispersing only hours apart. In the wet season, herds of up to thirty can be found.

Like many animals, early morning and late afternoon are feeding times for kudus. Many kinds of leaves, herbs, fallen fruits and melons, succulents, flowers and new grass make up their varied diet. They spend up to half of a twenty-four hour period foraging.

Greater kudus are very vocal animals. The bark of the bull kudu is one of the loudest sounds that antelopes make. It is used to advertise dominance along with body language and bristling of dorsal crest and tail. Courtship is occasion for whining, gasping, clucking, grunting and humming sounds by the male.

Successful courtship leads to mating and nine months later, in the rainy season, single calves are born to each mother. Tall grasses brought about by the heavy rains offer excellent concealment for the calves for two or three months. Cows and calves are close. Mothers and other females engage in social licking to calm and reassure the young. As they grow and begin to wander, the mothers call them back with smacking sounds. When distressed, calves call out – u-u-u-u. At six months, calves are independent but stay with the herd for up to three years.

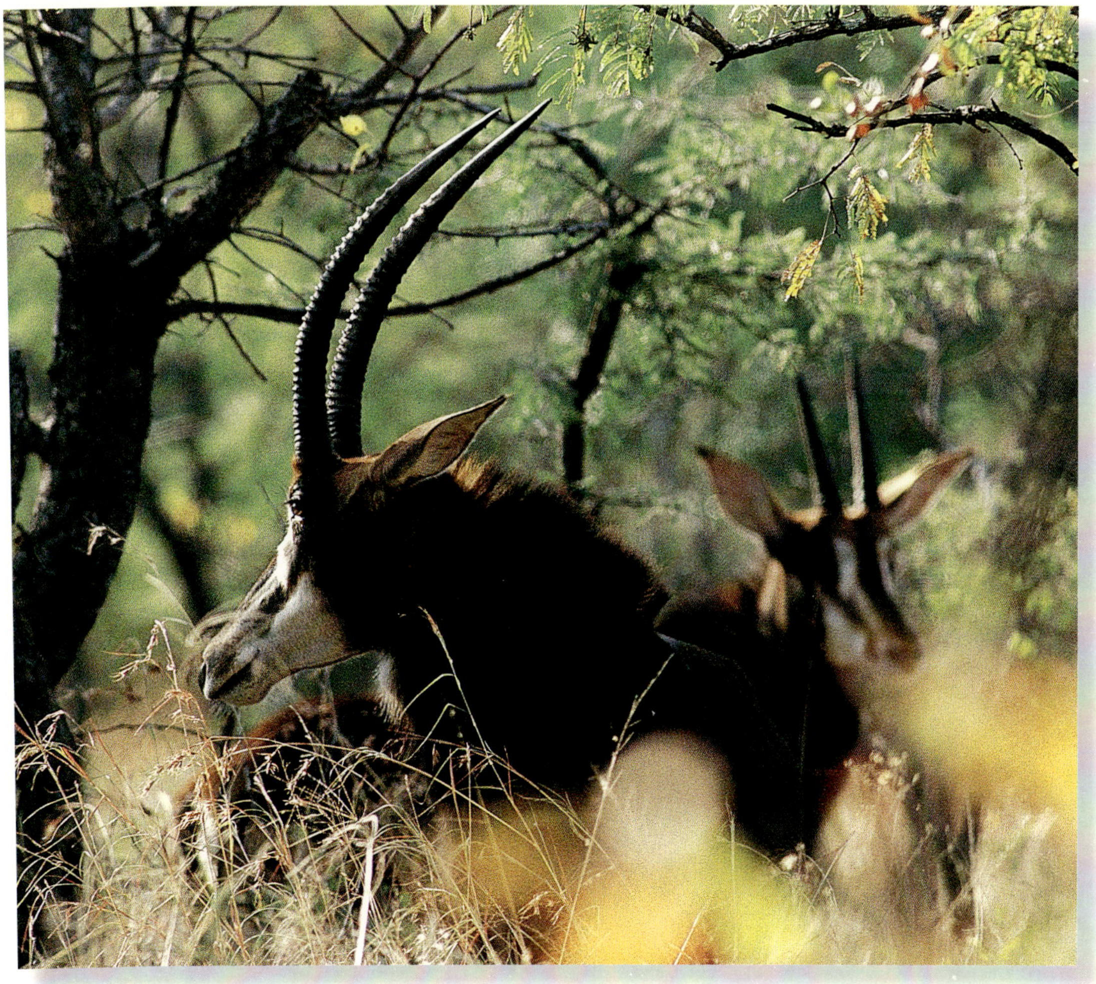

*A Sable cow and calf find refuge and shade in this heavy thicket.*

Kudus watch out for lions and Spotted hyenas. If they see a predator, but are not seen, they rely on the ability to sneak away into the bush rather than bolting. If they are pursued, they run, tail upturned, exposing its white underside so that others may easily follow through the underbrush. Males hold their horns flat along their backs to prevent snagging on branches. And they make spectacular leaps – clearing obstacles up to eight feet high.

Thickets and dense woodlands are home to another member of the Bushbuck tribe, the nyala. The spiral horns of the nyala twist one and a half or two and a half times, attaining a length of twenty-four to thirty-three inches. Male and female differ greatly from each other in size and coloration. The male weighs up to two-hundred seventy-five pounds and has a longish coat of charcoal grey with a long white spinal crest from head to tail. Below, hangs a dark fringe of hair from throat to hindquarters. The female weighs half that of the male, at about one-hundred thirty pounds. Her coat is shorter and bright chestnut in color with eight to thirteen white stripes across her torso, and white spots and chevrons on her chest and head. Males have fewer or no stripes at all.

Their coats aid in camouflage for these secretive antelopes. They spend daylight hours in dense cover – emerging at dusk to graze and browse. Deep in the night they rest and at dawn they again for-

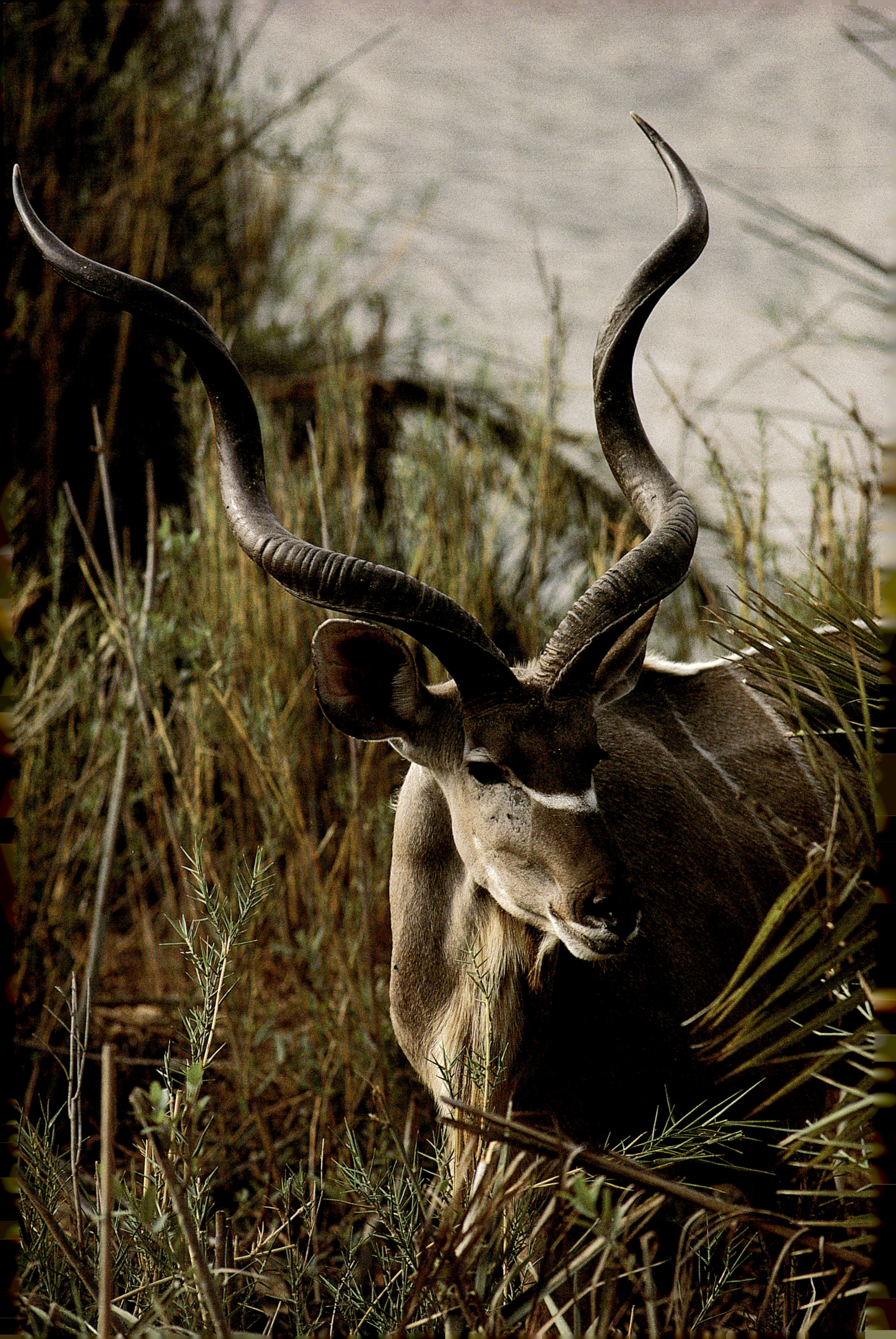

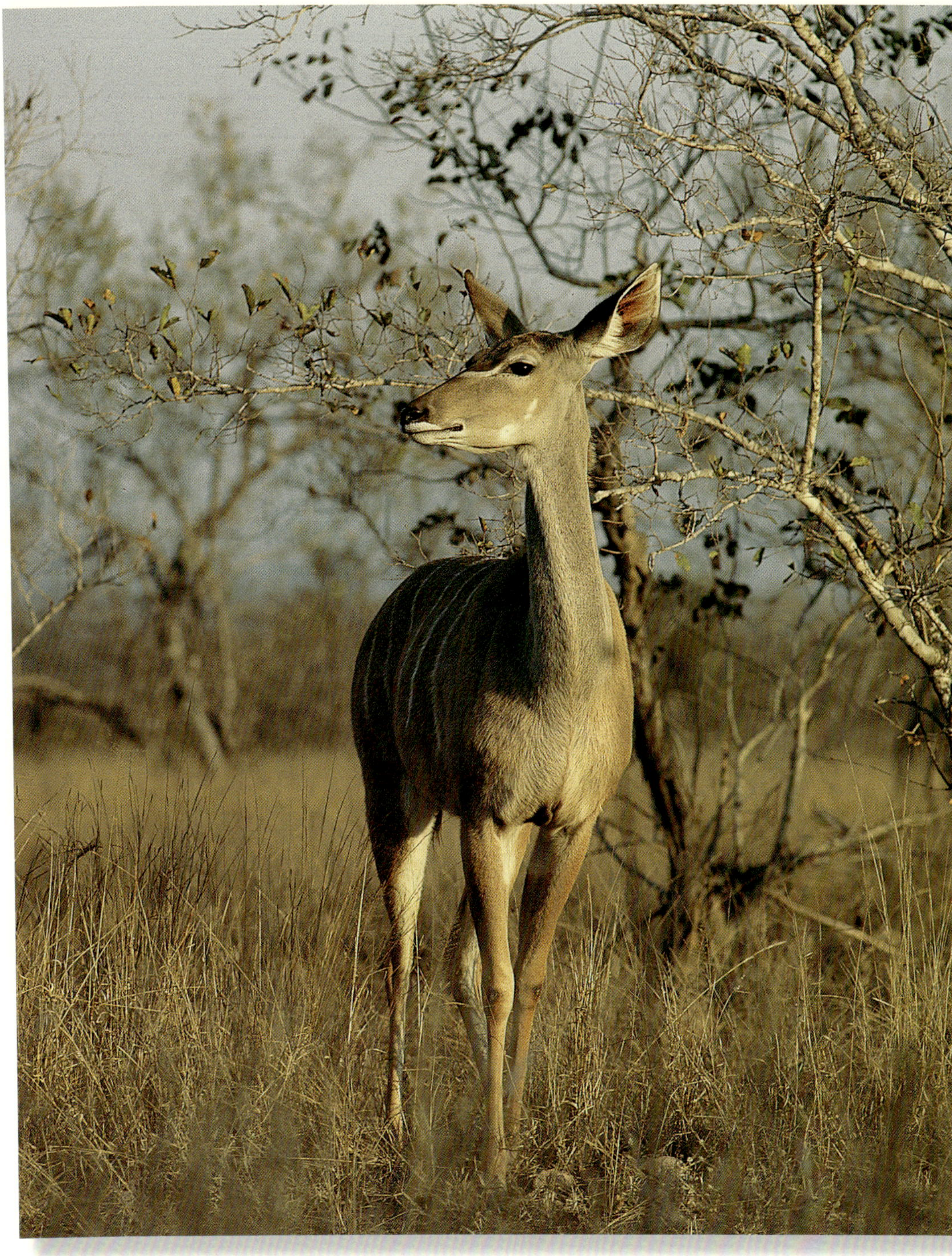

*The habitat of kudu must provide plenty of cover furnished by bushes and thickets.*

age. In the rainy season, nyala seek fresh green grass. In dryer months, leaves, pods, fruit and herbs make up their diet. They can be seen standing on their hind legs to browse. At times, nyalas gather below trees full of baboons, to munch the leaves and fruit dropped by the feeding monkeys. They make daily trips to water holes, as well, in small herds of five or six females and calves. There they may meet other herds where their territories overlap.

FACING PAGE

*This magnificent kudu bull is probably the handsomest antelope that I have ever photographed.*

Breeding occurs year round with peaks in early summer and fall. Gestation takes eight months for nyalas. After calves are born they are hidden for ten to eighteen days. During the time of

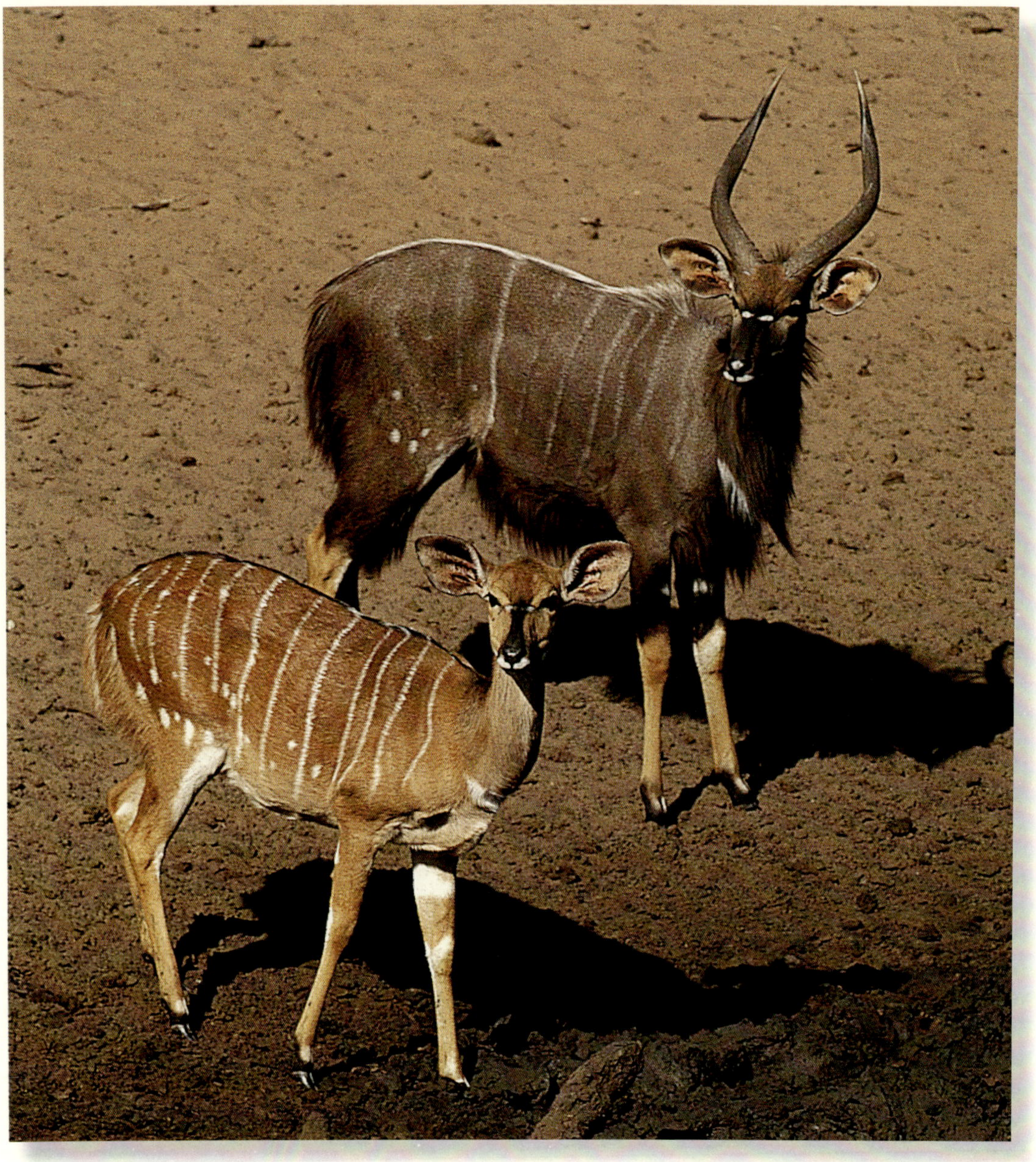

*Gender differences between male and female nyalas are extreme. A male can weigh up to two-hundred-seventy-five pounds. while a female can weigh up to one-hundred thirty pounds. Males have a dark charcoal gray or brown coat and the females have a light chestnut color. Only the males have horns that are between two to three feet long.*

hiding the mothers remain apart from the herd. One interesting fact about nyala females is that they are ready to mate again just one week after giving birth.

FACING PAGE

*I photographed this impressive nyala bull on the edge of a riverine forest. I found him at Pafuri, which is in the far north of the Kruger Park. They are related to the Greater kudu and are just as shy.*

Because they are smaller, nyalas need fear more predators. Lions, Spotted hyena, leopard and wild dogs hunt them. It is to their advantage that they react to the alarm calls of baboons, impalas and kudus.

In order to see another magnificently horned antelope, follow your nose. The waterbuck can be detected by smell long before it is seen, for its coarse greasy coat is redolent of a strong musk-like scent. Then, when spotted, the focus becomes the long crescent shaped horns; heavy, with deeply ridged rings – stacked like coins to a length of up to thirty-nine inches.

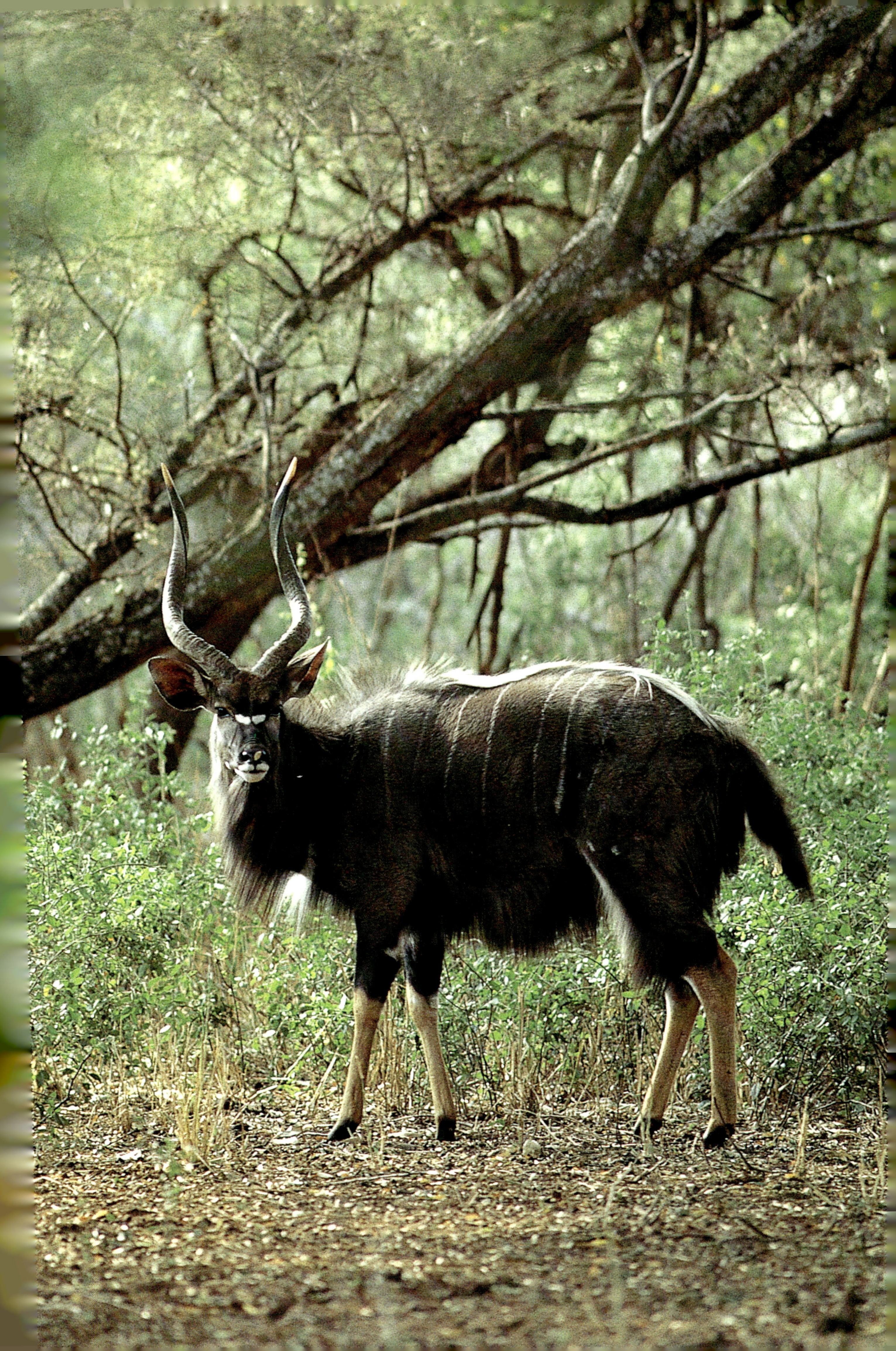

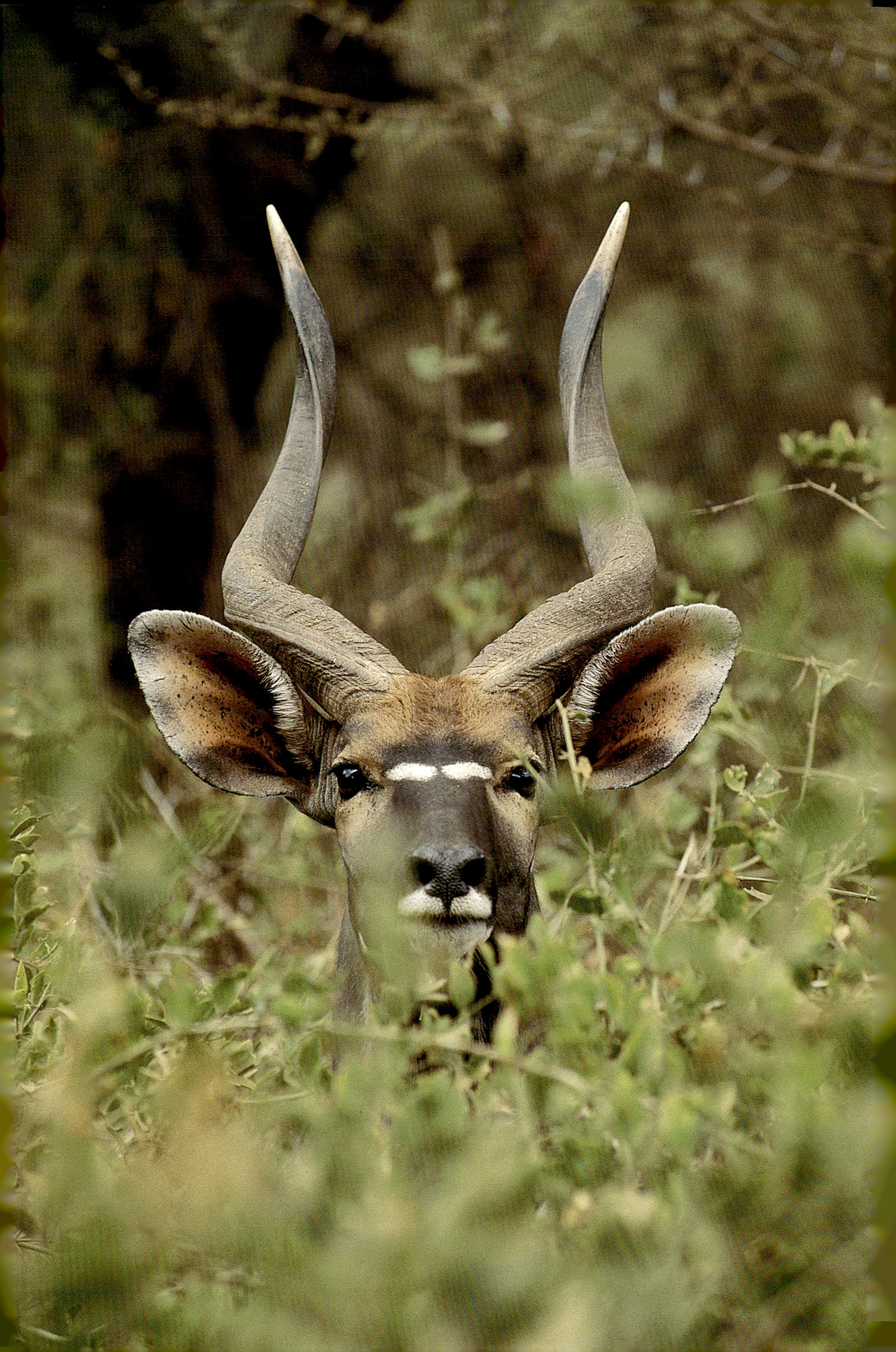

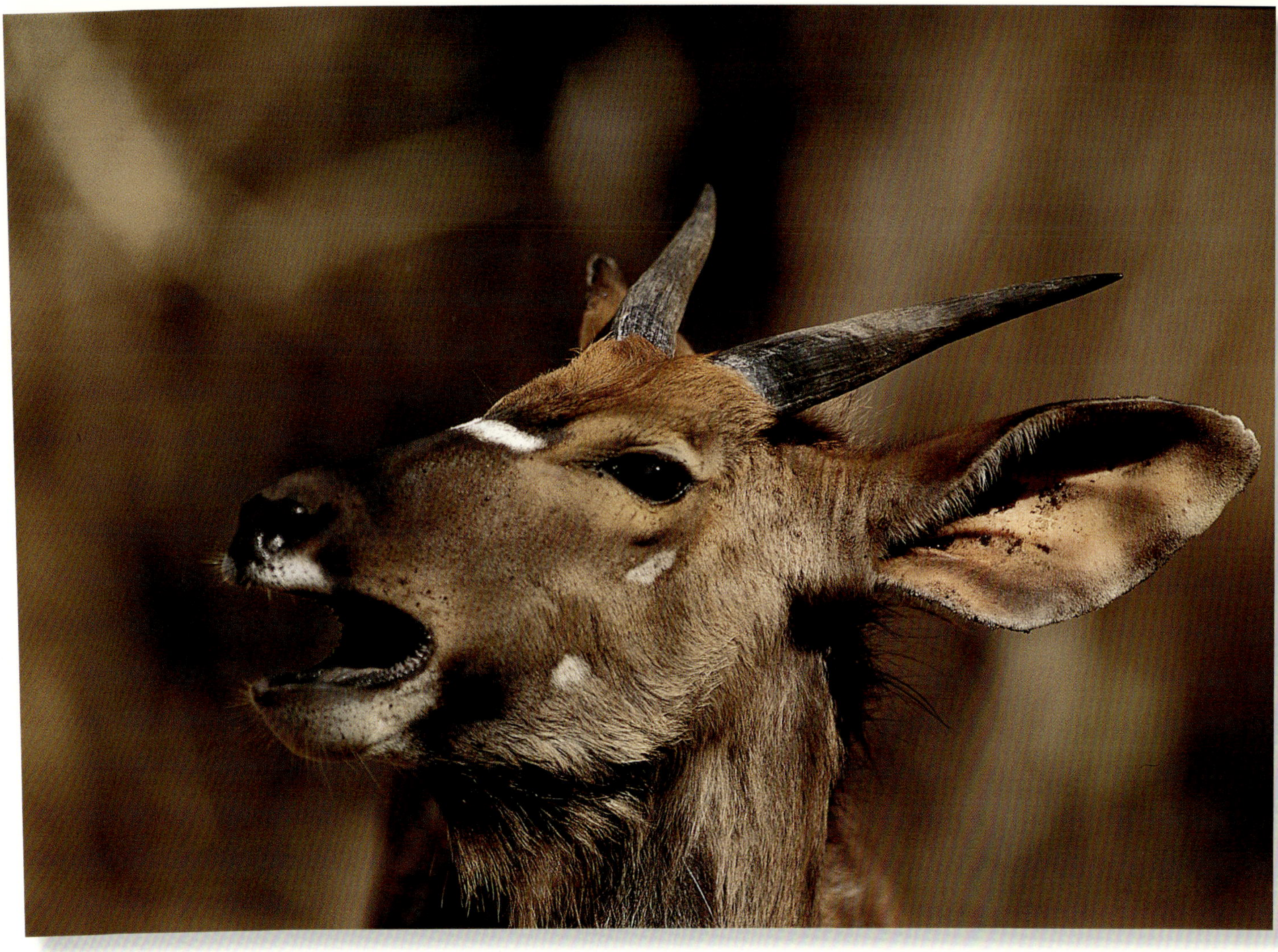

*We were able to get very close to this young nyala male, probably because he was more interested in his food than in us.*

The waterbuck is large, weighing up to five-hundred twenty-five pounds. It stands four to four and a half feet tall at the shoulder. The waterbuck's coat is yellowish-brown to red-brown, grizzled with white and grey hairs. It has black lower legs and a tail with a long black tuft. The neck is maned and the ears are large and hairy, white with black tips. White highlights the eyebrows, snout, throat and underparts. One distinctive marking is a broad white ring around its rump.

Waterbucks are an edge species, inhabiting grassland and woodland near water. Aptly named, they must drink often, every day or two.

Small herds of six to twelve members graze tender, protein rich grass that grows near water. In drier times they browse foliage and herbs. Waterbuck males are territorial but defend their land only against strong mature bulls. Younger males are tolerated if they exhibit submissive behavior like females. A master bull performs intimidation displays. He will stand proudly with his head held high, prominently showing off his white throat and patches near the eyes. Or, he may stand laterally, flaunting a thick strong neck. When waterbucks fight, the battles are more fierce than between other antelopes, sometimes resulting in death.

Facing Page

*It appears that this nyala decided to take a closer look at us. He gave me just enough time to get this shot, after which he disappeared into the thicket.*

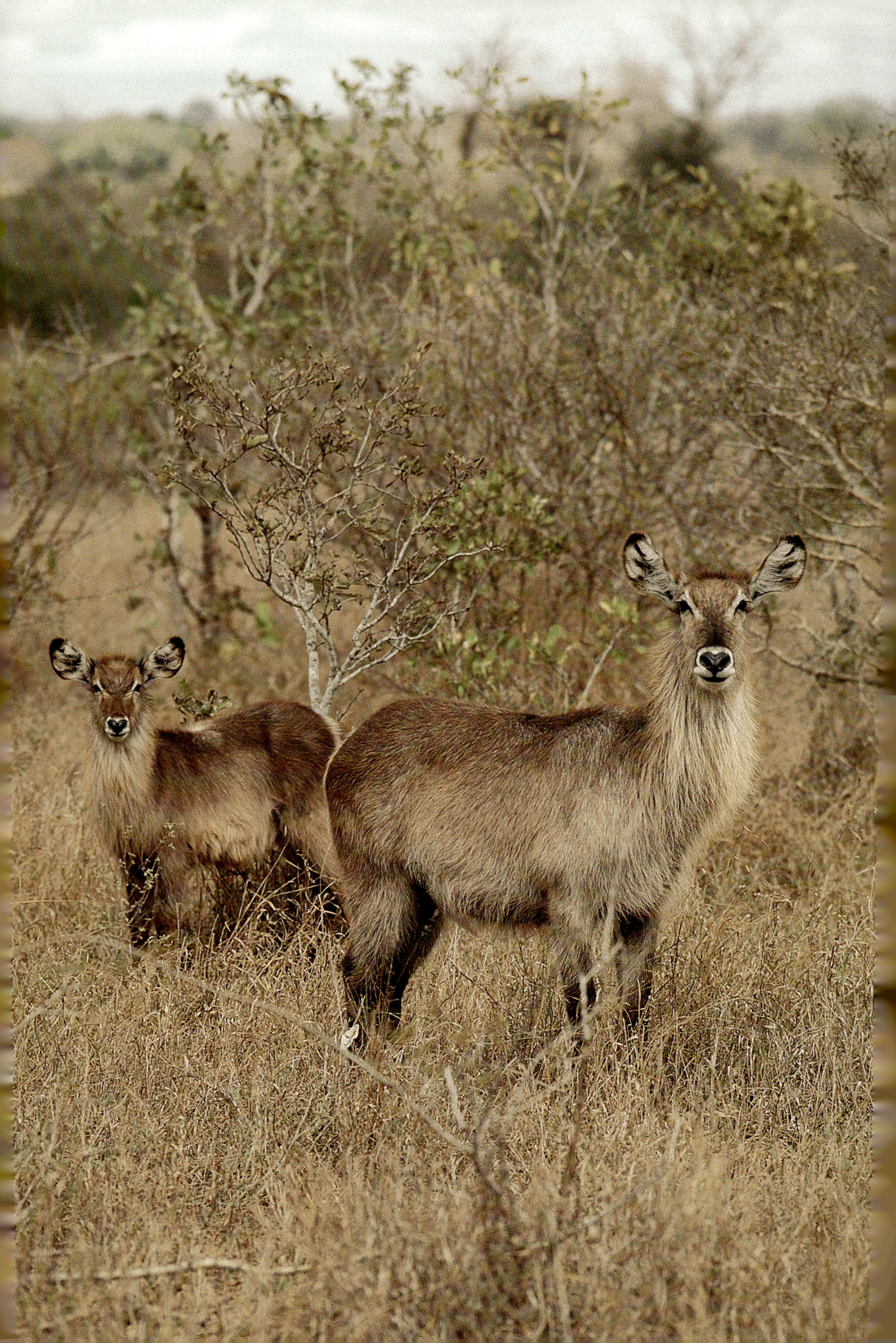

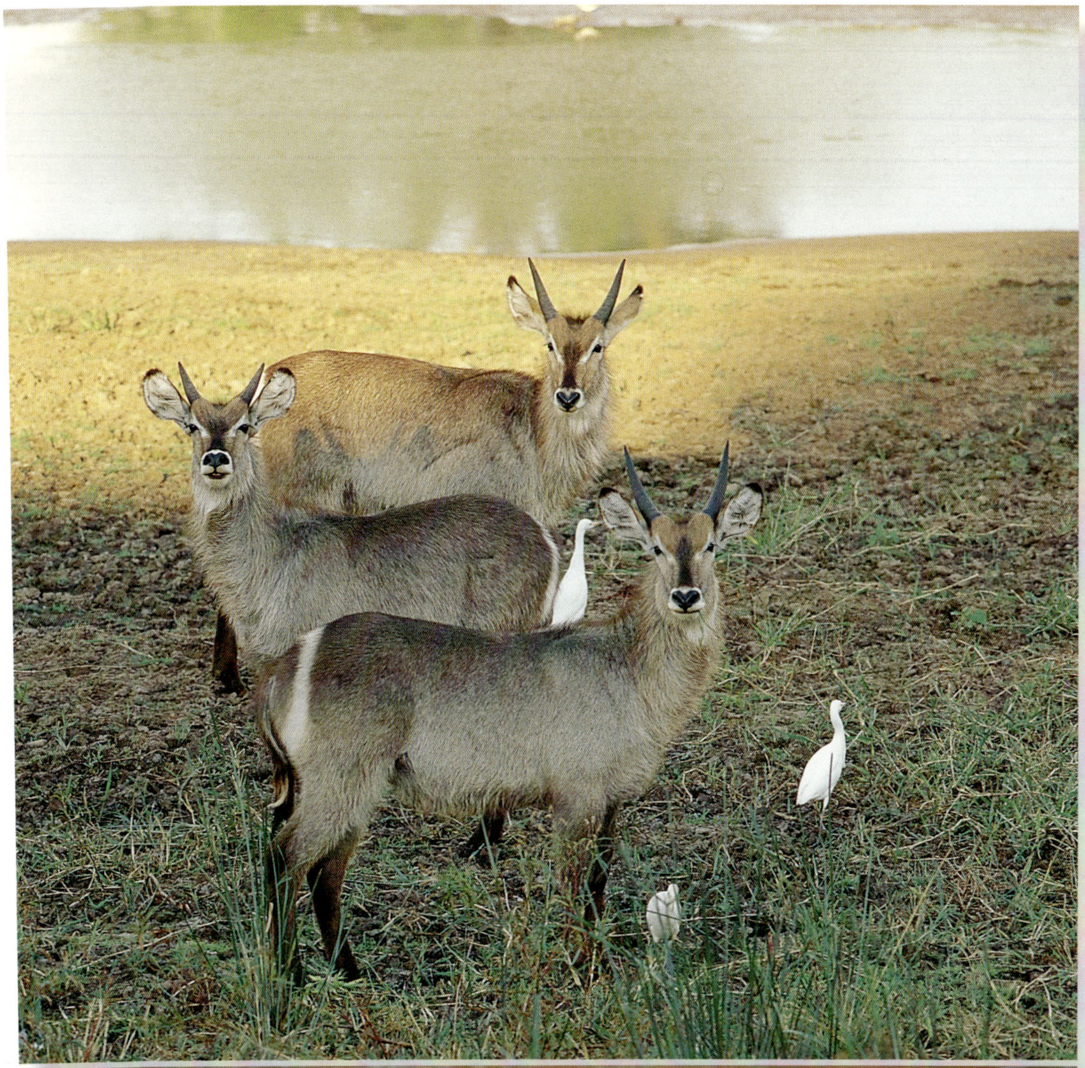

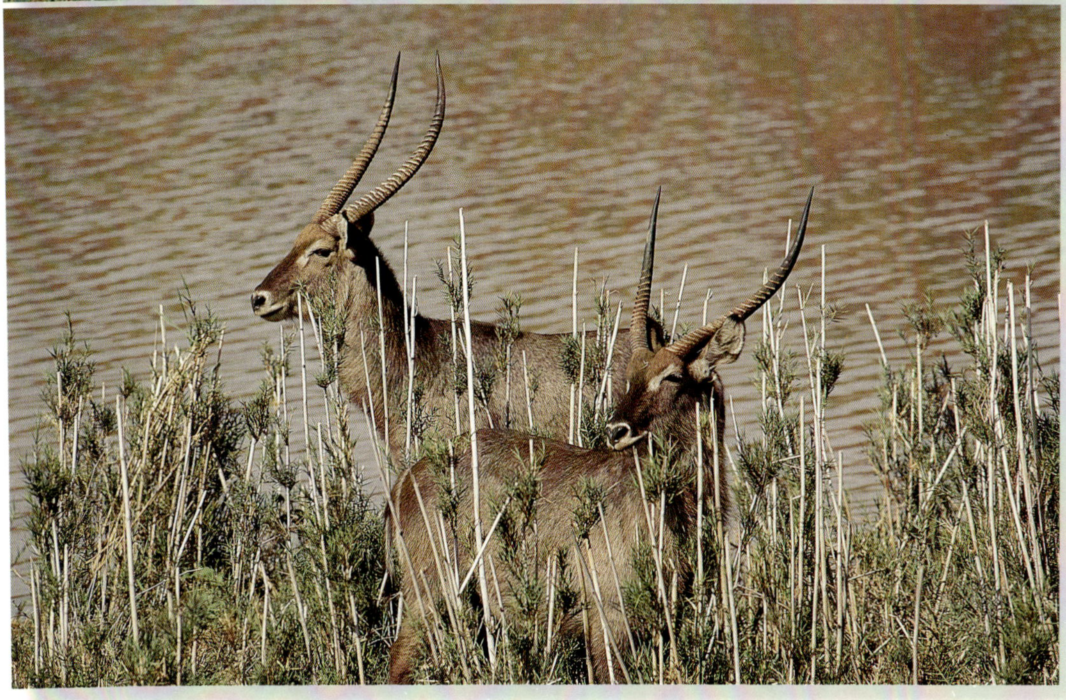

*The name "waterbuck" is a very appropriate one because these antelopes are extremely dependent on water.*

PRECEDING PAGES

*I have found from my own experience that waterbucks are more easily photographed than most other antelopes. Mother and son were kind enough to pose for this picture.*

FACING PAGE

*This huge waterbuck is awe-inspiring. I estimate that he weighed over five hundred pounds. And his horns are about three feet long.*

Cows are docile and have a loose social structure; no ranking order, no greeting ceremonies, no social grooming takes place. However, their breeding patterns match those of many other antelopes. Calves are born in the rainy season. A nine month gestation results in the birth of a single calf which is hidden for two to four weeks. Mothers vocalize to their calves with bleating and mooing sounds.

Like the other antelopes, waterbucks seek cover for protection from predators. A loud bull snort alerts the herd when lions, leopards, hyenas or Wild dogs are lurking. With refuge always close by, the waterbucks need not run, for they can quickly blend into thickets or reed beds, standing motionless, their magnificent horns intermingling and becoming one with the branches or tall thin reeds.

Only their musky scent gives a clue to their presence.

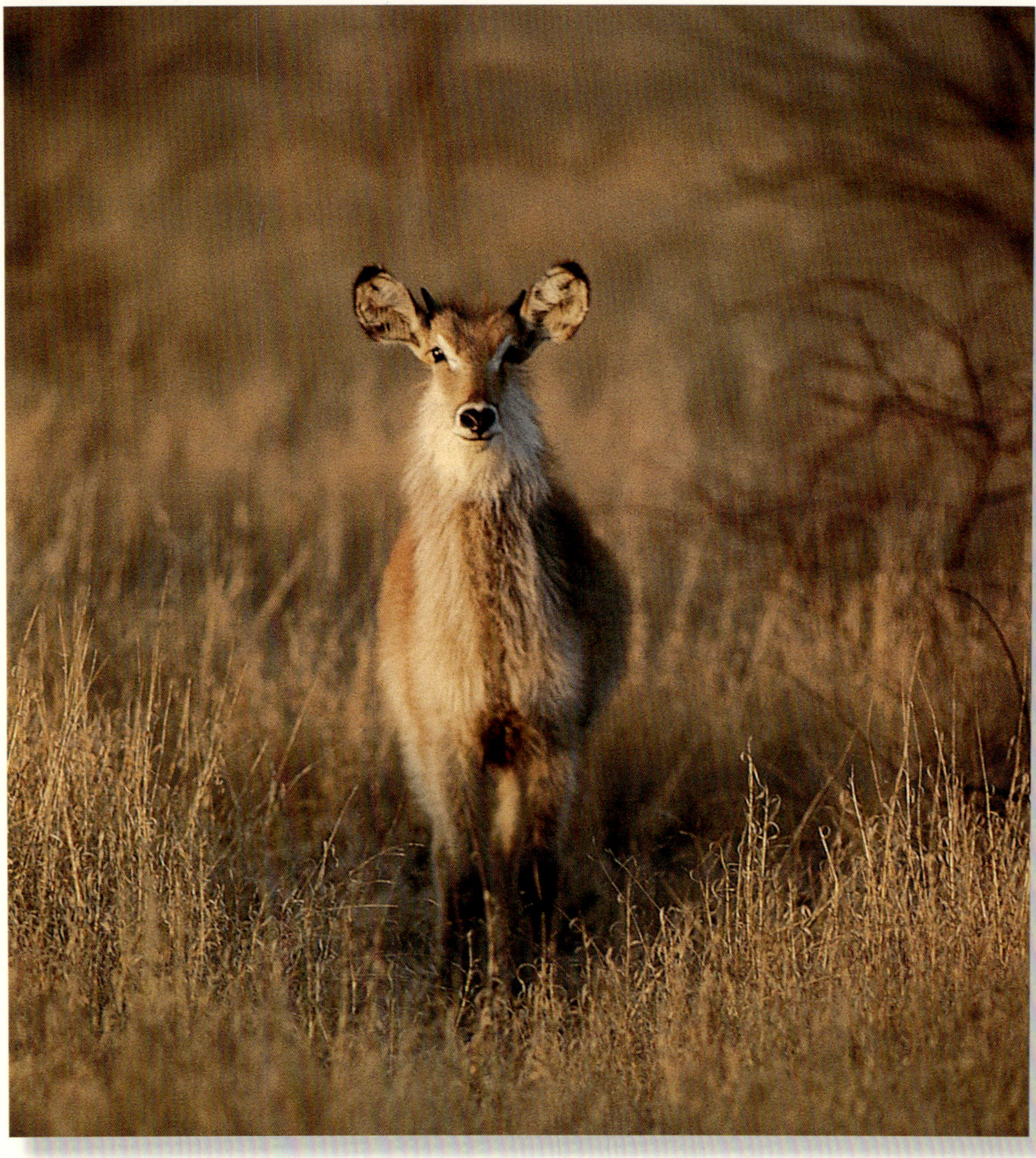

*This young waterbuck was probably chased from a nursery herd by a territorial bull. He will join a bachelor herd and remain until fully matured.*

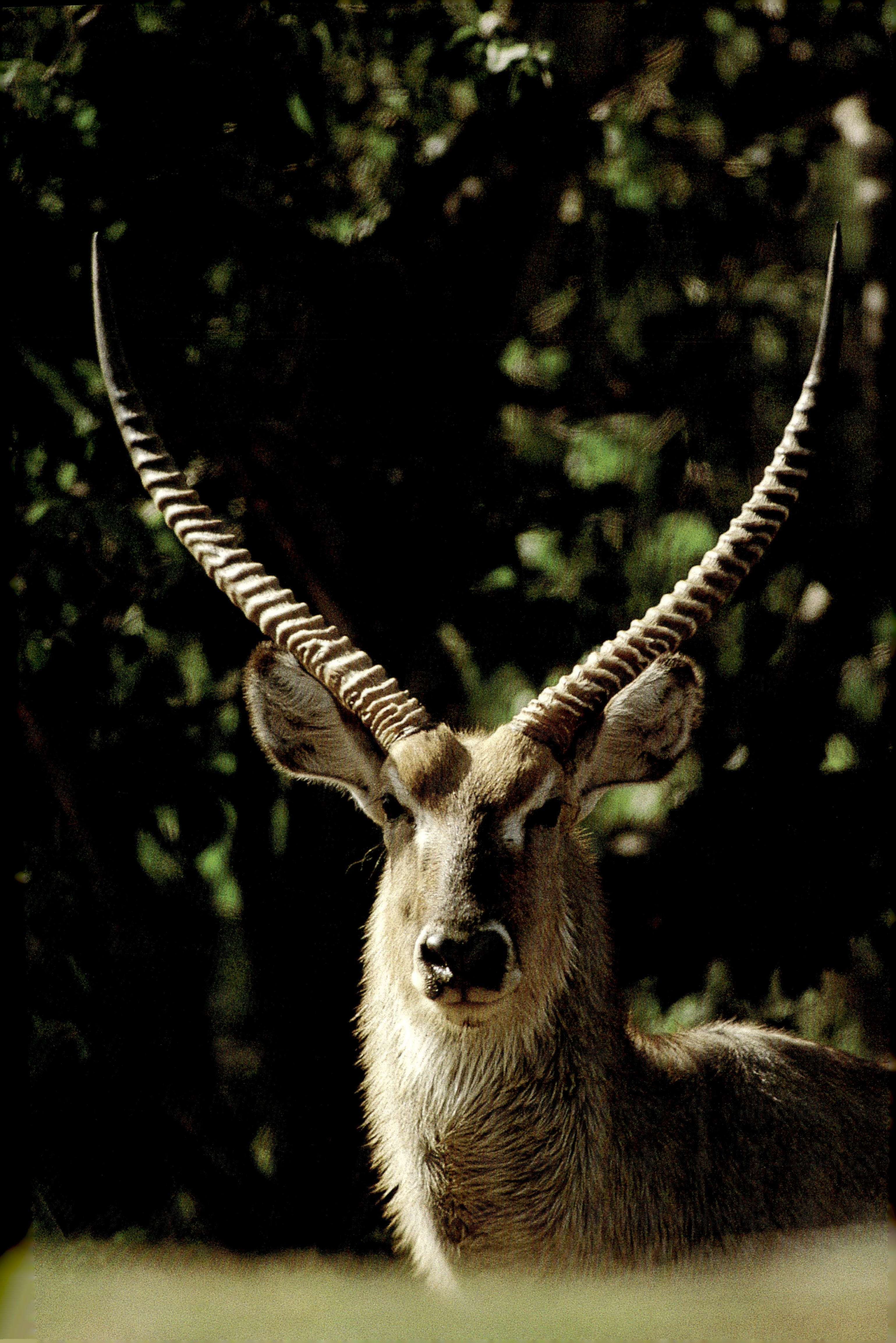

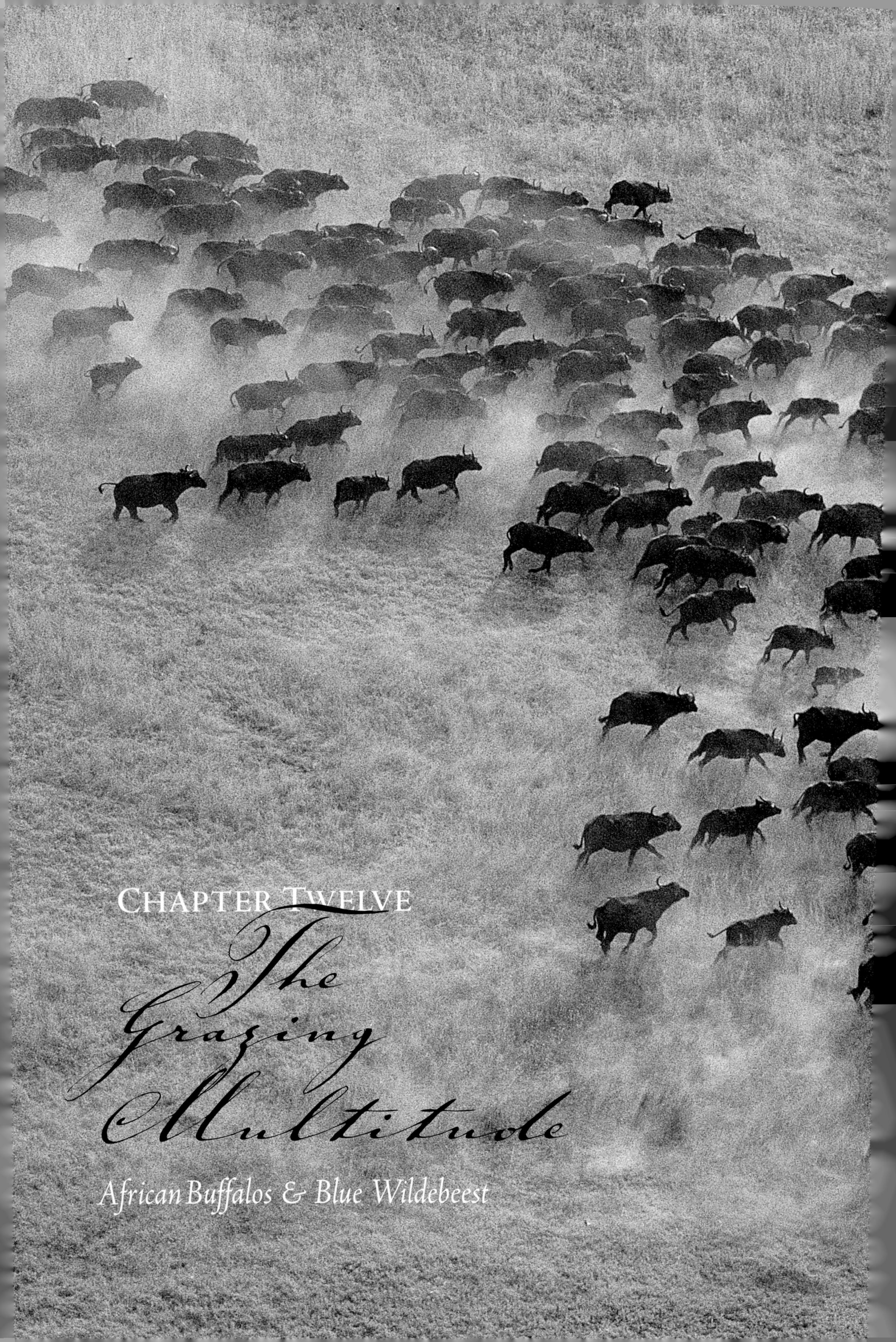

# Chapter Twelve

# The Grazing Multitude

## African Buffalos & Blue Wildebeest

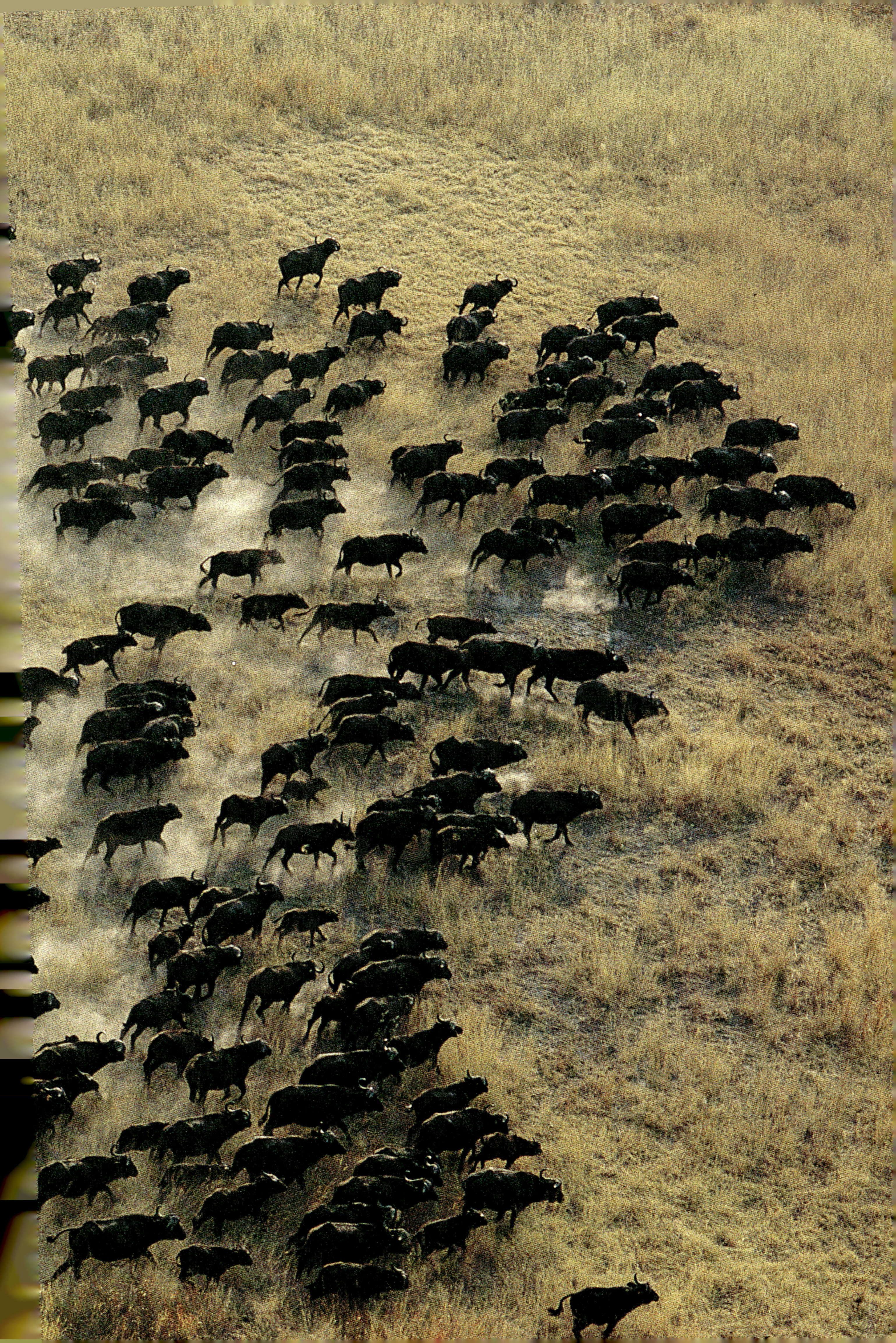

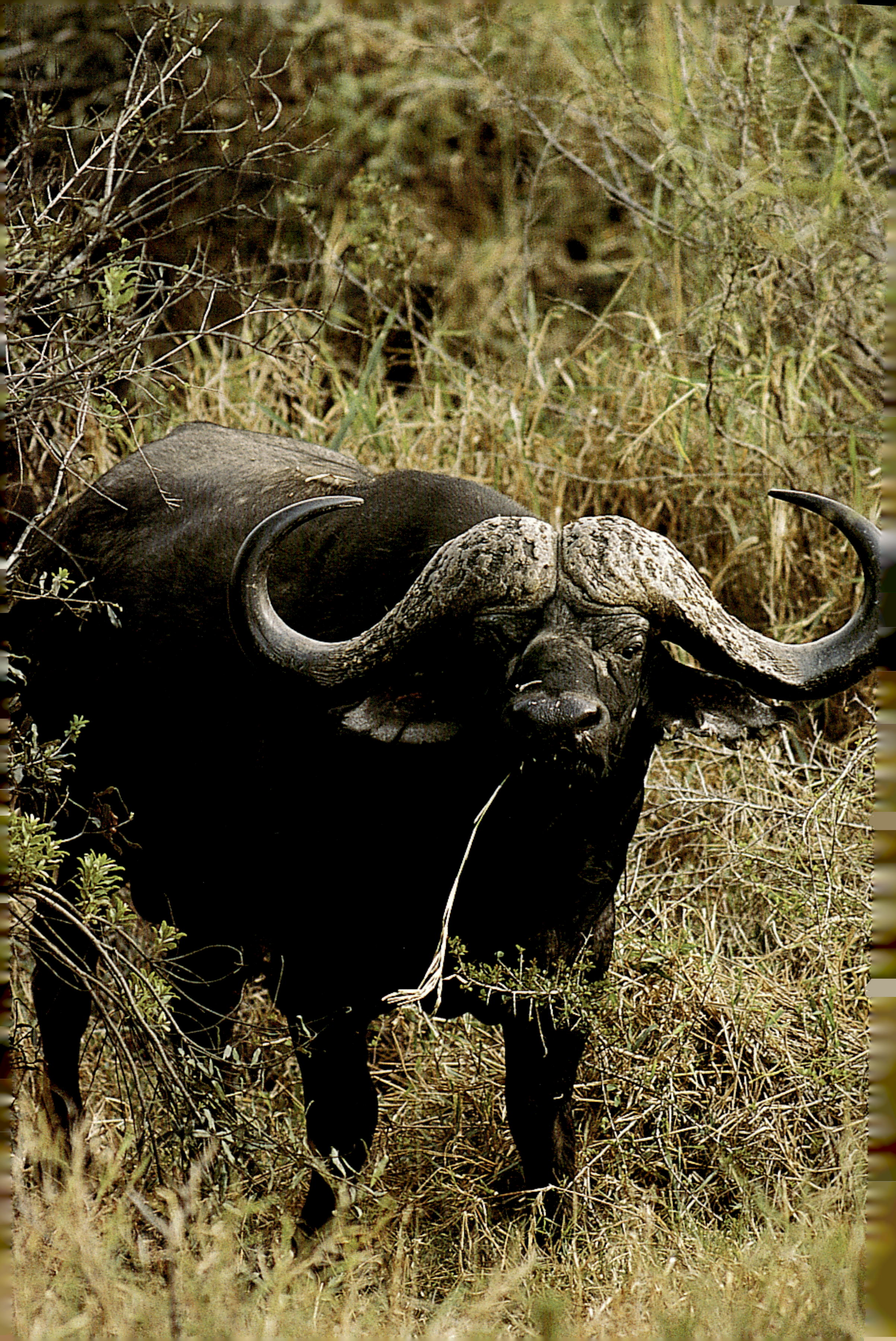

By day and by night, the vast African grasslands feel the trampling hooves and the hungry mouths of thousands of herding beasts. In times of drought, the land suffers along with the beasts for want of sweet fields of green. Then, when Mother Nature chooses to give rain to the thirsty land, the herds thrive and multiply. The land is the table of sustenance, and it is the bed of life and death for the grazing multitude.

This verdant land is where the African buffalo roam. Large herds of these formidable beasts live best where water is abundant, near swamps and flood plains. They seek reed beds and thickets for shade and cover.

The African buffalo is the only species of wild cattle in Africa. Also called the Cape buffalo, it is a massive animal weighing up to a ton, and standing sixty to seventy inches tall. It has a broad head with drooping ears fringed with hair.

It has a wide mouth and a bare nose that is perpetually moist. Atop the buffalo's head, a set of horns grow out from a wide boss, which covers the skull and forehead. The look is that of a Viking helmet. The curving length of the horns can be forty to fifty inches wide. Horn size determines gender and age; a broad heavy boss identifies a mature male. The coat is thin and black to dark brown, though young buffalos are a lighter brown and the heads of old bulls become grizzled. The front hooves are larger than the rear hooves to support the massive head and shoulders. On the other end, the buffalo's tail is much like a domestic cow's, with a tuft of long black hair.

The buffalo's sight is rather poor, and so it depends on its hearing and sense of smell to communicate and to detect enemies. Smell is used, also, to find food and water. The buffalo grazes on tall, coarse grass. It has a prehensile tongue which it uses to gather bundles of grass which it clips with wide incisor teeth and then grinds with large powerful molars. While feeding, it may grunt, honk, croak or briefly bellow – sounds of contentment. When grass is scarce, herbs and foliage round out the diet. Twice daily, in early morning and late evening, the buffalo seeks water for its great thirst. And where there's water there's mud. Buffalos like nothing better than to wallow in mud to cool themselves in the heat of the day.

African buffalos are not territorial, moving from range to range without challenge. Herd size depends on season, condition of the habitat and availability of food. In the dry season, when water is scarce, herds are larger. Five hundred or more will gather where there is water and good grazing. When water is plentiful, the large herds break up into smaller groups of around fifty and select their own range. The herds are made up of several clans; cows and their calves, bachelor herds of breeding males that stay close by the cows, young males and females which are not yet of breeding age, and older males in geriatric bachelor herds.

Constantly on the move, they roam up to seventeen miles to find water, but they usually graze near their home range. Individuals take turns as pathfinder, leading the clans to good grazing grounds and

PRECEDING PAGES

*I photographed this herd in the Okavango Delta. It reminded me of the old Roy Rogers and Hopalong Cassidy cowboy movies. Now I wonder if anyone knows who these western film stars were.*

FACING PAGE

*The African buffalo will graze on grasses that are too coarse for other animals to digest.*

FOLLOWING PAGES

*After coming upon a herd of buffalo, this bull came forward and raised his head in a threat display. In a layman's term it meant, "don't get any closer." We did not.*

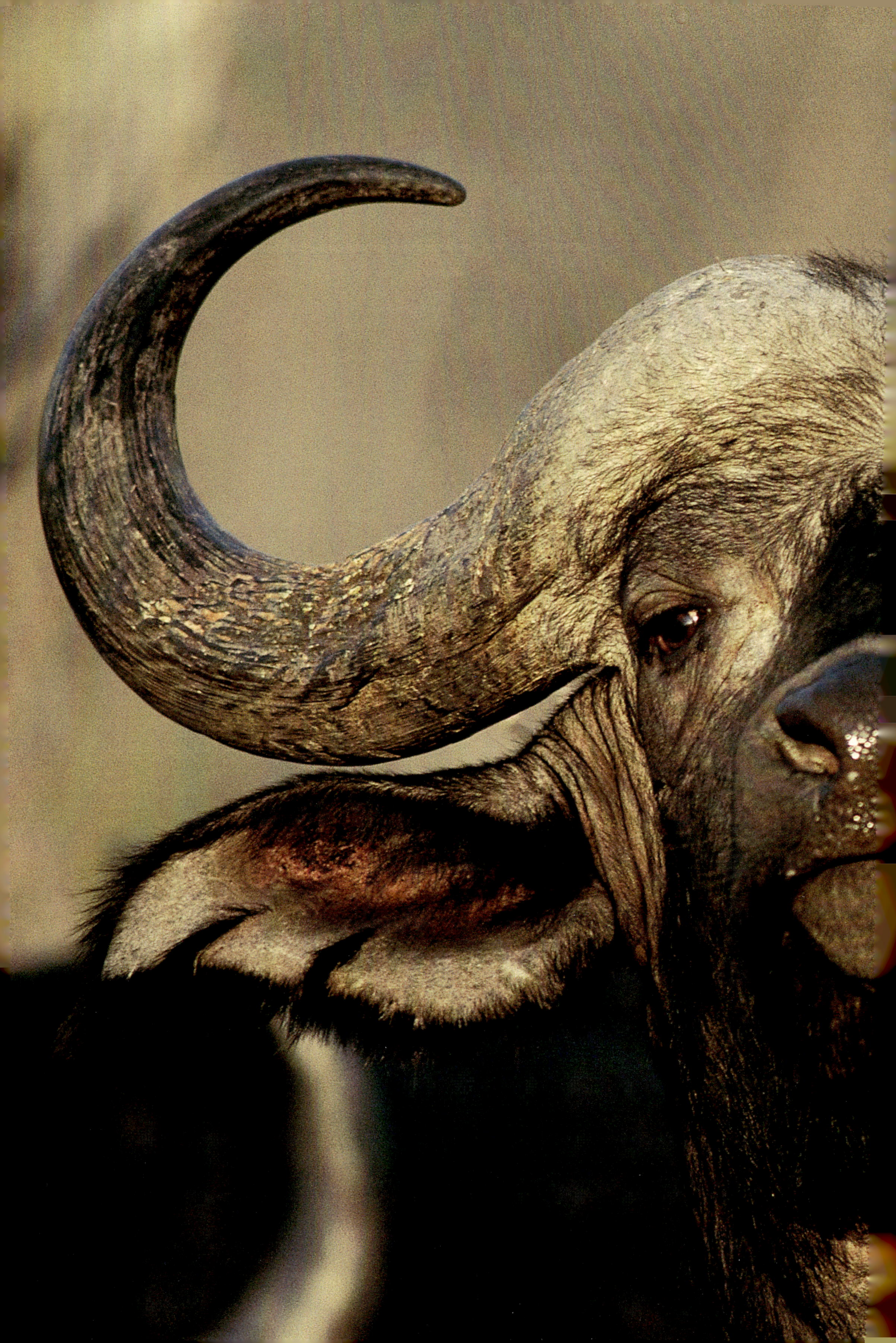

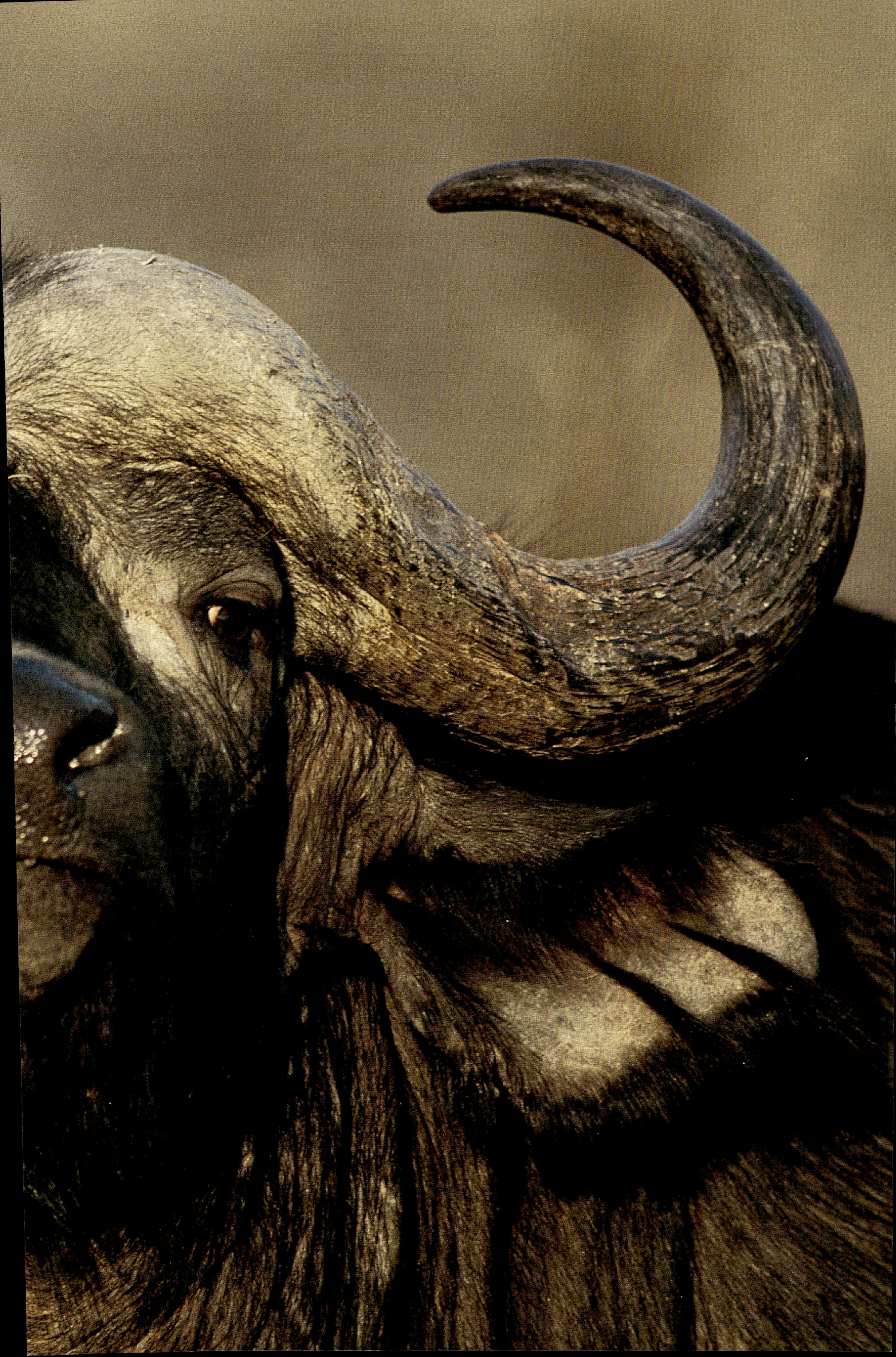

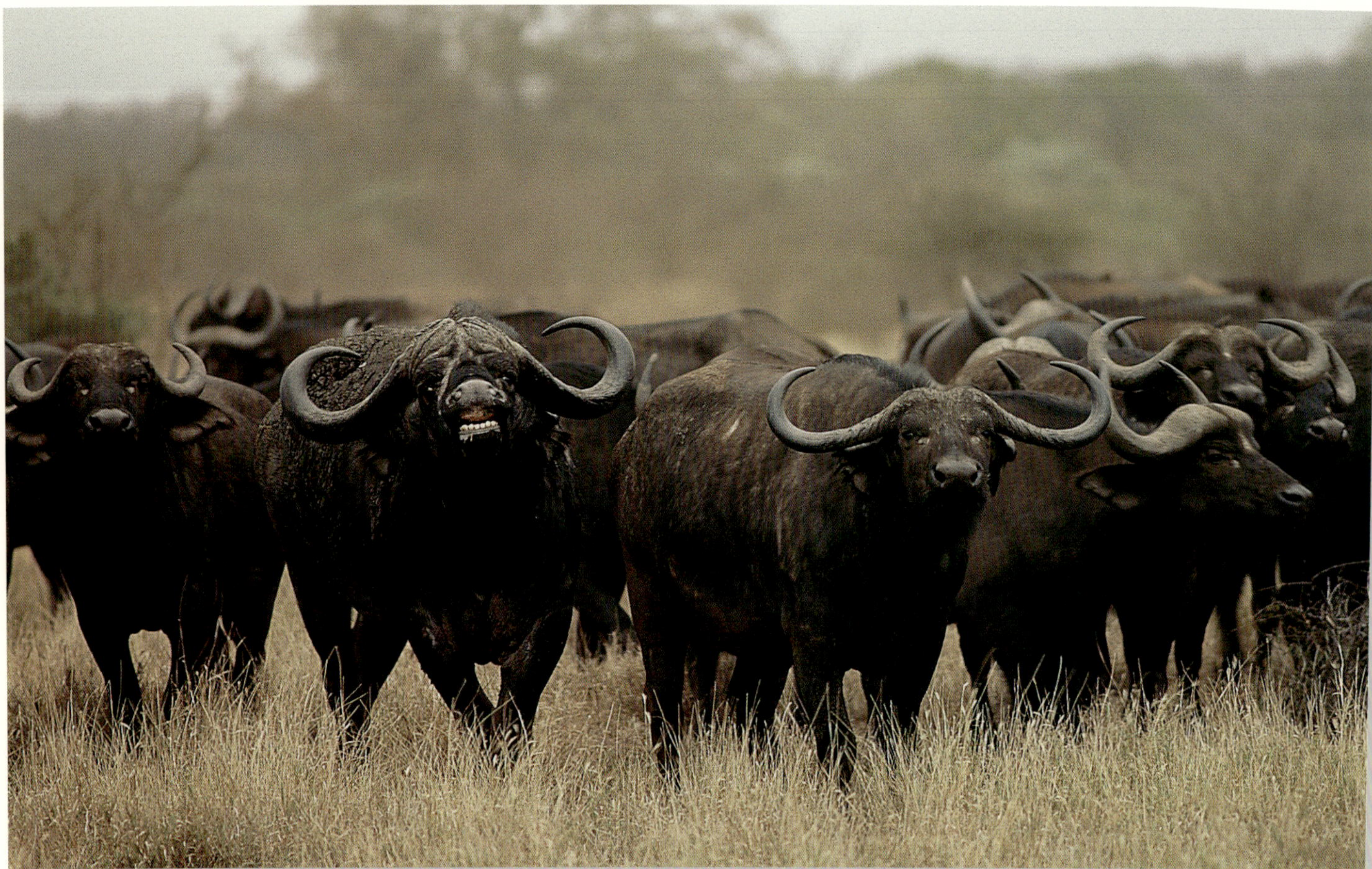

*My "chief scout" and I came face to face with this herd in the Kruger Park. The large bull with a grimace on his face is "tasting" the air. The term for this is "flehmen."*

water. With a vocalization that sounds like a creaking gate the pathfinder signals the herd to follow. Buffalos have such a strong herding instinct that they follow one another in all their activities. If one starts to eat, all will eat, if one lays down, all will lay down. And when they relax, they lie touching each other, often resting their chins on their friends. These extremely gregarious animals greet one another with nose touching. Social licking is a frequent activity – usually performed by lower ranking individuals for higher ones.

Both bulls and cows develop a hierarchy in the herd. Bulls dominate females and other bulls. A dominant male will display in various ways. He uses his horns to dig and toss dirt and vegetation. If he is provoked, he will present himself sideways, and with head held high, will walk toward the challenger, grunting, growling and tossing his horns until the subordinate gives up. The submissive bull may run away, bellowing with open mouth and curled tongue. Or, he may advance and put his head under the chin or belly of the victor. African buffalos' facial muscles are limited and so they are unable to communicate with expressions.

Dominance displays are prelude to mating as well. The bull seeks to impress the cow with the same digging, grunting and horn tossing actions. He will also lick her and rest his chin on her rump. If the courting behavior is effective, the cow will consent to mate, and eleven and a half months later, in the rainy season, she will give birth to a single calf. The calf is born within the safety of the herd. In only ten minutes the calf can stand, but for several hours it remains too weak to follow its mother. If the herd moves on, the mother will stay with her newborn until it is stronger. Several weeks pass before calves can keep up with the herd.

The mother-calf relationship is short, for by the time the calf is weaned at ten to fifteen months, the cow is again pregnant. She becomes hostile toward the growing calf when the new calf is born. Despite her rejection, the older calf remains with its mother for another year.

Young calves take to socializing early. They enjoy playing with one another, running and chasing with tails arched high behind them. Their play behaviors mimic those that will be necessary for future interactions with the herd; and, for dealing with predators. Mature adults fall prey mainly to lions, but sometimes are taken by crocodiles. Calves, the injured and the sick are vulnerable to spotted hyenas. When buffalos sense the presence of predators, they call out in distress with explosive snorts, croaks and bawls. They stand alert, toss their heads, then wheel and run. At a top speed of thirty-five miles per hour, they are comparatively slow runners. They can however outrun lions. But a slow start can be their downfall. It takes time to muster the power to get two-thousand pounds moving; time enough for lions laying in wait to ambush.

If a buffalo is attacked, the strong herding instinct comes into play. The herd comes to the rescue by bunching together and mobbing the lions. This strategy is so effective that weak individuals often thrive within the herd. Lions are often chased to seek refuge in trees. The buffalos often loll about below the trees for hours, sniffing and licking at the precariously perched felines. Lions that succeed in attacking an individual buffalo can be kicked with powerful legs or gored with sharp horns. A buffalo will fight off lions for hours, and sometimes, wins.

African buffalos are survivors, tough and often unpredictable. They have a history of being dangerous to humans – whether or not the buffalos feel threatened. Hunters have been killed by wounded, charging buffalos, as unstoppable as a freight train. Some people have even been killed by buffalos after being stalked through the bush for no apparent reason. It is wise to be wary and observe these powerful beasts from a safe distance.

Whereas the African buffalo is famed for its fierceness, another grassland resident is noted for its peculiarity. The Blue wildebeest has been described as an animal *'designed by a committee and put together with spare parts.'* It is a large antelope with a bison-like head, a black leonine mane and beard, and cow-like horns. It has a short neck, high humped shoulders and a deep chest. Bulky forequarters slope to a narrow rump and end with a black horse-like tail hanging nearly to the ground. It stands forty-six to fifty-seven inches tall and weighs four-hundred to six-hundred pounds, all supported on thin antelope legs. The front feet are larger to bear the weight of heavy forequarters. The coat is short and glossy, dark silver-grey to brown in color, and sports black vertical stripes of longer hair. The wildebeest has also been called a clown for its long black face and tufted beard, a gait like a rocking-horse and a vocabulary of frog-like grunts and snorts. A deep lowing *ga-noo, ga-noo,* is the signature call for its other name – Brindled gnu.

FOLLOWING PAGES
*The male wildebeest in this picture is well aware of the lions in the dunes; he also knows that there is a sufficient safe distance for him to escape from the lions should they charge.*

Whatever it may be called, the Blue wildebeest is a complex and fascinating animal. Territorial instinct is strong in wildebeest society. Mature males stake out territory on the grassy plains – conspicu-

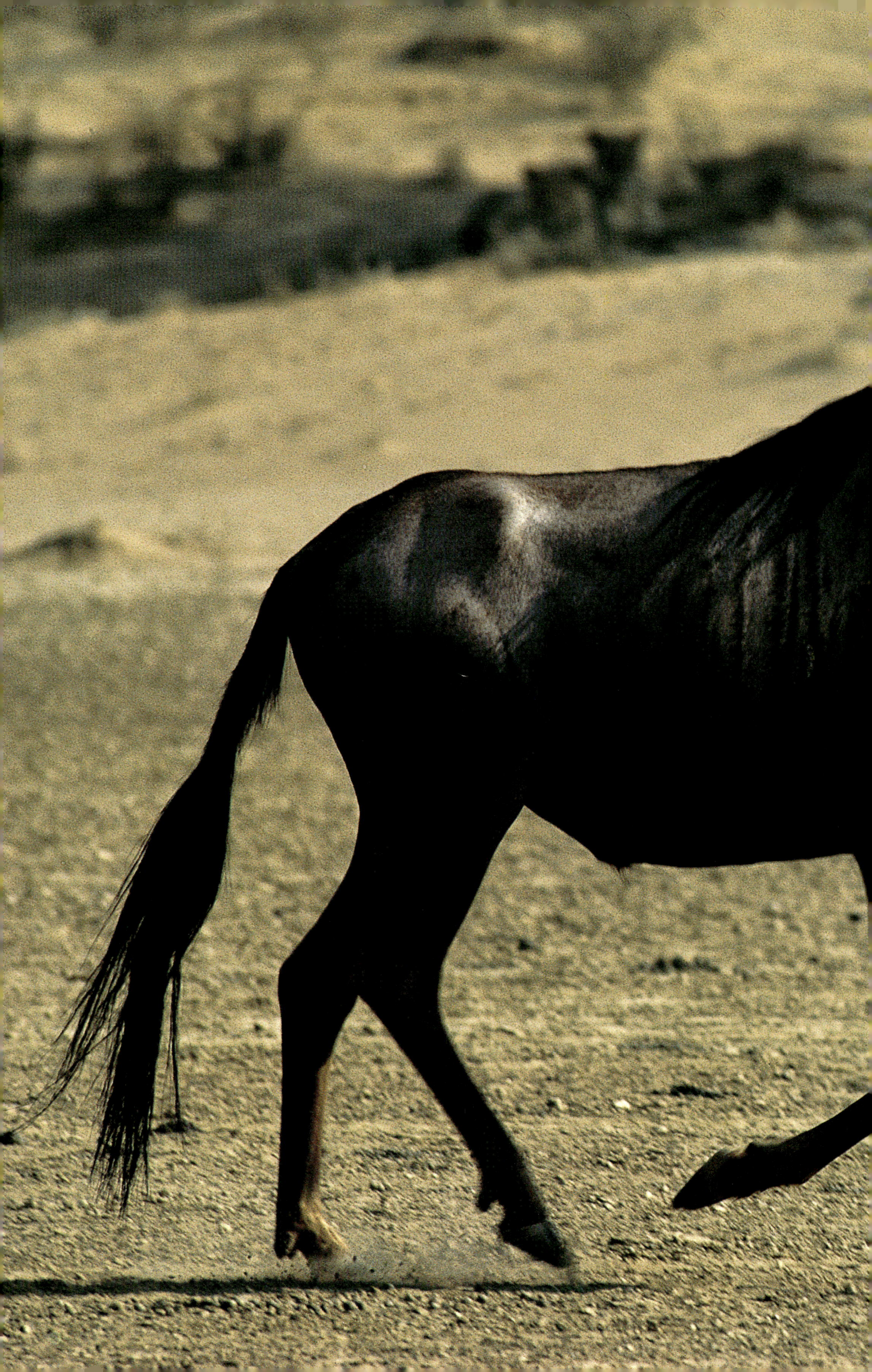

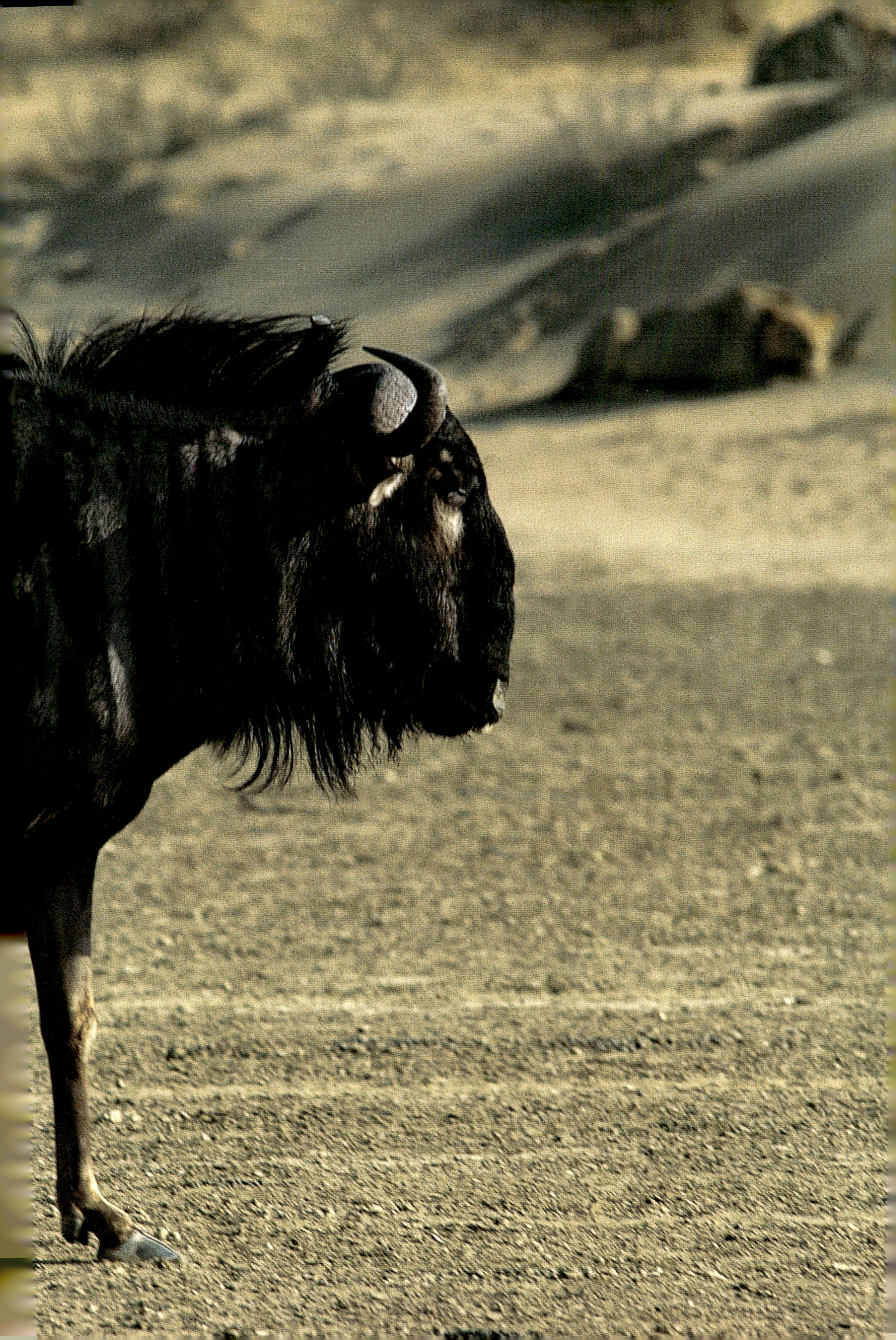

ous as solitary sentinels in a sea of grazing cows and calves. They engage in rituals that advertise their control over the small square of real estate they call their own. The landlord stands erect and calls, *ga-noo, ga-noo,* up to eight times in a row. He will paw the ground to deposit his tar-like scent, kneel and horn the soil, or roll around on his back. Any male that dares to trespass will be witness to a wild cavorting frenzy of jumping, kicking and reeling, punctuated with manic tail swishing. Usually the explosive display is enough to make the intruder back off. Sometimes, on bended knees, horn sparring follows – but rarely do injuries occur.

The defensive bouts continue daily, as a steady stream of bachelors vie for a chance to claim territory of their own. Competition is most fierce in the breeding season. Some succeed, and some must try another day. Those that give up trying are relegated to the fringes of the herd, where they are vulnerable to predators, especially lions. Males on the periphery become lookouts that alert the herd with alarm postures and loud snorts when lions are near. Unfortunately for them, they are often the first victims when lions attack.

Blue wildebeest herds are comprised of clans and individuals; cows with young calves and yearlings, bachelor herds, and lone territorial males. Great herds in the thousands are present in Africa's Serengeti Plain where arduous migrations occur. Whereas smaller herds inhabit some of the Southern African National Parks where migration is limited to short seasonal movements to find good grazing and water. They are active by day and on moonlit nights, grazing, moving and socializing. In the heat of the day and part of the night they bed down, closely positioned in rows, so that should an emergency occur, they can mobilize quickly and run without confusion. They are fast and enduring runners when the need arises.

Always seeking greener pastures, wildebeest herds follow rain. They watch storm clouds, listen for thunder and sniff the air for water content. That ability enables them to find the water which they drink daily and the fresh green grass they prefer. Their wide lips are adapted for clipping big bites of short grass, newly sprouted after rains or in burnt-out areas. Wildebeest are often found grazing with zebra, gazelles and other species. The different feeding habits of each allows for plenty of food for all. Wildebeest have a habit of thrashing small saplings with their horns which helps clear land for grasses to thrive.

The tree thrashing is also part of territorial display. The displays increase at mating time which some studies suggest coincides with the lunar cycle. During courtship, the males approaches the female with his rocking-horse gait. With ears lowered and chin up, he will rear up on his hind legs and make grunting sounds. Only the territorial bulls mate – but the cows will wander from area to area, mating with many bulls.

Gestation takes eight and a half months and ends just as the rainy season begins – in a time of plenty. Ninety percent of calves are born within a three week period, usually between dawn and midday, for danger is more imminent in the night. Lions and hyena are always nearby when wildebeest are born, looking for an easy meal. A Wildebeest mother has the ability to interrupt a birth if she feels threatened. She may also delay delivery of the afterbirth until her calf is up and able to run. A calf can stand within three minutes and in five minutes can run. When it is safe for both to move, the mother will release the afterbirth and hastily depart, her new calf in tow.

Calves try to stay with their mothers at all times. If separated, the mother will call – mooing. The calf will answer – bleating. They have no trouble recognizing each other and are soon reunited. Within two days calves can run with the herd. Older calves sometimes gather together while their mothers graze. In early morning they enjoy frolicking together and butting heads. Long term bonds are often formed. Within eight months, the calves are weaned. Females stay with the herd. At about one year the males leave to join or form a bachelor herd.

And soon, the young bachelors will feel the primal urge to prove themselves; to challenge the territorial bulls, to mate, to travel with the herd and the multitude of grazing beasts – to survive many seasons on the vast African grasslands.

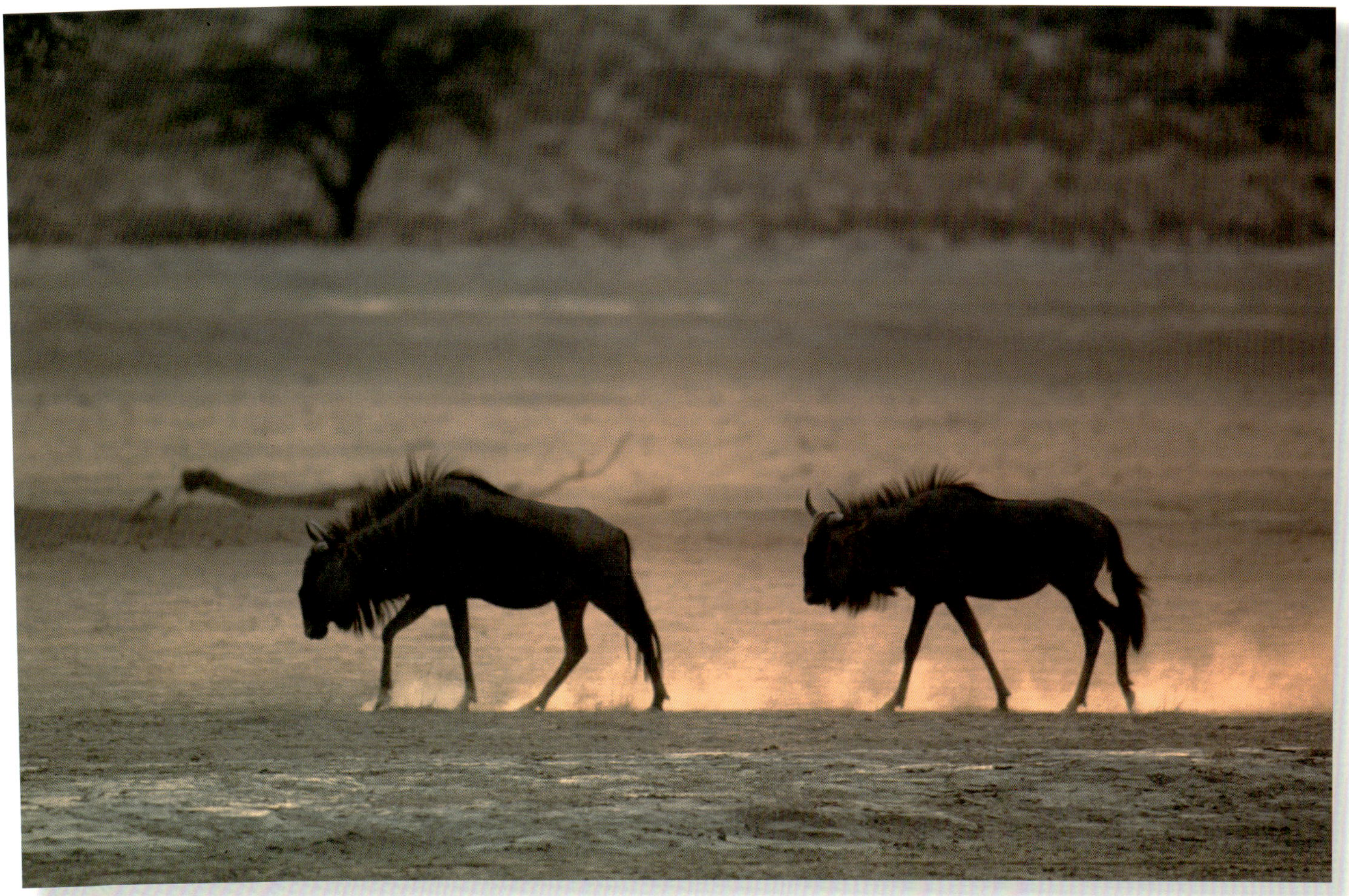

*The Kalahari Desert was hot, dry and dusty this early morning. Even the wildebeest seem affected by the weather.*

FOLLOWING PAGES
*Waking up before sunrise may be the toughest part of wildlife photography. However, a scene such as this makes it very worthwhile.*

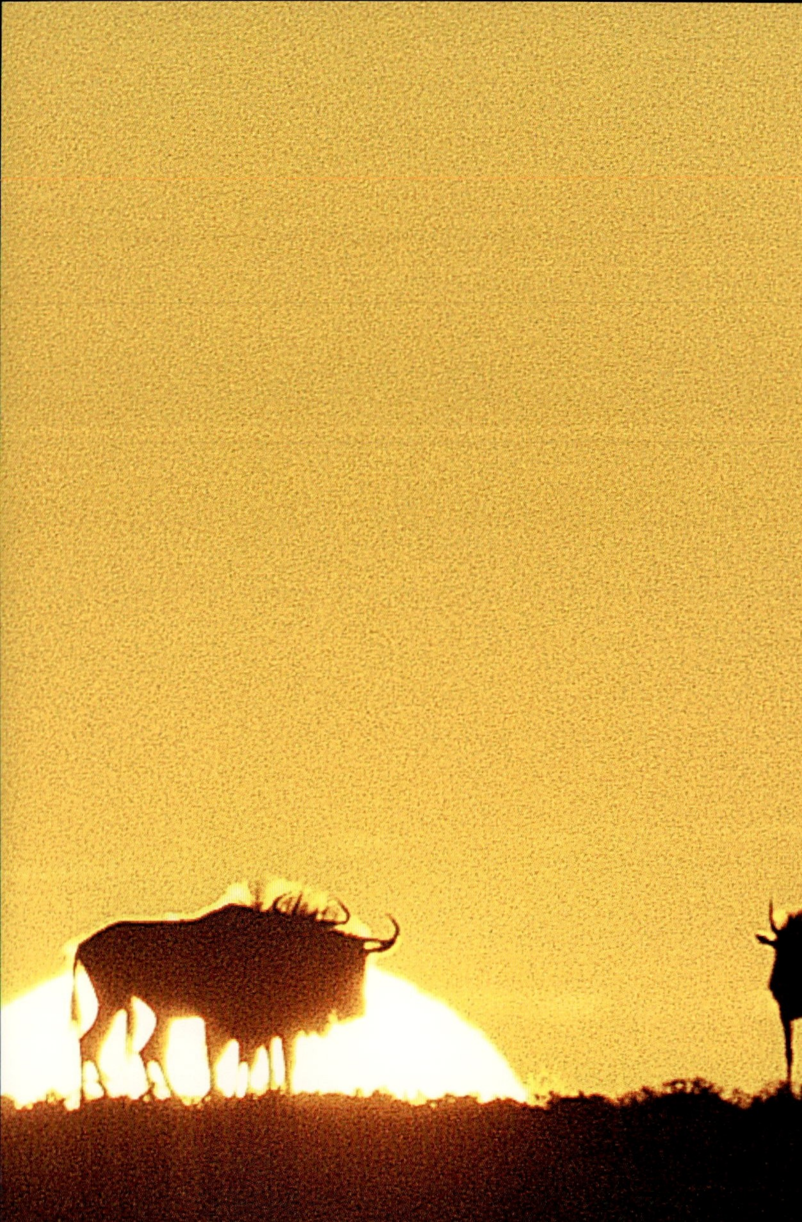

CHAPTER THIRTEEN

# Wet, Wild & Deadly

*Hippos & Nile Crocodiles*

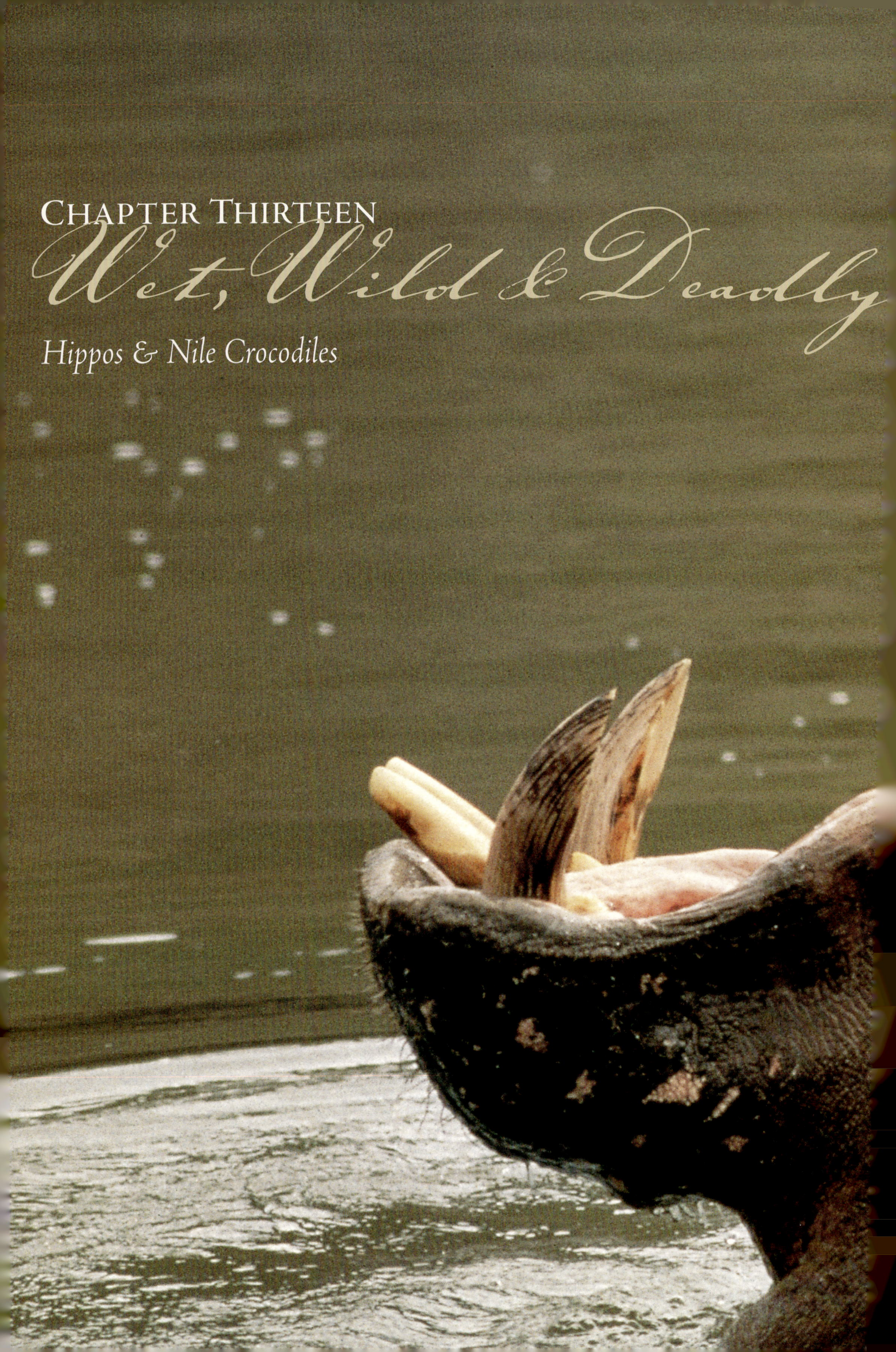

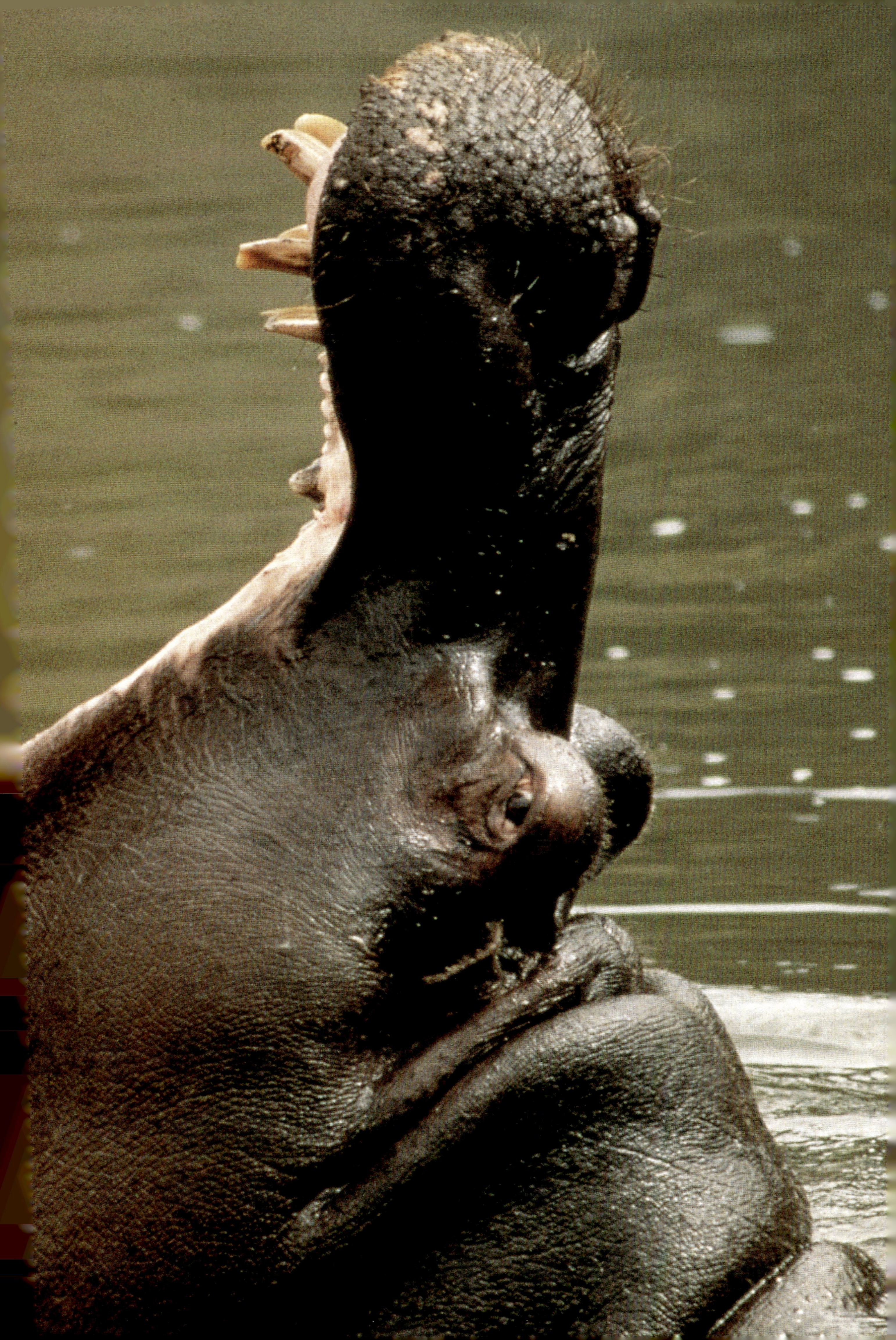

It is a fact, water is life. We all must drink it to live. But for many creatures it is more. It is a place to eat, a place to hide and a place to mate. It is their world. Aside from the obvious inhabitants – fishes, frogs and such, Africa's streams, lakes and rivers are home to aquatic giants that live near the surface, their bodies submerged, exposing only their ears, nostrils and eyes. And just below, are mouths full of sharp teeth. These are hippopotamus and crocodiles, Africa's most dangerous animals. They lie still and alert, looking, listening and waiting; in the case of hippos, to attack hapless trespassers, or for crocodiles, to grab a tasty meal. In their world, it is also true that water is death.

PRECEDING PAGES

*A male hippo with a gaping mouth is seen as a threat display. A warning such as this is usually enough; however combat can often ensue when two bulls are evenly matched.*

FACING PAGE

*The hippo's timing was perfect as I took this photograph of a beautiful sunset over the Letaba River.*

Upon first encountering hippopotamus, it seems absurd that such placid beasts could be dangerous. Much of their day is spent lounging in shallow water or wallowing in mud. Groups of fifteen or so snuggle closely, often with their heads resting on one another's backs, nearly motionless. From a casual glance, they look much like boulders or floating logs.

The hippopotamus, which in Greek means *river horse*, is the world's third largest land mammal after the white rhino and the elephant. Barrel shaped, with short thick legs, it can weigh three-thousand to seven-thousand pounds. One third of its ten to fifteen foot length is its massive head which can weigh up to half a ton alone. It has a huge muzzle that opens into an intimidating maw with long sharp teeth. The bottom canine teeth are tusks that grow to twenty inches long

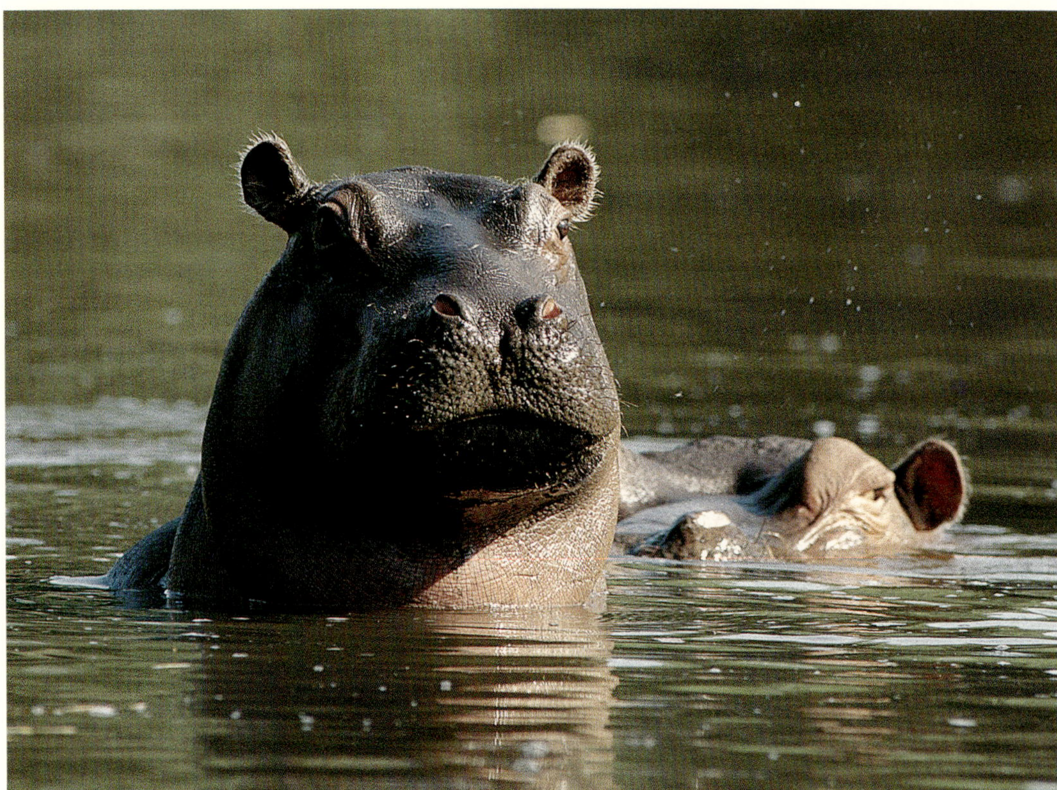

*A young hippo seemed curious and got a closer look at me while I photographed him. The adult in the background kept an eye on both of us.*

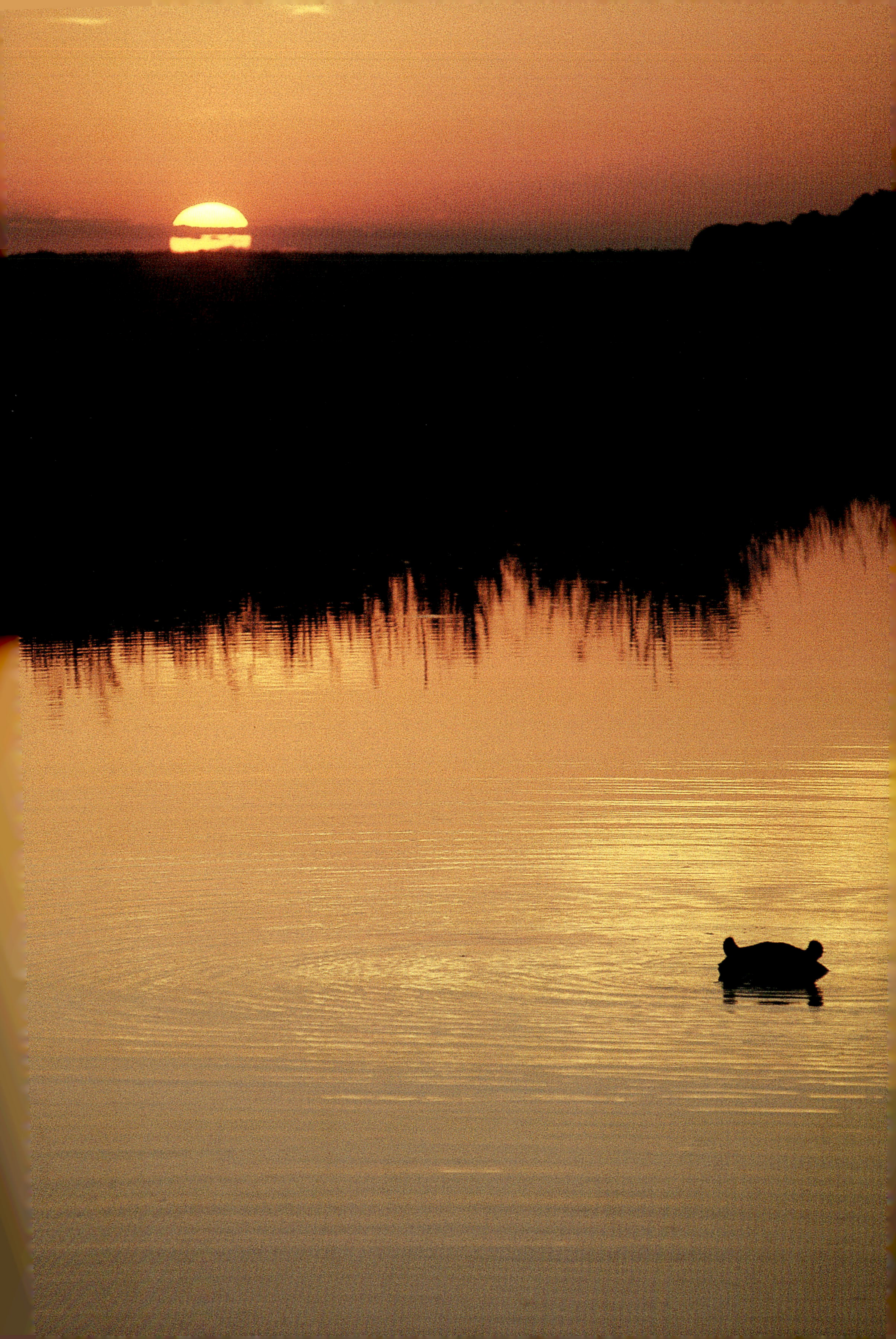

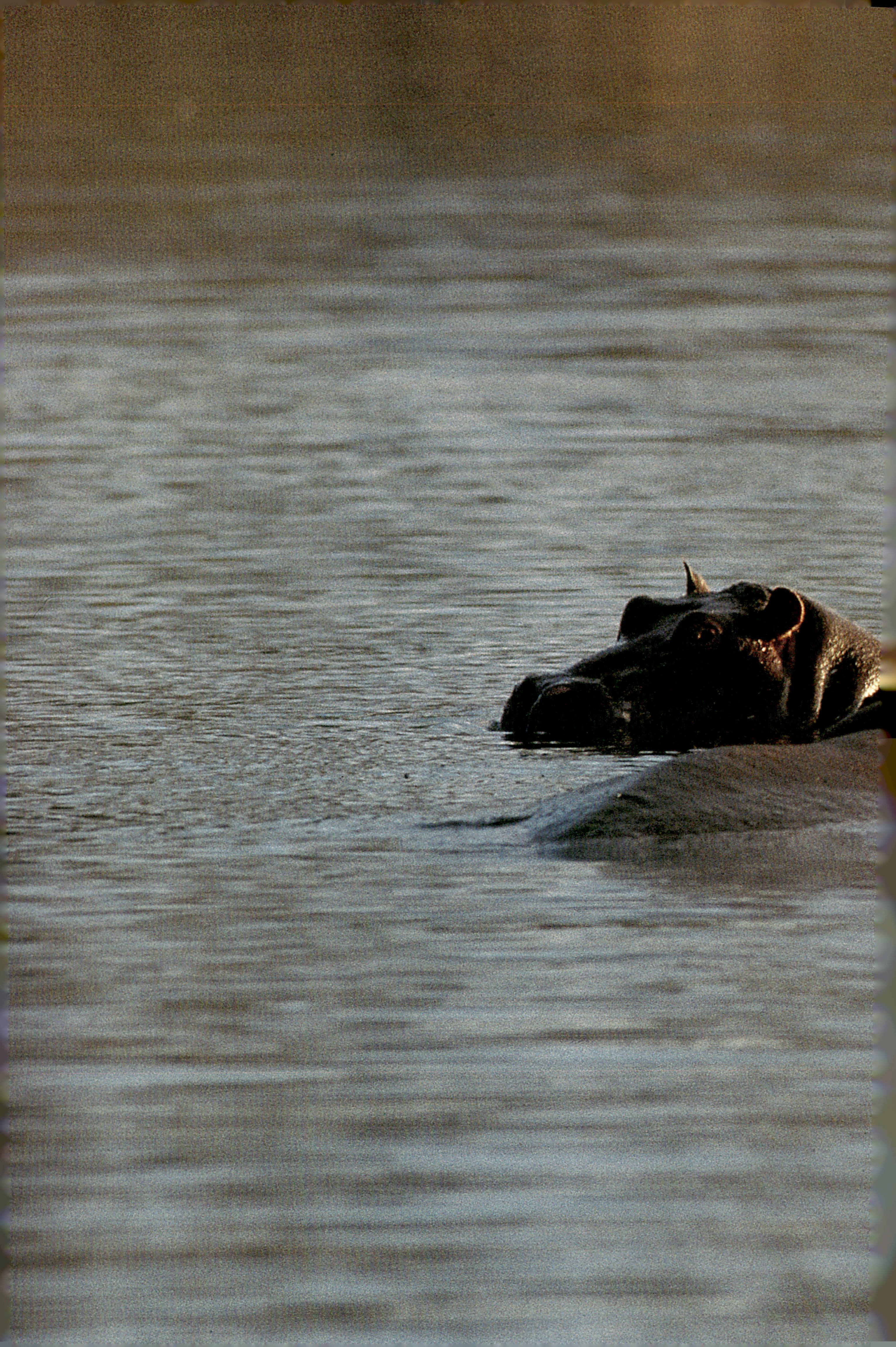

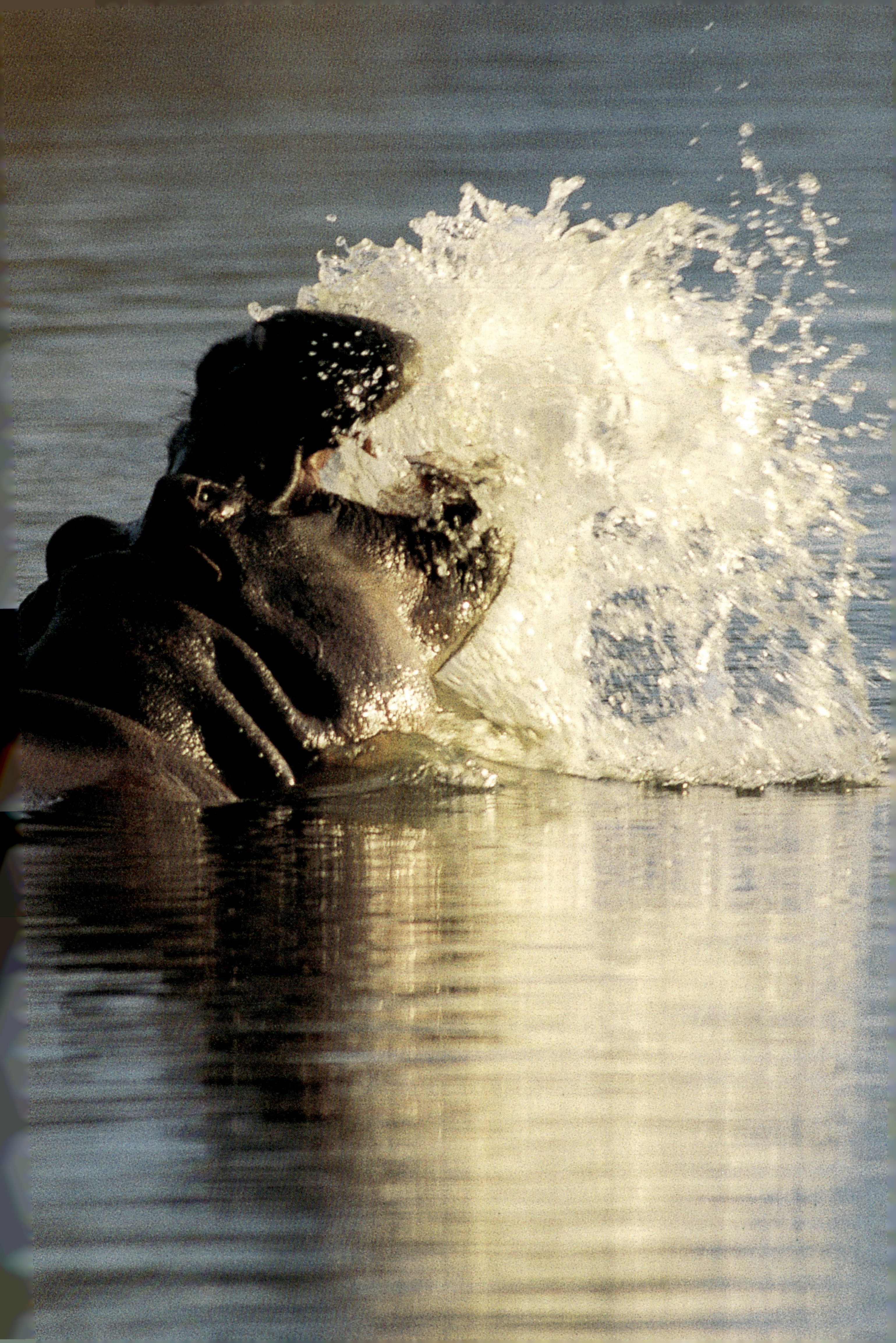

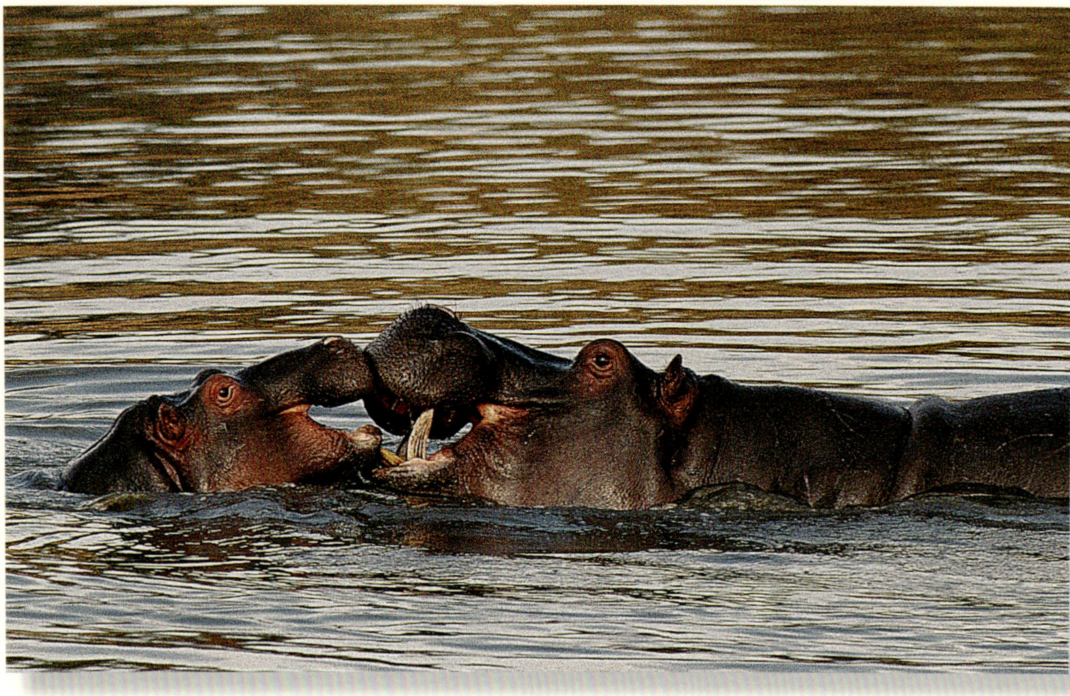

*The female hippo accepted the courtship of the water scooping bull.*

and weigh four pounds each. Its nostrils, eyes and tiny ears sit atop the face, and conveniently, above the water's surface while the animal soaks.

In deeper water the hippo must swim, for its solid, heavy body does not float. Just before it dives, the hippo takes a deep breath, then closes its nostrils tightly and folds its ears into creases, flat against its head. It swims by kicking the rear legs, or it may sink to the bottom and walk or run. Five or six minutes later the hippo must come up to breathe. As its head breaks the surface, it exhales with a blast of water droplets and unfolds its ears – wiggling them to shake out any moisture.

The hippopotamus must stay in cool water or mud wallows during the heat of day to regulate its body temperature and, because its skin is very sensitive to dryness. The smooth skin is about two inches thick, and brown to grey-purple with pink underparts and creases. It is hairless except for short bristles on the animal's nose, ears, eyebrows and tail. The outer layer of the skin loses moisture rapidly even though it has no sweat glands, and so, to prevent dehydration, the hippo must stay wet under the sun. It was once believed, that hippos sweated blood; an erroneous observation that derived from the reality that the skin secretes a viscous pink fluid. The secretion dries to a lacquer and protects the skin from sunburn, drying or getting waterlogged. It is also thought to have antiseptic properties. This is deduced from the fact that hippos often have open gash wounds from fighting, which remain free from infection even when exposed to the unsanitary conditions of their environment.

Fighting is a big part of hippopotamus society. Males fight for dominance over the herd. They also fight for territory. Dominant males have been known to maintain sections of lake or river, shallows and shoreline as their territory for many years. They will charge at any intruder that comes too close, animal or man. When hippos challenge or threaten one another, they yawn widely, showing their giant tusks, and bellow loudly. They also rear and charge, scoop water with their bucket-like mouths to throw it, and defecate – scattering the dung with their short paddle-like tails. If one does not submit to the display by retreating or lying down prone, fighting will ensue. Hippo fights are lengthy, often up to two hours, and bloody, as they bite and slash each other with the sharp tusks. Their thick hides protect them during battles, but sometimes, it is not enough. Fatalities do occur. Most hippos wear the scars of many a duel.

Hippopotamus herds are made up of fifteen to thirty cows and calves, led by a single territorial bull. At mating time he will fight off all males of breeding age that trespass. Males of non-breeding age are tolerated as long as they behave submissively. When it is time to mate, the bull will court the cow with water scooping displays. Mating is timed so that the birth of calves will take place at the beginning of the rainy season. Hippos mate in the water, with the female submerged. She periodically lifts her head to breathe during the process.

Gestation lasts eight months. When the cow is ready to calve, she moves to a secluded place in the shallows where she gives birth to a single young. At birth the calf weighs only about one-hundred pounds. Calves can swim before they can walk, for sometimes they are born in deep water and must immediately swim to the surface to take their first breath. They must surface often thereafter, for they can hold their breath for only twenty seconds. Calves instinctively close their nostrils and fold in their ears while nursing – whether under water or on land. The mother stays alone with her calf for ten to fourteen days before rejoining the herd. During that time, the calf is imprinted to recognize its mother. The mother hippo is fiercely protective of her young, keeping it close at all times. If a youngster strays, the mother will discipline it with nudged, sideswipes or bites. The calf responds by lying prone. There are many dangers to young hippos. Lions, hyenas, and crocodiles prey on hippo calves when they get a chance. A mother hippo will threaten predators with a *wheeze-honk* sound, water scooping, head shaking and charging. If the predator doesn't retreat, she will attack. A bite from her powerful jaws is capable of killing a lion or cutting a crocodile in half. In the water, a very young hippo will ride on mother's back to be safe from cruising crocodiles. Mothers with calves have been known to overturn boats and attack the floundering humans, often killing them. Adult hippos are rarely killed by predators. On land, calves are vulnerable to being trampled by adults that are fighting or stampeding. Groups of females will mob any bull that may possibly endanger the calves. The bull's response is to lay down prone in submission.

PRECEDING PAGES

*A hippo male was water scooping while courting a female. It was his way of showing her that he was strong, protective and worthy of her.*

FOLLOWNG PAGES

*This calf, which is just several hours old, is bonding with its mother. Woe to any creature that tries to get too close to them during this early bonding period.*

Cows and calves live in nursery herds that associate on the banks of the waterways. One or two cows will baby-sit all the calves while their mothers attend to other activities. Calves are weaned at eight months, although they begin grazing at five months. Young sons eventually leave and form bachelor herds while daughters stay with their mothers for years. A mother may have calves from four successive years following her about, led by the most recently born.

When evening approaches, hungry hippos prepare to leave the water to find food. As the sun sets over the watery domicile, deep roaring grunts can be heard for miles. Hippos are one of Africa's noisiest animals. No, it is not the sound of their stomachs growling; rather, the vocalizations are a prelude to

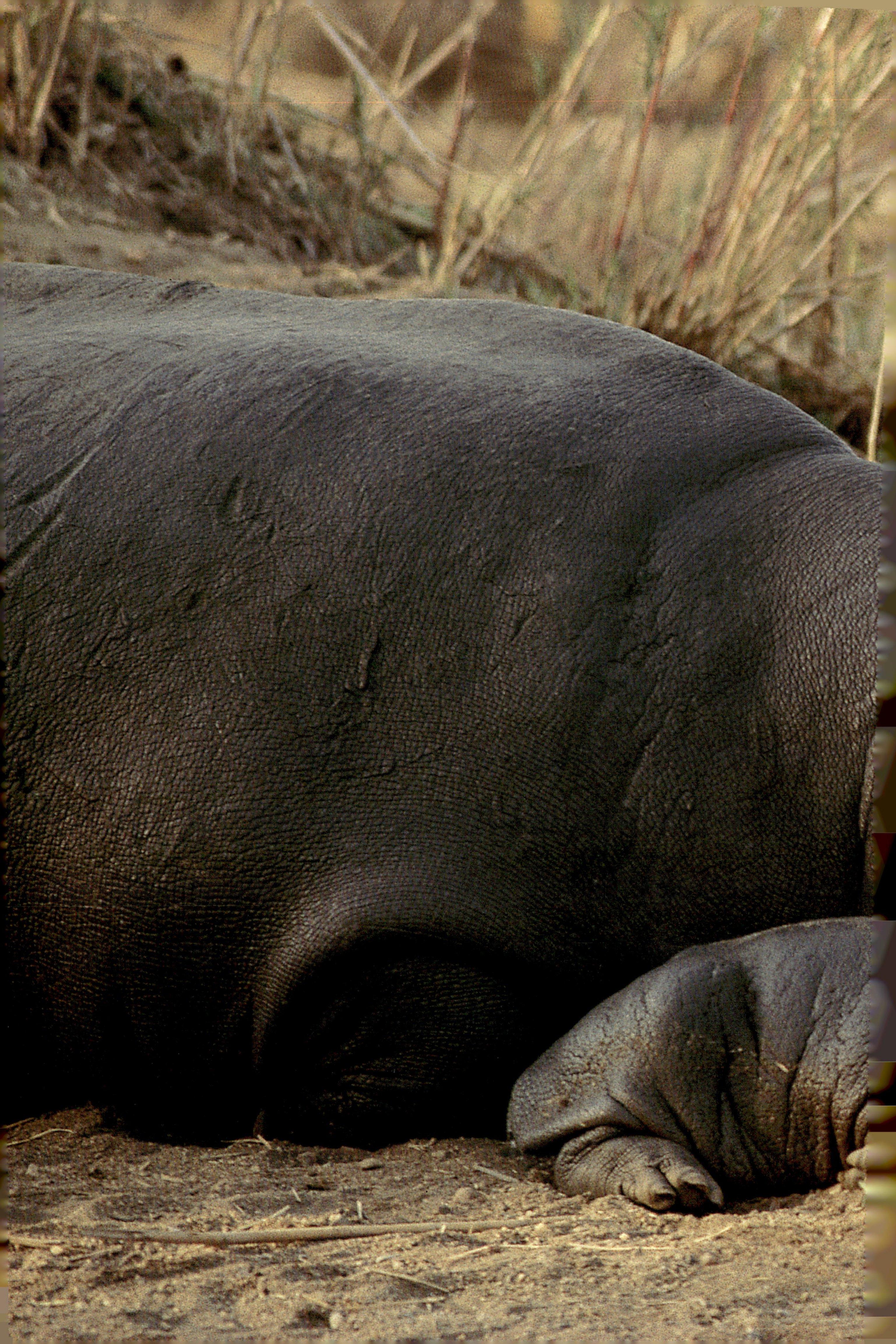

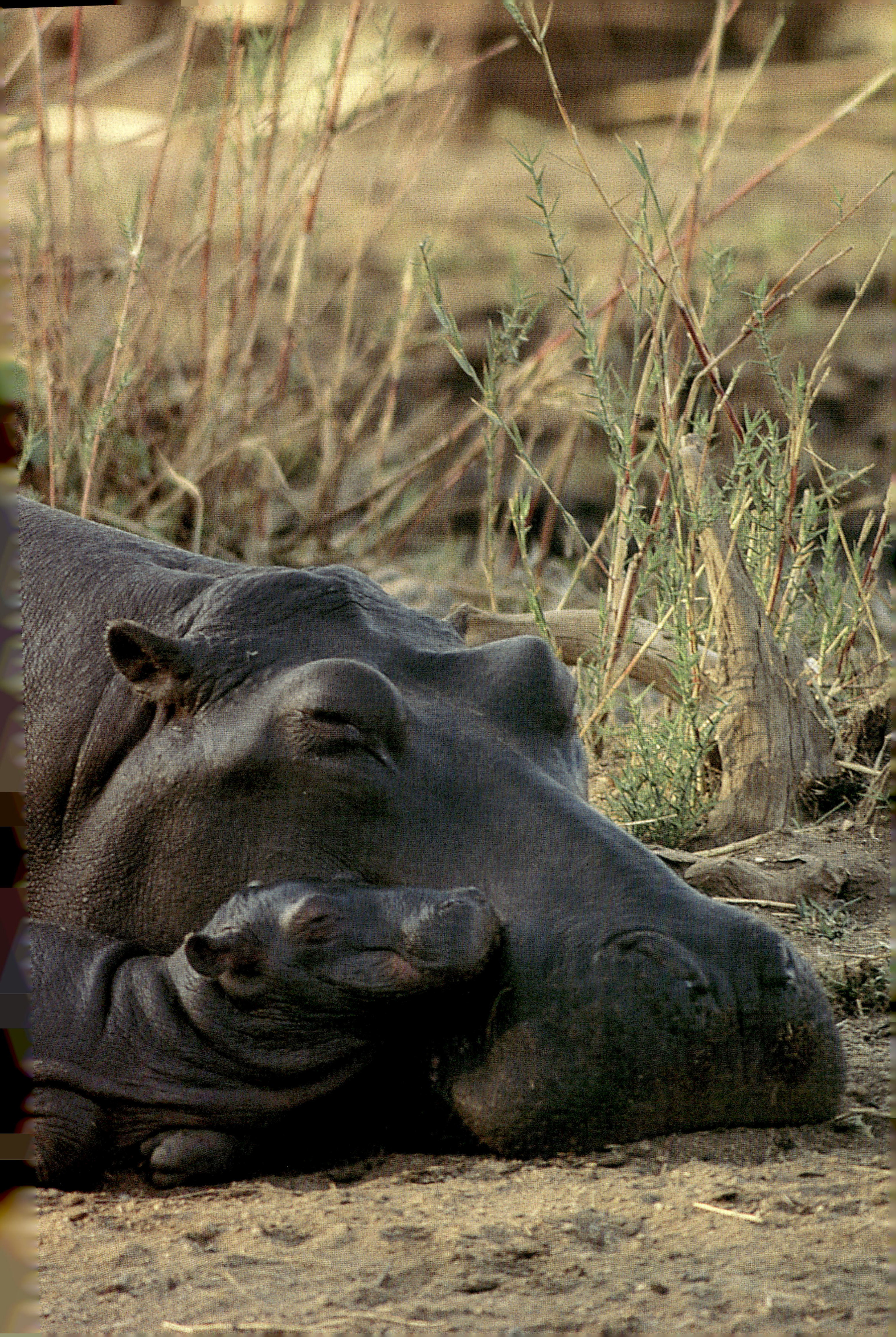

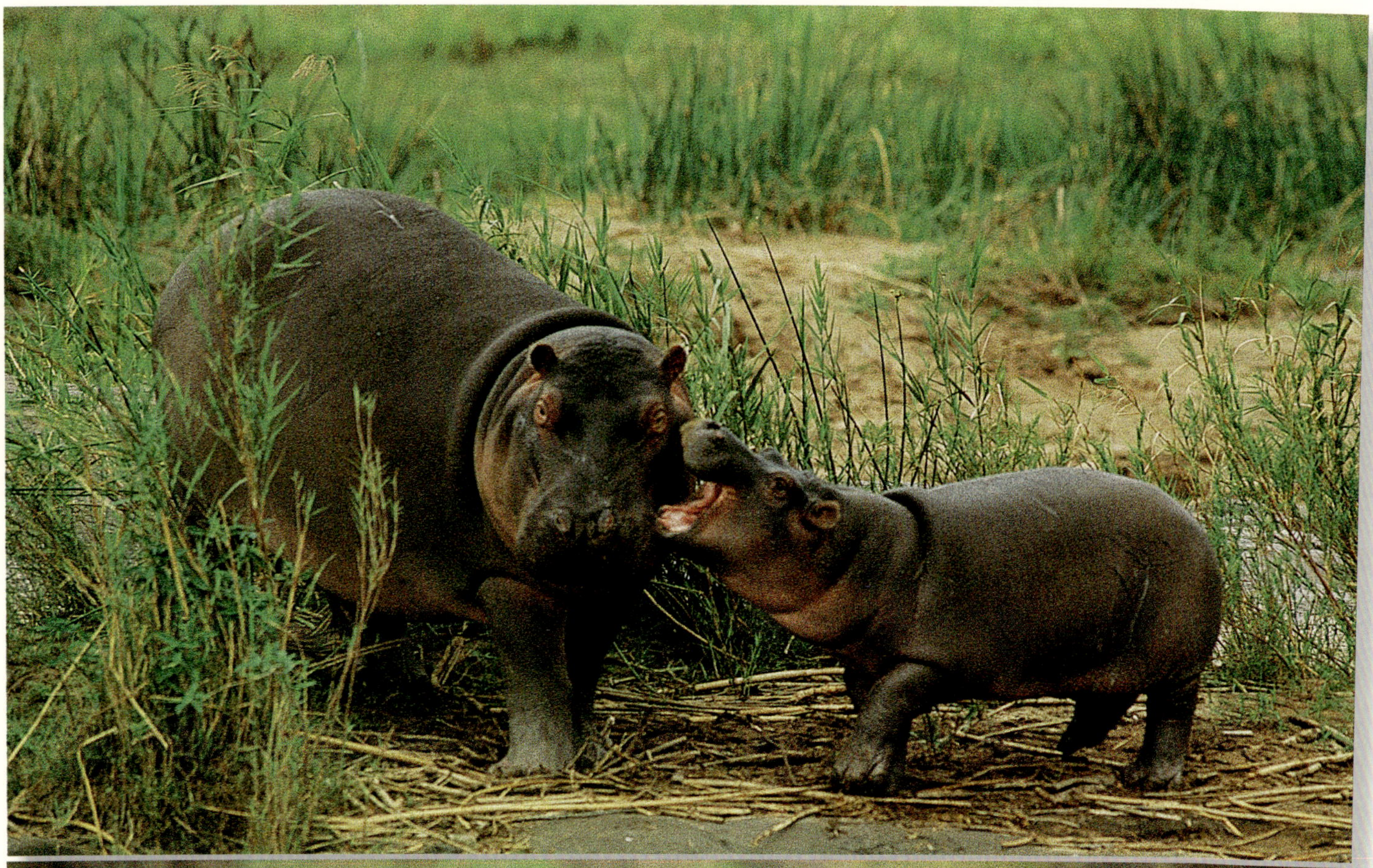

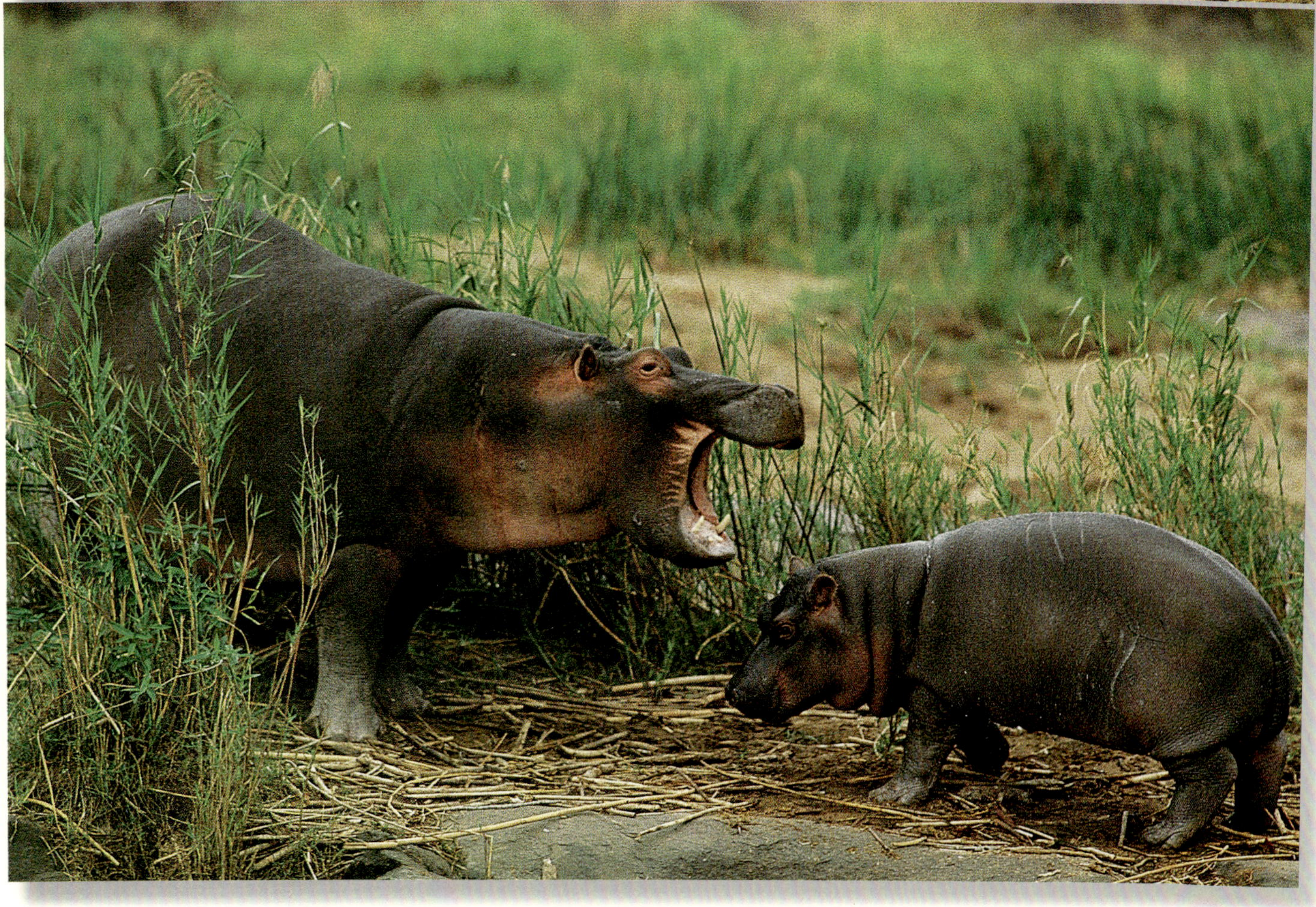

grazing time. Individuals slip out of their aqueous environment and into the night to follow familiar trails leading to the 'hippo lawns' where they graze. The trails are marked with dung heaps – scent markers that keep the hippos going in the right direction. They walk single file and stay close together, protecting calves.

Hippopotamus are as agile on land as they are in water. They climb steep banks and trot in a bouncy gait at about ten miles per hour. They cannot jump, and won't even step over objects in their way, preferring to go around them. When they stop to graze, they are usually within two or three miles of the water. They eat short grass, clipping it with two-foot wide horny lip edges. Grazing for about five hours, they can eat up to one-hundred pounds of plants each night, which is not a lot considering their size. Since most of their daytime is spent in leisurely pursuit, they conserve much energy and don't need a lot of food. At dawn they begin returning to the water. If something disturbs them along the way, they may stampede, galloping up to eighteen miles per hour. It is unwise to get between a herd of hippopotamus and the water.

FACING PAGE

*All youngsters will try to sass a mother from time to time and this hippo is no exception. Not to worry. Mother hippos know how to handle a situation like this in short order.*

As day begins once again, the hippos slip back into their liquid home to ruminate and sleep. Those that sleep underwater automatically rise to breathe without waking. Their presence is beneficial to the water's ecology. Herds release tons of manure into the water which feeds plankton and invertebrates. Those, in turn, are food for fish. Waterways with hippopotamus have a bigger fish population than those without. Some fish stay near the inert hippos, nibbling algae from their bodies.

These slow moving animals also serve as mobile islands for other aquatic wildlife. Wading birds use them as perches from which to fish. Turtles and young crocodiles bask on their broad backs. The hippos don't seem to mind, they're just doing what comes naturally.

---

*Why Hippo Eats Grass* (A Bushman story)

*In the beginning, when each animal was given its place in the world, Hippo humbly asked Creator if he could live in the cool African waterways. Creator hesitated, for He had already granted Crocodile a home in the water. Creator felt that two giant water beasts would quickly deplete the supply of fish – and so he denied Hippo's request. Hippo was distraught and tearfully appealed once more, for he loved the water so. Compassion moved Creator to acquiesce – under one condition. Hippo had to leave the water at night and eat nothing but grass. Hippo quickly agreed and yawned with pleasure. Hippo has ever since kept his promise, and to prove it, he scatters his dung to show Creator it contains no fish bones. And each evening, Hippo roars in thanks for the joy of a wet and wild life.*

---

As the bushman story tells, the hippopotamus must share its habitat with another huge aquatic creature. The African waterways are also home to the Nile crocodile, one of the world's largest crocodiles. Two-thousand years ago, Greek adventurers named the giant reptile for its resemblance to a small Greek lizard, the *krokodeilos*. Little changed in sixty-five million years, the crocodile is a vestige of ancient creatures from prehistory. Still resembling its ancient ancestors, this behemoth wears a scaly hide of earthy green. Its head is mostly mouth with about sixty teeth that remain visible when the mouth is shut, giving it the appearance of smiling. The teeth are used for tearing flesh, not for chewing. The eyes and ears are positioned on top of the head and the nostrils atop the snout so that the crocodile can be fully aware of its surroundings as it lays submerged and camouflaged. Underwater, a transparent third eyelid protects the eyes while the nose and ears close to keep out the water. More than half of the body's length is a powerful tail – adapted for swimming. Short legs and feet are built for walking on land. The back legs also act as rudders when the crocodile is swimming. Crocodiles continue to grow throughout their lives, which can stretch to one-hundred years. The Nile Crocodile averages sixteen to eighteen feet long and can weigh up to two-thousand pounds.

BELOW
*Hippos graze most of the night and dawn is normally a time to sleep.*

It is the most intelligent of reptiles, having a larger brain than most, which enables it to think as well as act instinctively. The thinking brain aids in planning the attack when the crocodile is hunting. Out in the water, lying still, submerged except for nostrils, watching eyes and listening ears above the surface, the hungry crocodile waits, not a ripple betraying its presence. At dusk, when an unsuspecting animal steps to

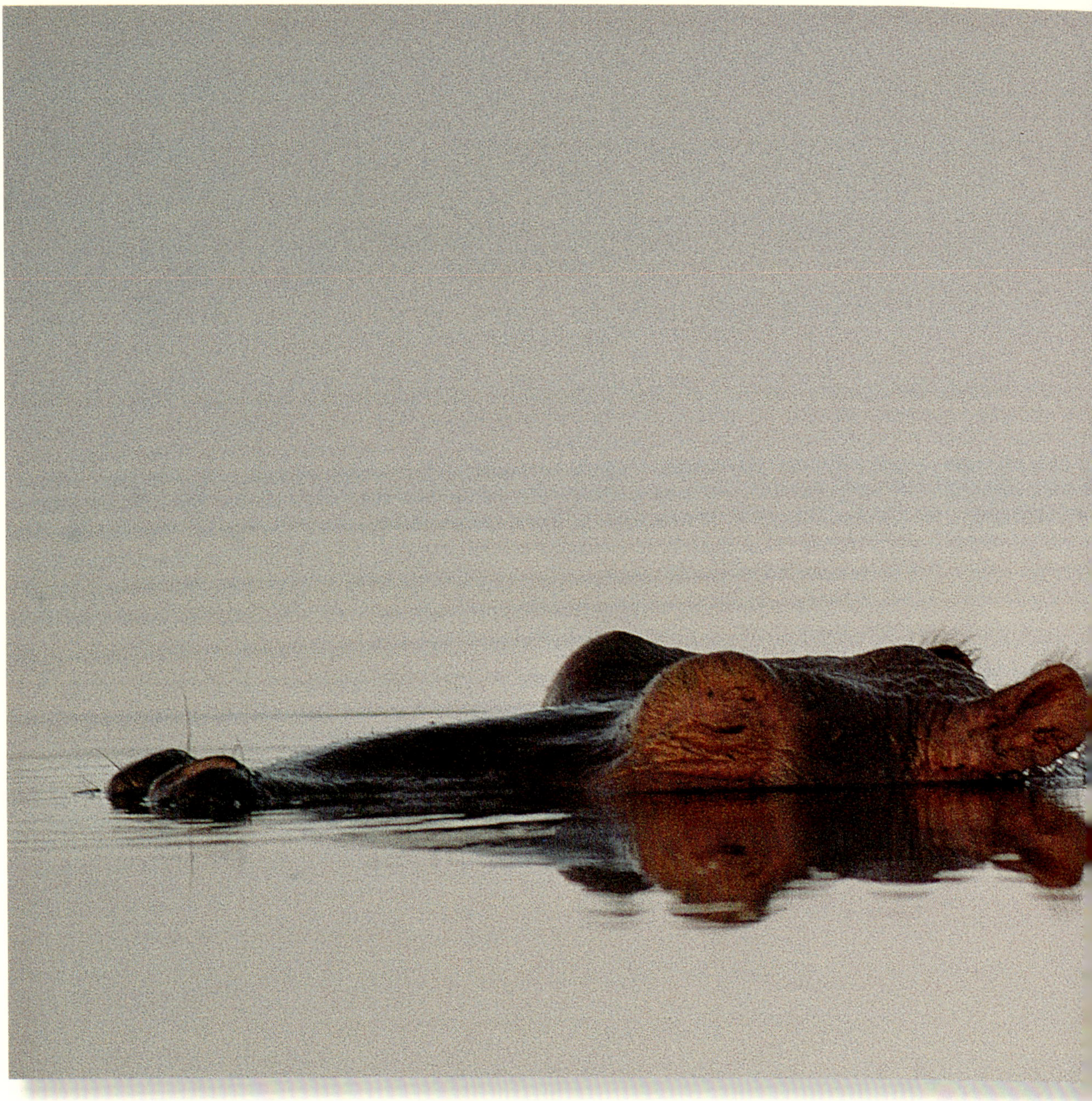

the water's edge to drink, the crocodile slips silently below and swims to shore. In a wink, it leaps out of the water and grabs the surprised victim and drags it under - where it soon drowns. The crocodile's jaws close with one ton of force – so strong that most animals are unable to escape. Once the victim is dead, the crocodile twists and turns to tear the lifeless animal into bite-size chunks. Sometimes, several crocodiles will feed together on one kill. The territorial bull will get the largest share. At other times, food is hidden beneath underwater ledges to be eaten later.

FOLLOWING PAGES
*Since the crocodile is cold-blooded, opening and closing its mouth is another way in which body temperature is regulated.*

During the violent process of eating, the crocodile often loses some teeth, but they rapidly grow back. Under each tooth is a tiny new tooth ready to replace it when lost. A crocodile can use and lose up to three-thousand teeth in a lifetime. It often swallows small stones, which, aided by the stomach's contractions, help digest food by grinding. When catching prey under water, a muscle in the throat seals the opening to keep water out. Although the adult Nile crocodile hunts mammals that

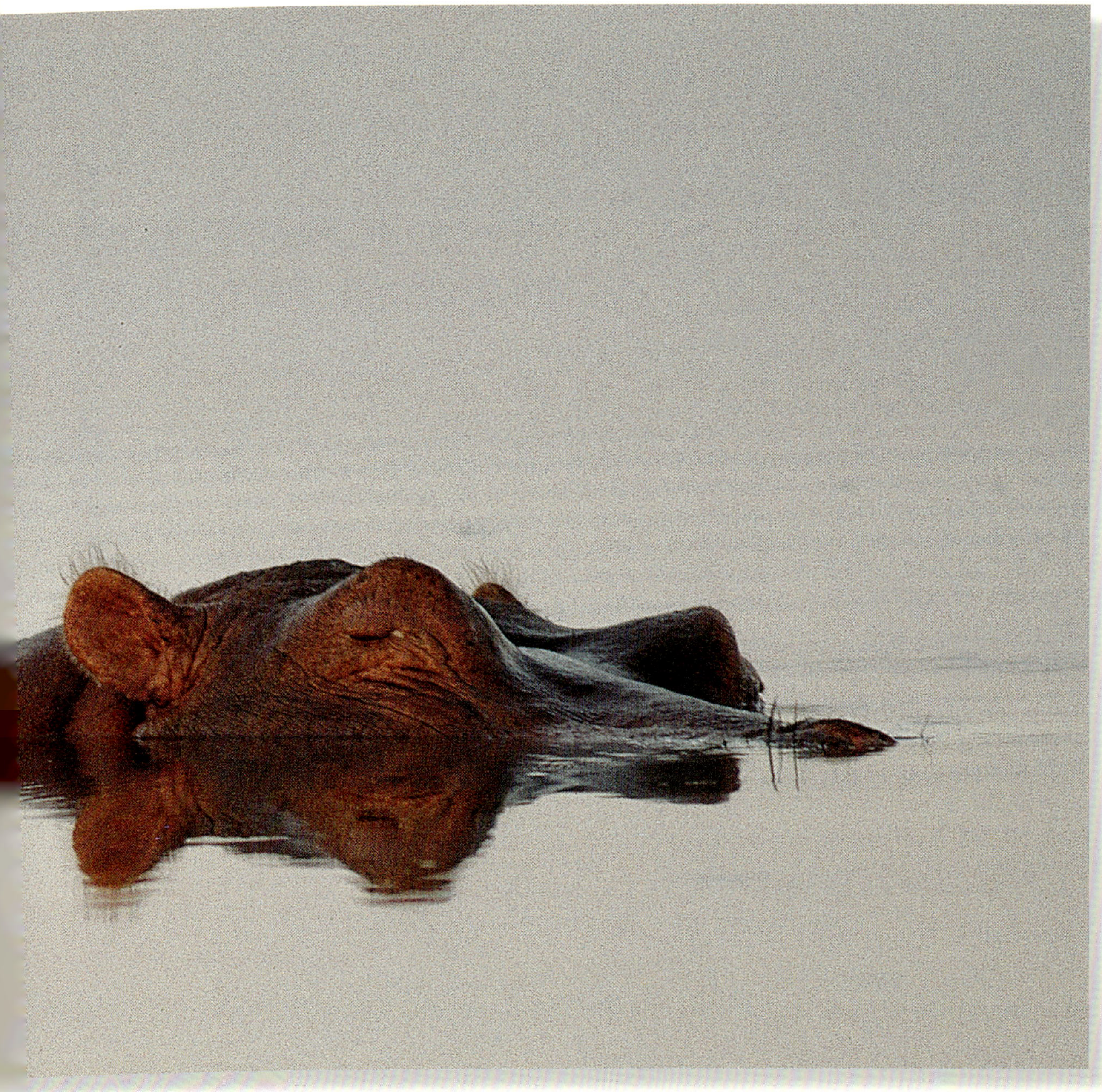

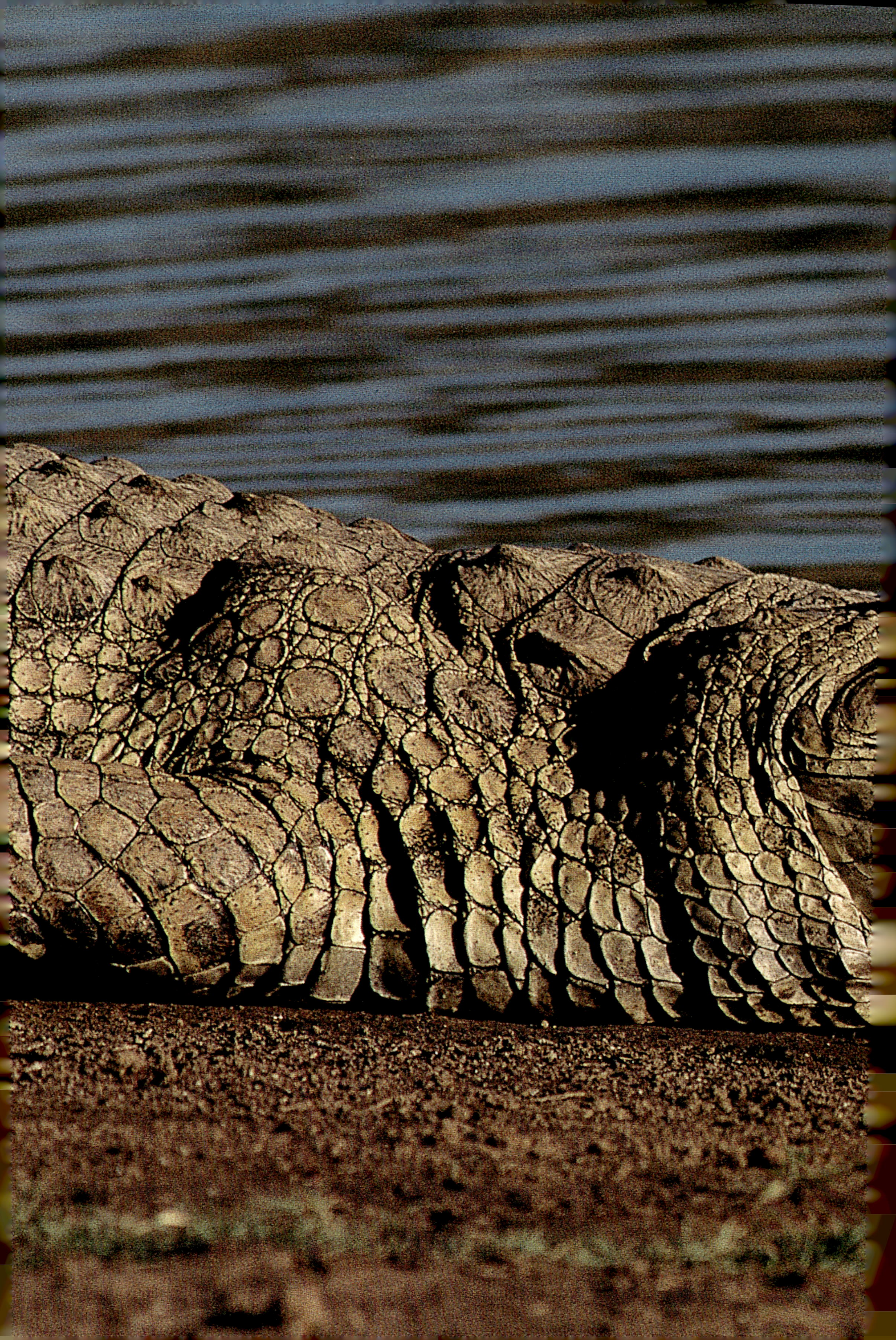

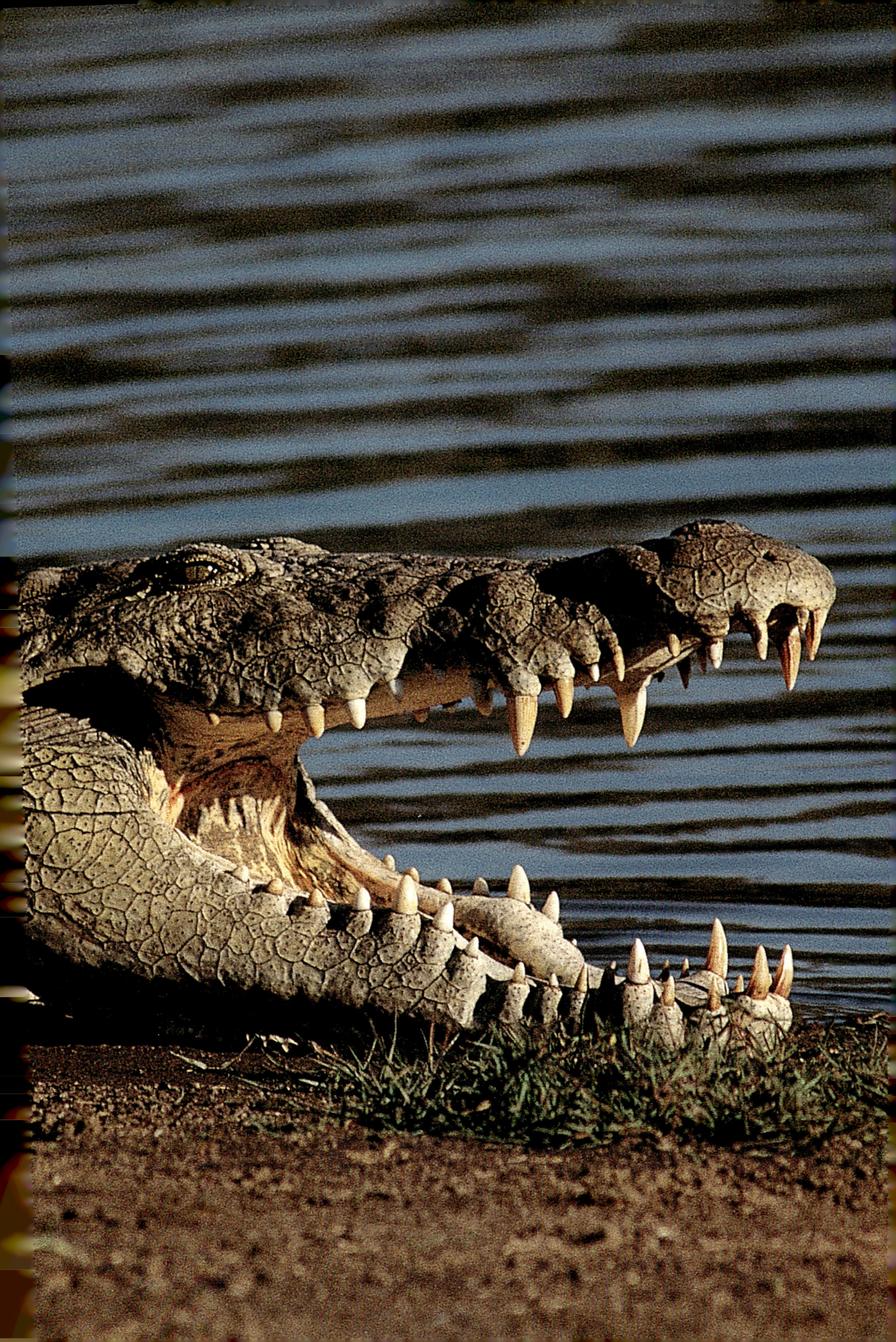

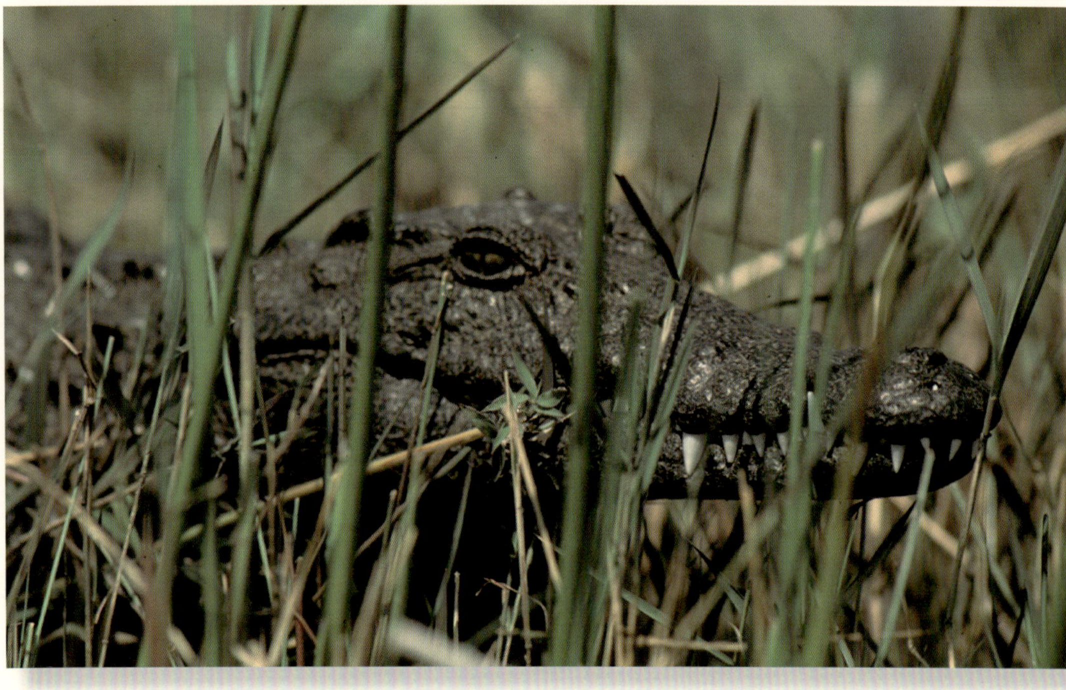

*The Okavango Delta is the setting for this crocodile lurking in reeds.*

come to drink or that live nearby, such as antelopes, impalas, turtles, pythons, birds and sometimes young hippos – its staple food is fish. Scavenging is not beneath a crocodile. One or more can sometimes be seen sharing a kill with other animals.

For the most part, unless involved in the activities of hunting and eating, the crocodile lives a leisurely life. Since it is a cold-blooded animal, it requires no energy to maintain body temperature and therefore does not need to eat every day.

To keep heated, the crocodile stays in its warm watery environment by night. At sunrise, it crawls onto the shore as the warming rays arrive. There it spends the day basking in the hot sun. If it becomes overheated, the crocodile will open its mouth to let evaporation aid in cooling. It will lay for hours in that position. Other cooling methods include seeking shade or re-entering the water, which is cooler by day. In cold weather or time of drought, it may dig a den for warmth and protection.

Although Nile crocodiles are solitary hunters, they can be seen in large assemblies, floating peacefully together. Males are territorial. They communicate with hissing and coughing sounds to warn intruders, to indicate distress and to initiate mating. In addition to vocalizations, they display with open jaws while splashing and thrashing about.

The male courts the female by swimming in tandem with her, his inside fore and hind legs across her body. Mating takes place under water as the couple sinks to the bottom. After two months, the female lays twenty-five to seventy-five hard-shelled eggs in a pit near the water. As they are laid, she gently places them into the pit with one of her back feet. When all the eggs are laid, she covers them well with soil and guards them as best she can for three months. During that time, if they get too hot, she will splash water onto them. When she must leave to hunt, they are subject to predation by lizards or wild pigs. When the young chicks, are ready to hatch, they call to their mother with peeping sounds. She uncovers them and waits for them to hatch. Each chick has a tiny 'egg tooth' that aids it in breaking out of the shell. If any fail to do so, the mother will gently crack the eggs with her teeth. When all the chicks have emerged, the mother gathers them in her mouth and carries them to a safe nursery area in the water. At this size, they can fall prey to foxes and crows, but the mother tries to guard them well. In the water, large fish and water birds eat any who stray. Sometimes, several mother crocodiles will form a créche so that most of the mothers can leave to hunt while one or two watch the brood. They must be protected from bull crocodiles as well, for they often eat young chicks.

Beyond that, the young are independent. They must find their own food – at first feeding on insects and snails. The size of their prey grows as they do; frogs and crabs, then fish and birds. Not until they are adult size, in two or three years, do they hunt larger animals. Full-sized adult crocodiles have few enemies. Angry hippo mothers attack when their calves are in danger. Lions sometimes kill crocodiles. But man has traditionally been the chief cause of their demise. The shoe and handbag trade has long had a taste for crocodile skin. In many places today, they are protected, and hunting is limited.

Ironically, turnabout is fair play. Crocodiles are dangerous to people as well. Many people have been killed in waters inhabited by crocodiles or attacked on the shoreline just as any hapless antelope might be.

Although Nile crocodiles pose a danger to humans and to many an animal, it will spare the life of the small Egyptian plover which bravely walks right into the yawning crocodile's mouth. The plover acts as a dental hygienist, keeping clean those pearly whites. And like their neighbors, the hippos, crocodiles make wonderful water taxis for various small creatures and wading birds. All together, reptile, mammal and bird co-exist in a fragile balance, in Africa's lakes and rivers and streams – where water is life.

FOLLOWING PAGES

*This is a situation in which I photographed a drifting crocodile. It is also the time Johann (my trusted friend) said, "I have never been able to find somebody stupid enough to follow me here."*

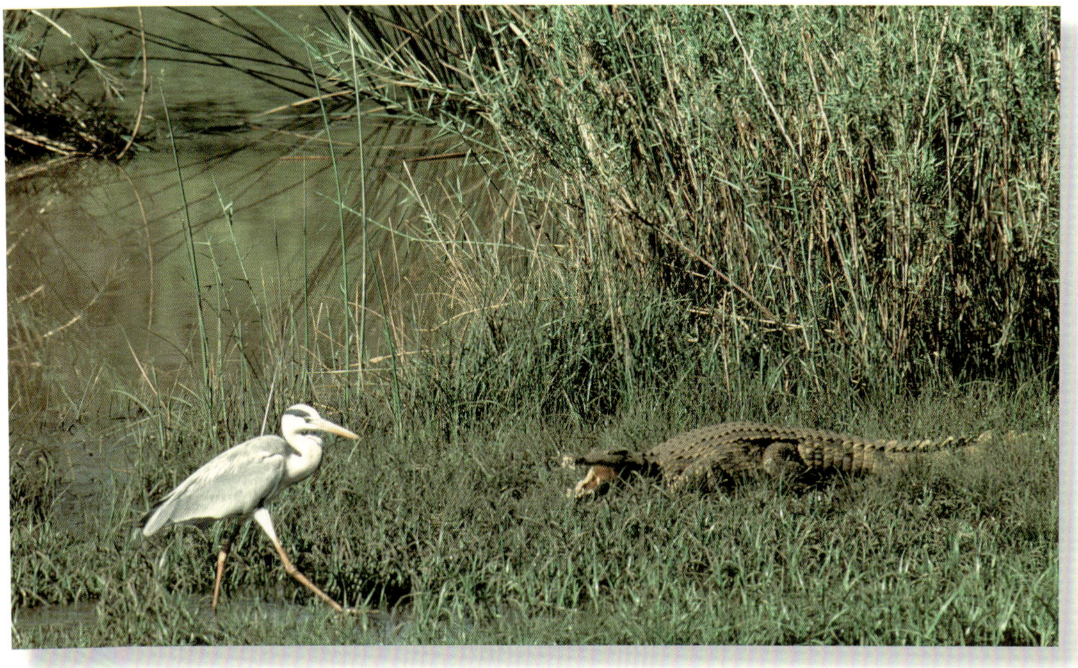

*Although this crocodile appears to be ready to attack the Gray Heron, appearances are deceiving. The Crocodile is simply cooling himself by lying with open mouth – while the heron instinctively knows the safe distance to leave between them.*

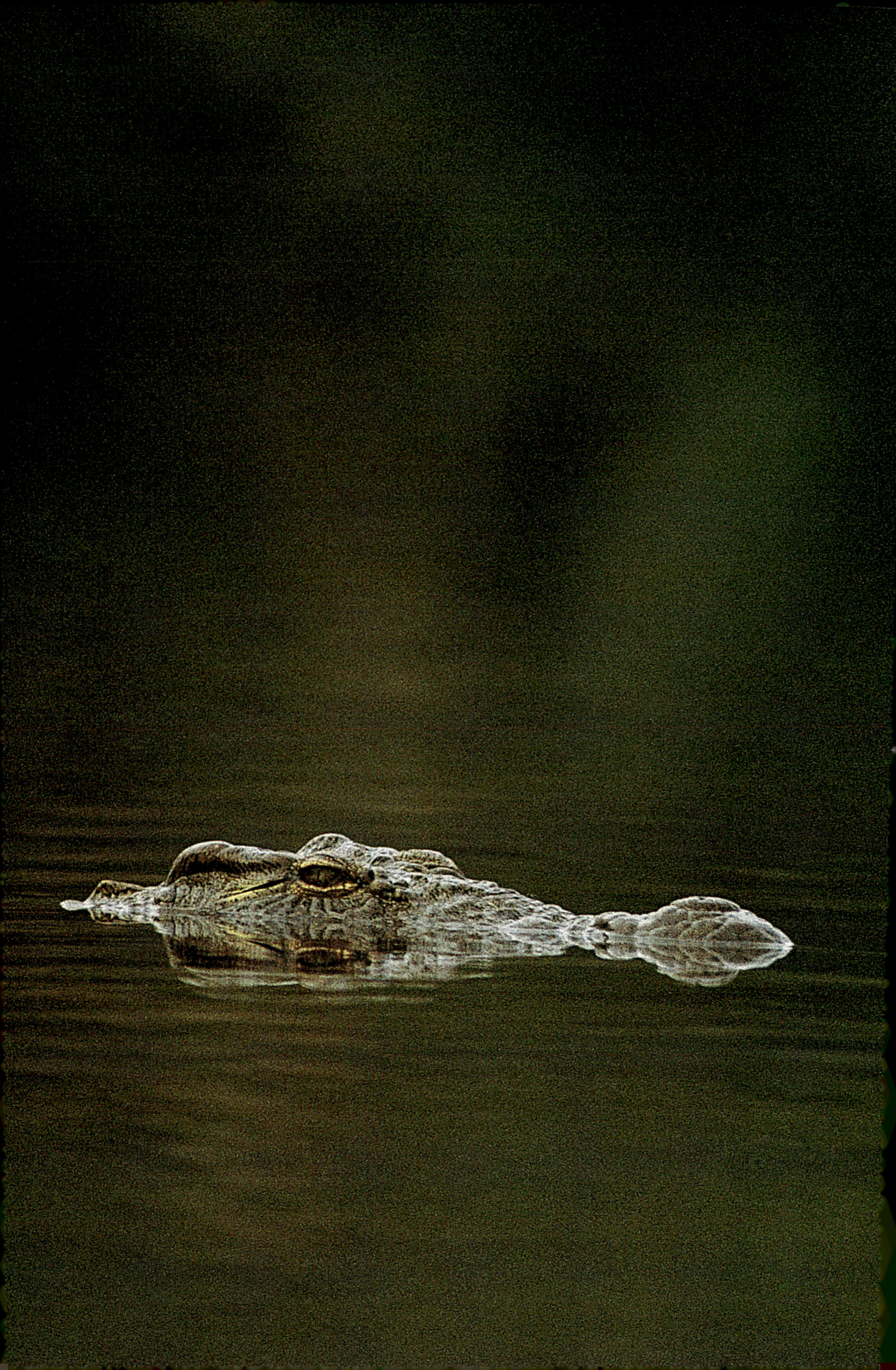

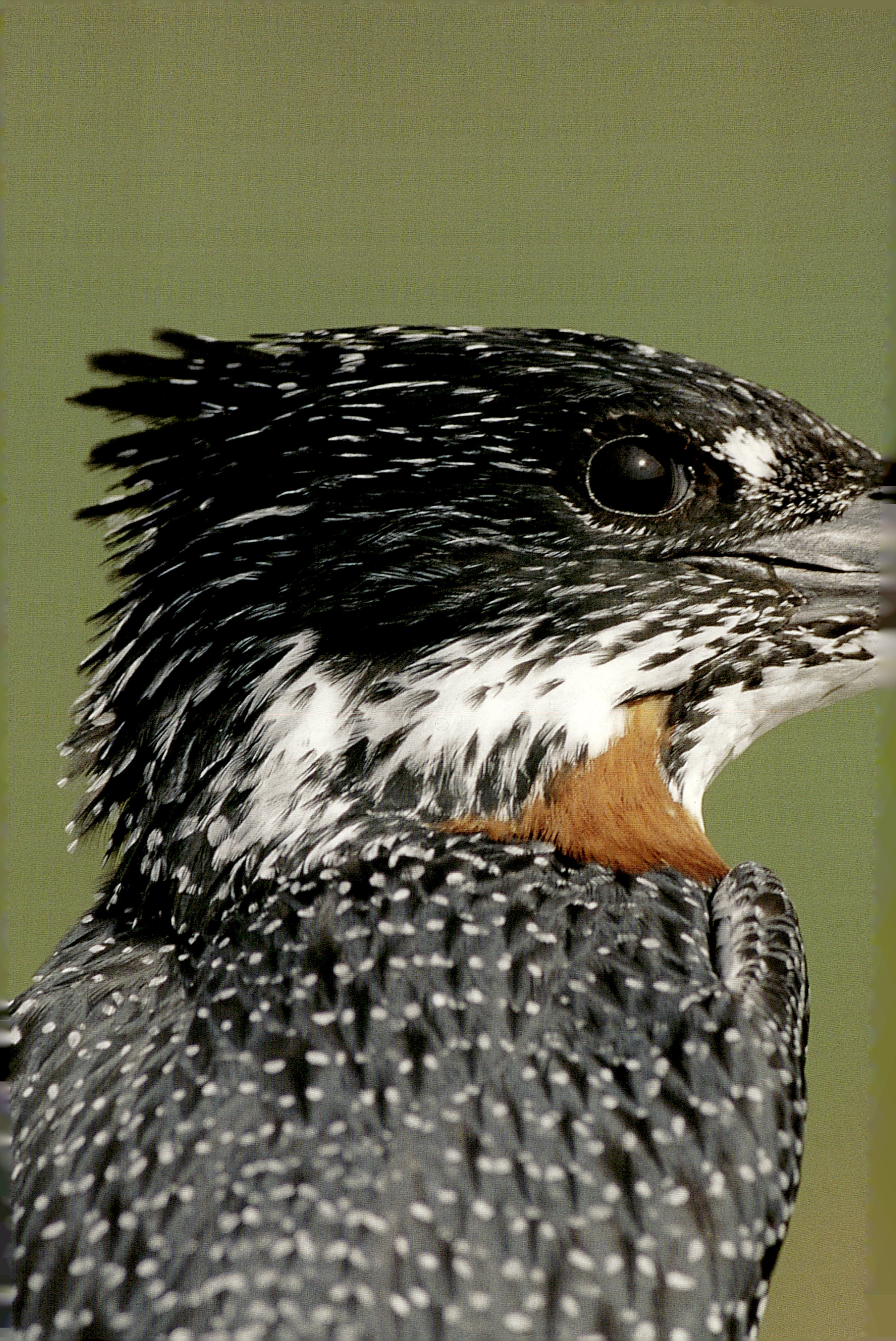

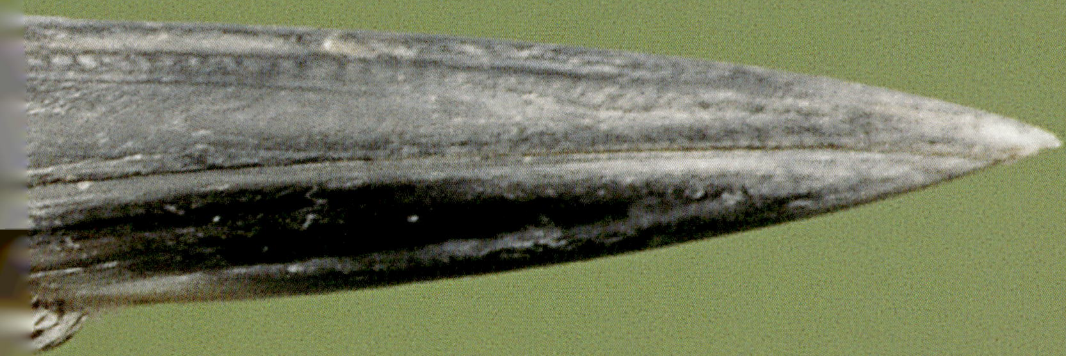

# CHAPTER FOURTEEN
# Birds of the Waterways

*Kingfishers, Hammerkops, Herons, Egyptian Geese, & Storks*

PRECEDING PAGES

*Getting this close and seeing all of the beautiful markings and textures on a Giant Kingfisher is always a big thrill for a photographer.*

FACING PAGE

*A Pied Kingfisher is hunting from a perch.*

BELOW

*A Giant Kingfisher will hunt from a perch or will sometimes hover to capture prey.*

Along the shorelines of Africa's lakes and rivers, pans and estuaries, dwell a diversity of feathered anglers - equipped with features perfectly adapted for catching wet, slithering, aquatic prey; long legs and toes for stalking through the shallows – stirring up the muddy bottom to expose crustaceans and amphibians, and rapier-like bills for spearing fish. Some dive headlong into the water, and with precision accuracy, seize an unsuspecting fish from its liquid world. These are herons, storks and kingfishers, which, along with their herbivorous kin, waterfowl, live abundantly by the waterside.

Several species of kingfishers are the diving birds. Most dive for fish and other aquatic creatures, yet, some use their skill to dive into grass for insects, lizards or small rodents. Kingfishers are stocky birds with short legs, short necks and large heads. They have large, straight beaks with dagger-like points. Plumage varies from black and white to multicolored – often iridescent blue, green, purple or red. Their wings are short and rounded. Tails vary as to species. Kingfishers nest in riverbanks, tree cavities or earthen mounds, where most raise three to five young in a season. They are monogamous and territorial. Some are gregarious; others are solitary. All species are carnivorous.

Watching a kingfisher hunt is fascinating. From a perch it will spot a fish, dive at a precise forty-five degree angle, close its eyes as it enters the water, and blindly seize the prey. A second later, it emerges and

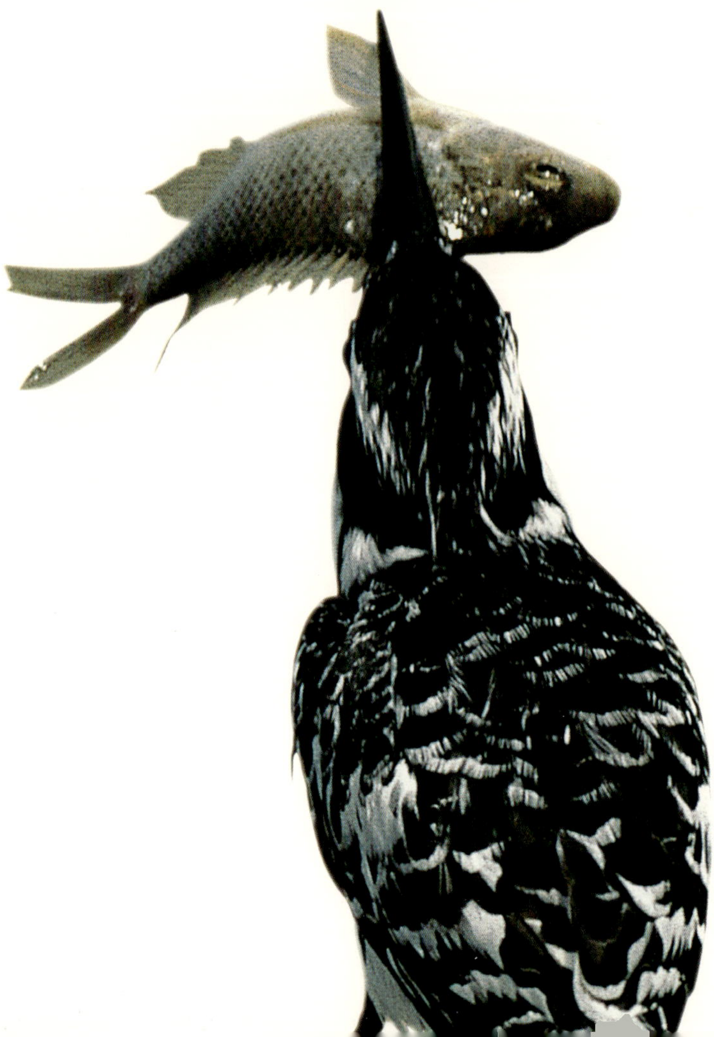

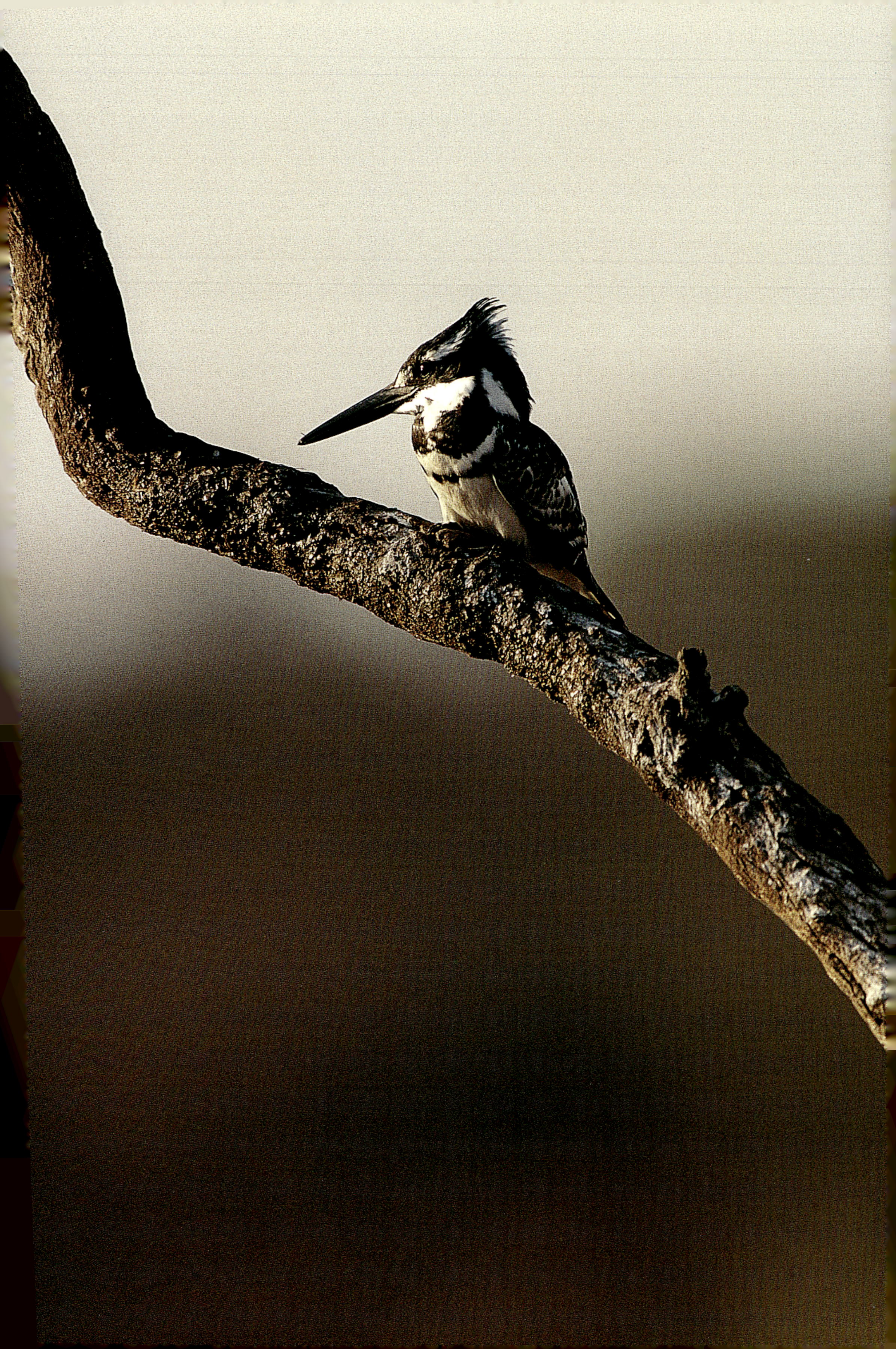

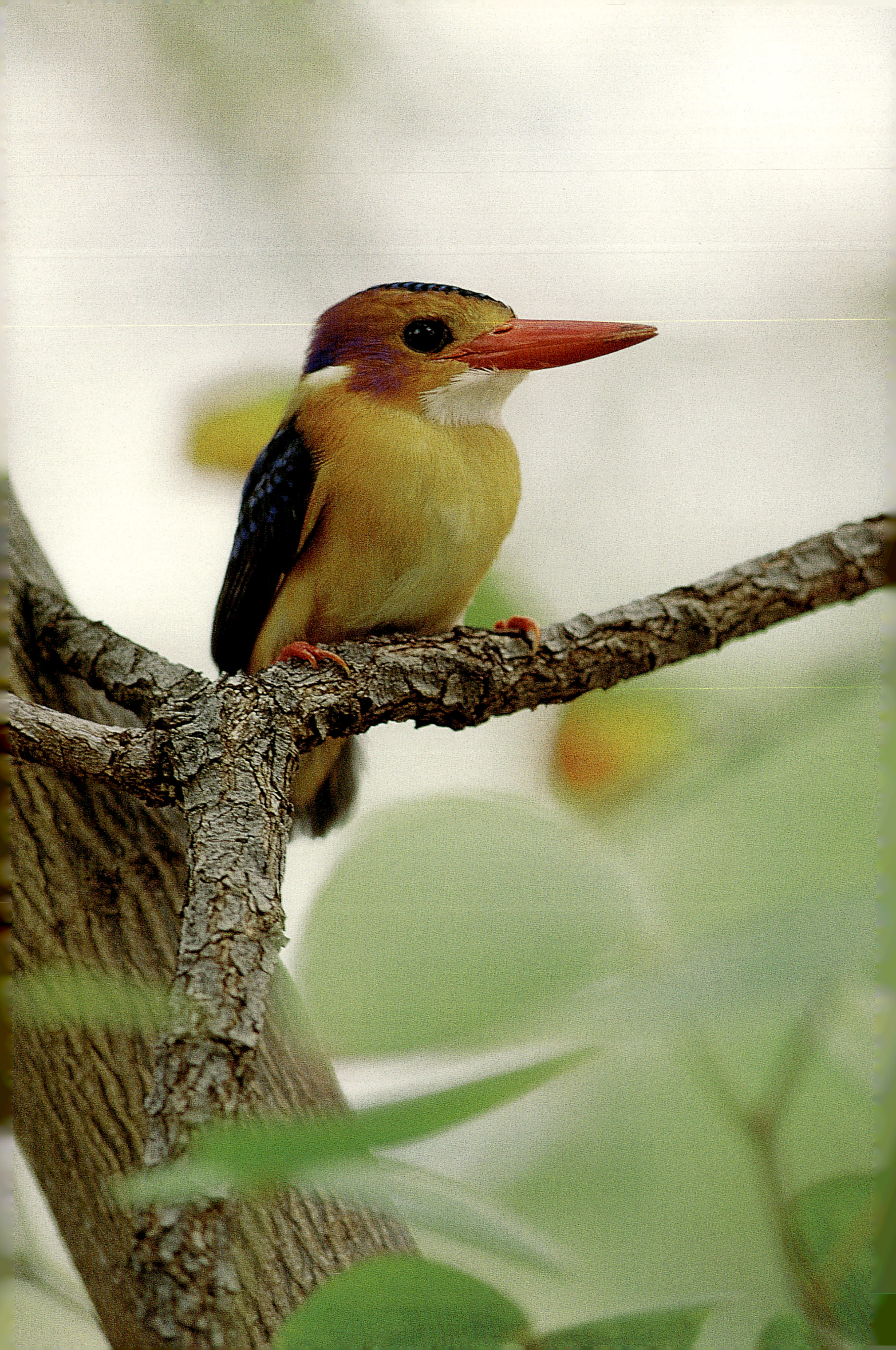

flies to a perch, where first it beats the fish on a branch to immobilize it, and then swallows it head first.

The largest is the Giant Kingfisher at eighteen inches long. It is a black speckled bird with a chestnut breast and a heavy black bill. This angler eats mainly fish, but also has a taste for frogs and crabs. The Giant Kingfisher prefers a habitat near wooded rivers and lagoons. As it flies, it laughs a loud ringing – *wak-wak-wak-wak-wak-wakririririri.*

The smaller Pied Kingfisher is ten to twelve inches long, with black and white patterned plumage. The male has a pair of black bands across its breast. The female has only one band. Pied Kingfishers are often found in pairs and are the only species to live in loose colonies when breeding. Their nests are burrows, excavated three to six feet into riverbanks. Within a chamber at the end of the burrow, they lay three to five round white eggs. After hatching, they are cared for by both parents and by a helper, which may or may not be related to the pair. The helper assists in defending the nest and in feeding the fledglings. The burrow soon becomes lined with fish scales and crab carapaces.

FACING PAGE

*This colorful Pygmy Kingfisher will easily fit into the palm of your hand. He was photographed in the camp at Shingwedzi.*

Another special feature of the Pied Kingfisher is its fishing technique. In addition to hunting from a perch, it will fly low over water to a chosen spot where it hovers high above the water, hummingbird style, for several seconds, body vertical, bill pointed down. Then, it dives – hitting the water with an audible splash – and traveling up to six and a half feet down to snatch its prey. It often catches many fish at once, or sometimes comes up with a crunchy crustacean. After a meal, it may perch, squeaking and twittering or calling sharply – *kwik-kwik-kwik.*

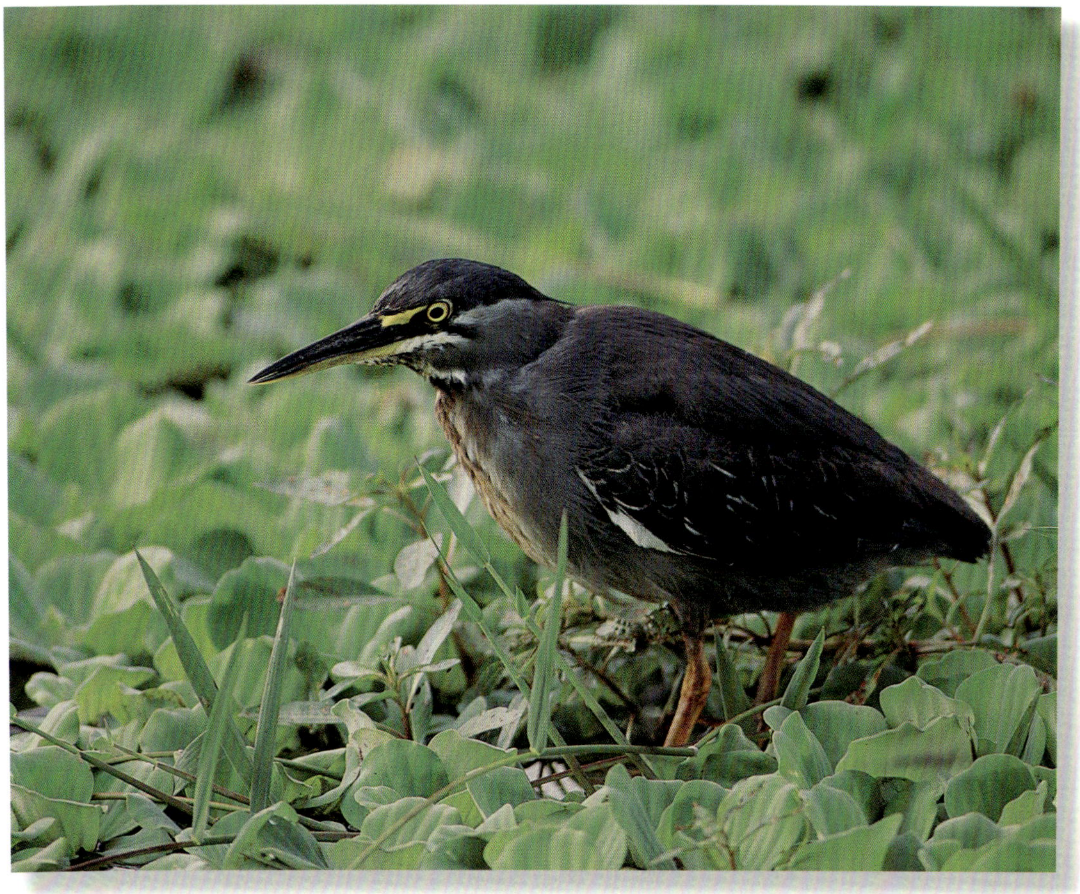

*Water lettuce supports a Greenbacked Heron and allows it to walk on water.*

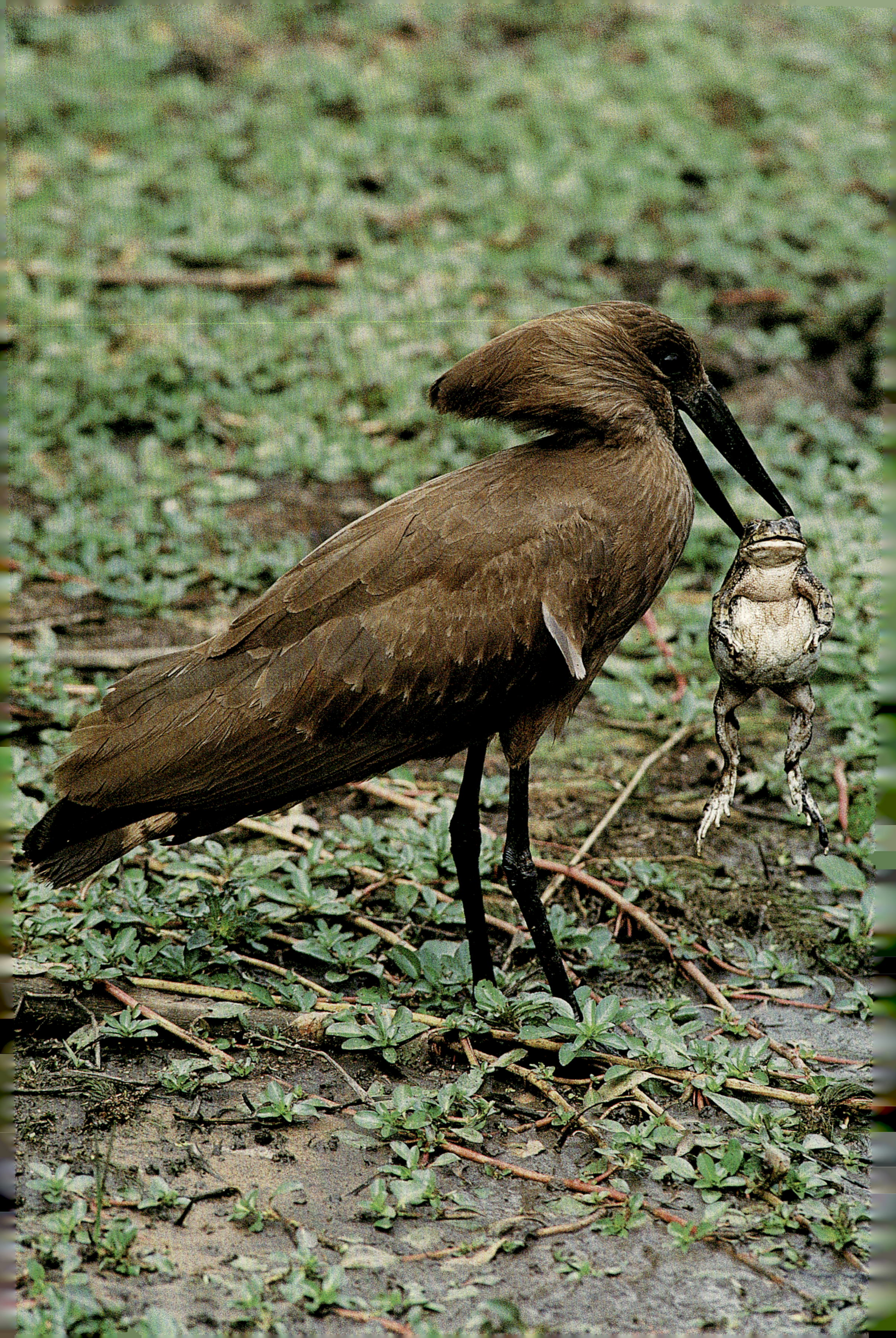

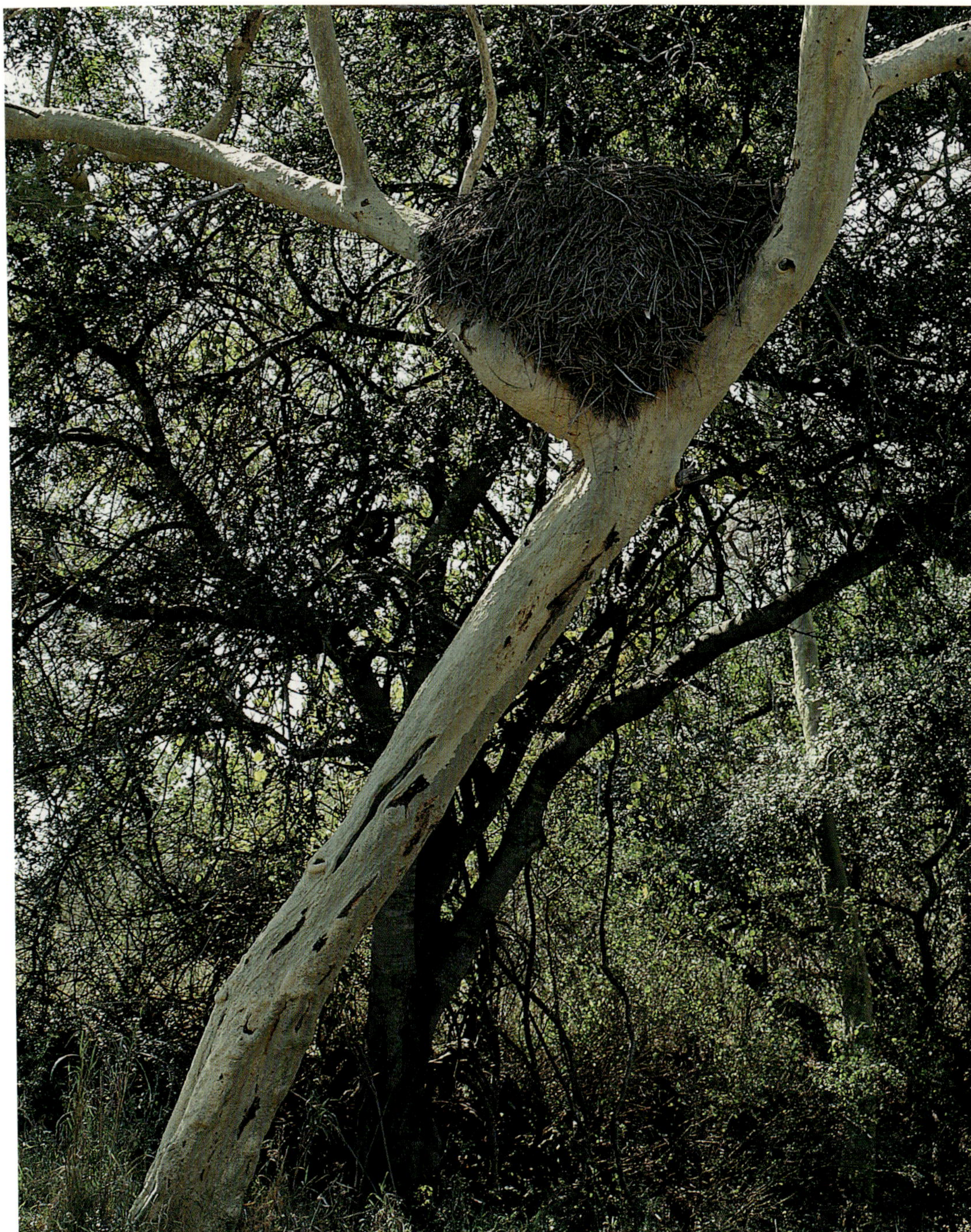

*A Hammerkop nest is indeed a wondrous architectural gem. It is strong enough to hold a human and weighs up to one-hundred ten pounds.*

In the deep woodland and riverine forests lives the diminutive Pygmy Kingfisher. At five inches long it is the family's smallest. What it lacks in size it makes up for in coloration. The Pygmy Kingfisher is a deep, bright blue above, including the crown of its head. It has a mauve patch behind the ear and chestnut-orange underparts. The bill, legs and feet are bright red. If the Pygmy Kingfisher's brilliant coloration doesn't give it away, perhaps the sound it makes in flight – a sharp *chip-chip* – will.

It is usually a solitary bird, but can sometimes be found in pairs or small groups. Unlike

FACING PAGE

*A Hammerkop eating a frog is not a pretty sight. The frog was beaten to incapacitate him and then swallowed whole. What is incredible is that the Hammerkop ate another frog just after he ate the first; the second frog was just as big as the first.*

its fishing cousins, the Pygmy Kingfisher hunts mainly insects; grasshoppers, caterpillars, beetles and such. It shares the characteristic diving technique of other kingfishers, but instead of into water, it dives from a low perch into high grasses to catch its prey. On occasion, it will venture to water and dine on small frogs and crustaceans. Instead of a riverbank, it burrows into earthen mounds built by termites or antbears to create a nest. Like other kingfishers, Pygmys raise a small brood of three to five young.

FACING PAGE

*The Gray Heron is conspicuous throughout the Kruger Park. The first time I saw one it reminded me of the Great Blue Heron we find here in the U.S.A.*

BELOW

*Cattle Egrets will forage from field to river in their search for food. They can often be seen taking a pause on top of hippos.*

The grandest nest of all belongs to a curious bird called the Hammerkop. This bird defies classification. Often called a stork, it is not. Neither is it a heron, or part of the Flamingo family where it is sometimes placed. It is a distinctive species with some distinctive characteristics. It is named for its uniquely shaped head, which has, on one end, a heavy, conical, flattened bill, and on the other end, a backward projecting crest. These two features, in profile, give it a hammer-head look. The plumage is dark brown with a purplish sheen and the legs and feet are black. At twenty-four inches tall it is about the size of a thin chicken.

The Hammerkop lives near inland water where it builds giant nests in the crooks of trees, on cliffs or on rocks in rivers. It takes four or more birds four weeks to six months to build a nest. The Hammerkop's grand home is a thing of wonder. It is a huge domed assemblage of sticks, reeds, leaves and debris, as well as man-made objects and trash. It can be up to six feet in diameter and weigh up to

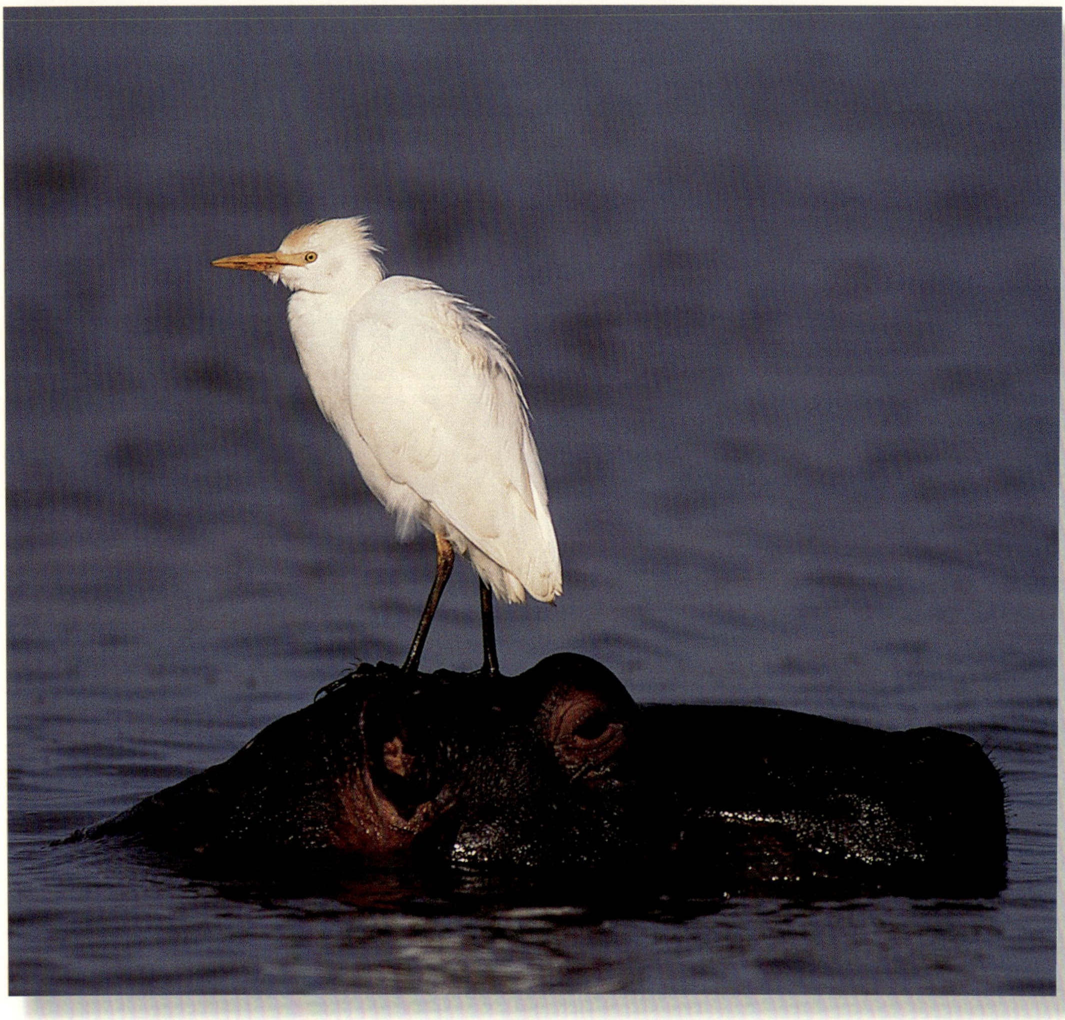

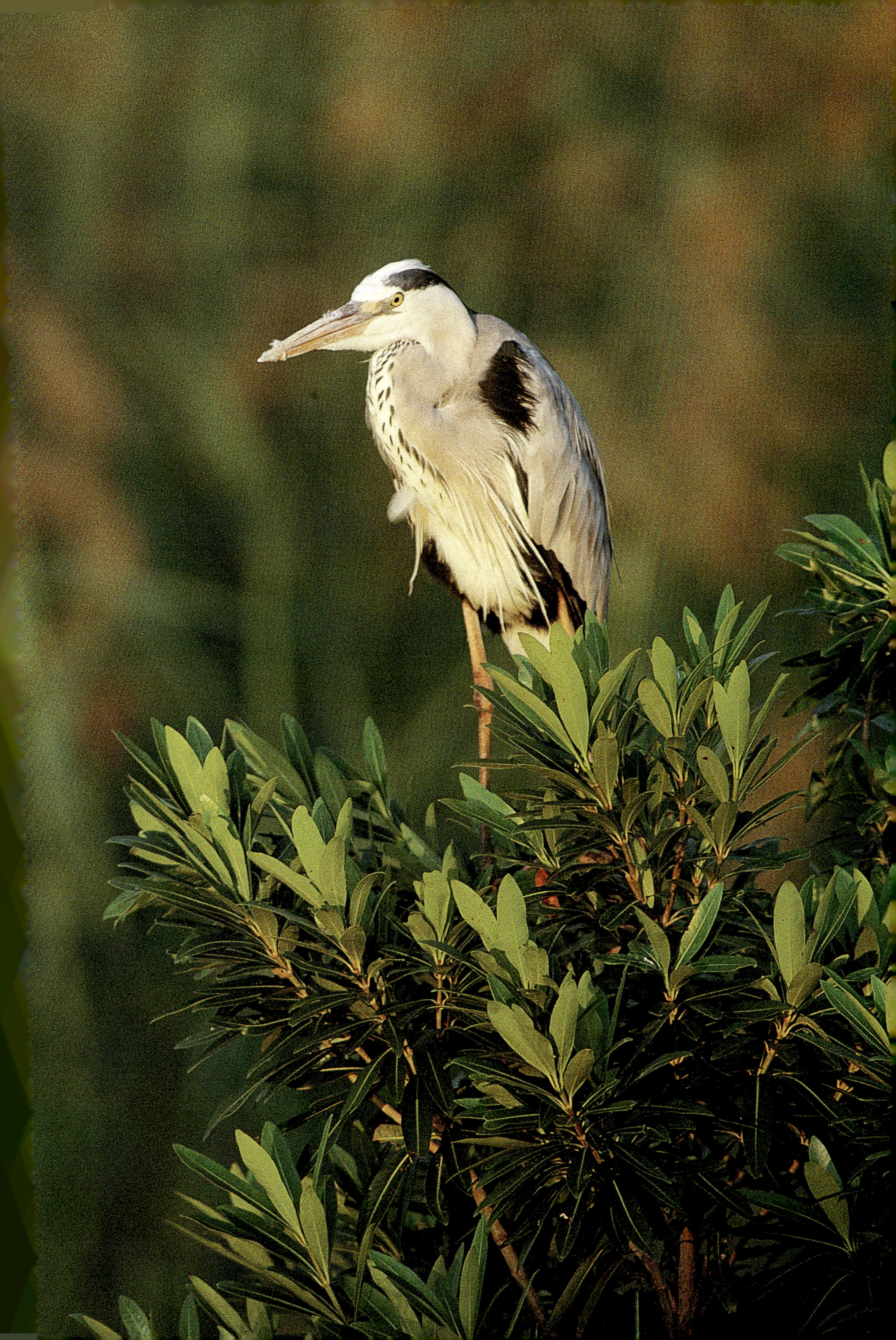

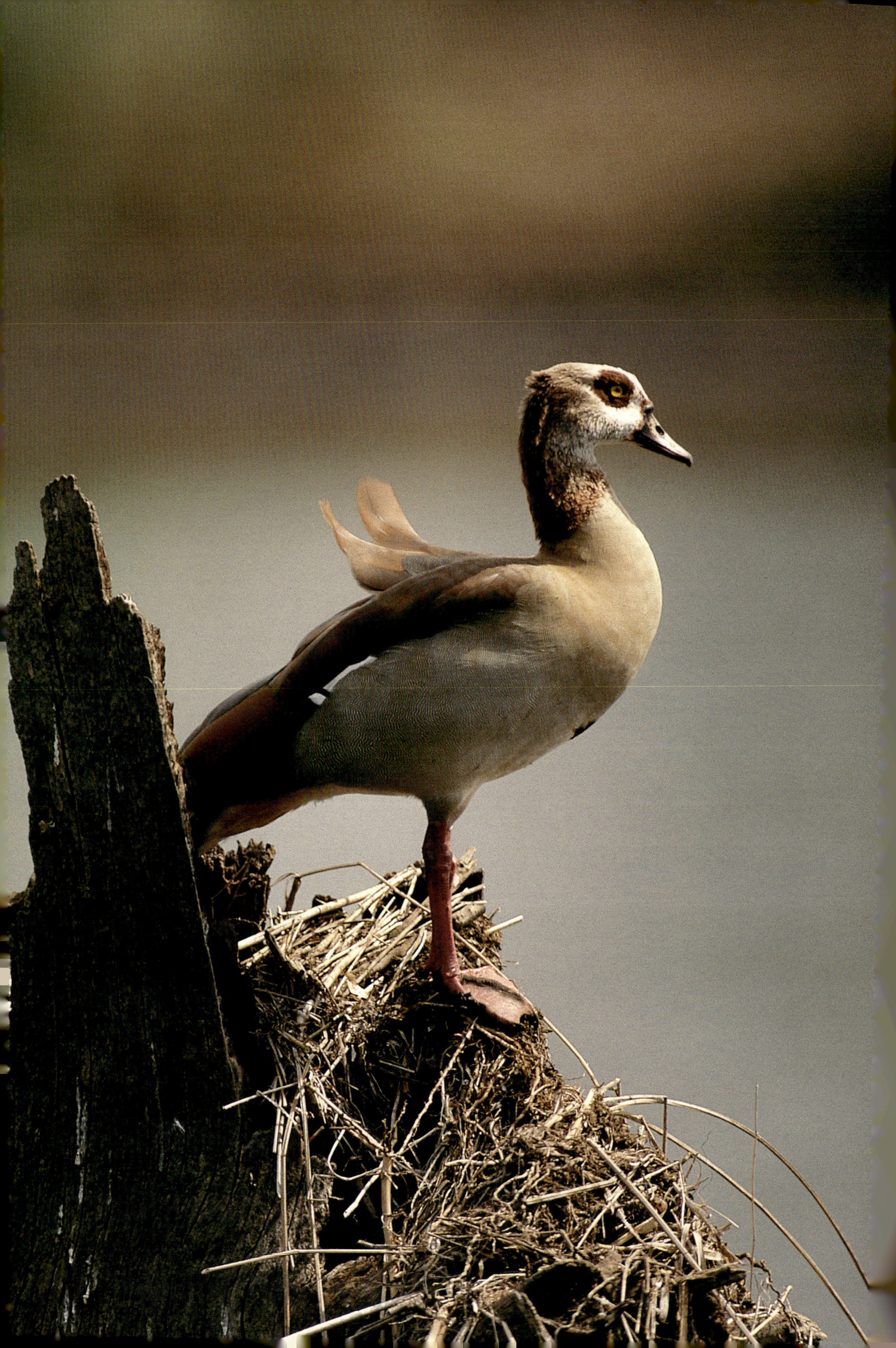

onehundred-ten pounds. A sixteen to twenty inch long entrance tunnel opens into a mud plastered interior chamber twelve to twenty inches in diameter. The three foot thick roof is strong enough to hold a fully grown man – but one might not like to venture that close – for the nest is often home to more than the Hammerkops. Bees, snakes or small mammals sometimes share residence. Weaver birds may attach their own nests. Monitor lizards like to invade to steal eggs, and birds such as owls and kestrels will sometimes evict the rightful owners and take over. But the Hammerkops don't go homeless, for they often build several nests in a season.

When the nests are completed, groups of Hammerkops gather nearby and display together, bowing with wings outspread, calling loudly – *yip-purrr-yik-yik-yik-purrr-pur-yik-yik* – in chorus, running in circles with raised crests and fluttering wings. Sometimes they simulate mating or stand on one another's backs, totem pole style. This behavior carries on for several minutes. Later, when actual mating takes place, a pair will display in a similar fashion as a prelude. This time, though, they vocalize with squawks and frog-like croaks.

FACING PAGE

*One morning an Egyptian Goose allowed me to get close. I photographed him along the Shingwedzi River in the Kruger Park.*

The female lays one to five eggs in the secure nest chamber and both parents incubate them. When hatched, chicks are downy and grey. Within seventeen days they have feathers, and are fully fledged within one month. They stay for one more month before venturing out on their own and joining the adults to hunt. They fly off in the hammerkop's characteristic jerky motion – off to the waterside, where they may fly low over the water, snatching fish or invertebrates from the surface. But mostly, they wade in the shallows and probe the muddy bottom for frogs – a favorite food.

On the shores, in the shallows, the large conspicuous wading birds balance on their long legs and search the water for a meal. These birds have long, unfeathered legs with long toes for wading, and long, straight, pointed bills which are specially adapted for stabbing live aquatic prey. Wading birds breed in colonies – often several species mingling together. They build large platform nests of sticks, usually in trees or on cliffs, but sometimes on the ground. They have special feathers that turn to powder when the birds preen themselves. They rub the powder into the other feathers and then comb it out with a flattened claw on their toes. Their varied diet of fish, crustaceans, amphibians, snakes, birds, insects and small mammals, have helped them prosper.

Among the wading birds are several species of day herons. They vary in appearance, but share many characteristics. At breeding time, their eyes, bills, feet and plumage change color – often to bright hues. They fly with their long necks folded and their long legs held out behind. All species vocalize with harsh croaks and squawks – often in flight.

With all the similarities, each species has its differences as well. One type of day heron, the Cattle Egret, divides its time between grassland/savanna areas, where it follows large game animals, collecting insects, – and shallow waters, where it forages for frogs and fish. The Cattle Egret often perches on the backs or heads of large mammals such as hippopotamus. It is an average sized bird at twenty four inches tall. When not breeding it is all white with a yellow bill and eyes. Breeding plumage is pinkish on the bird's crown, back and breast. The eyes and bill change to red and the legs, usually olive-brown, turn yellow-orange when in the breeding stage.

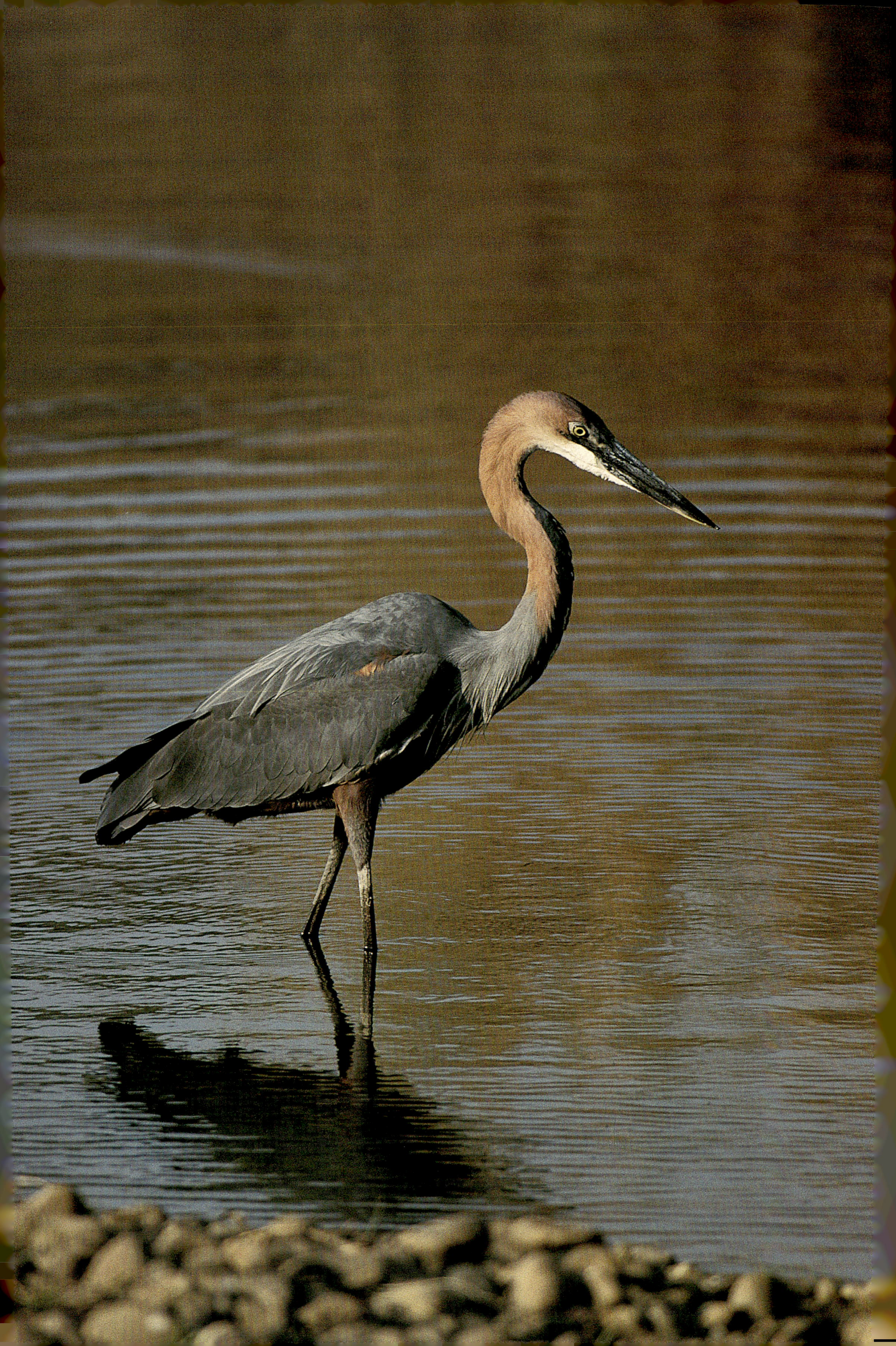

Cattle Egrets breed on small platform nests made of sticks or reeds. The nests take about eleven days to build. The male gathers the materials and the female builds the nest. She lays two to five pale blue eggs. After hatching, both parents feed the chicks which quickly fledge and leave the nest. They can fly within thirty days and are independent in forty-five days.

FACING PAGE
*Goliath Herons will wade in the rivers until they discover a good fishing hole.*

Egrets are gregarious. They gather in large flocks to roost in trees over water. When going to and from roosts they fly in a 'V' formation, chattering and croaking – *rik-rak-kraaa – thonk-thonk-thonk.*

The Greenbacked Heron lives consistently near water in dense vegetation near rivers, streams, ponds, lakes or swamps. It prefers cover, as it is a shy, solitary bird. The Greenbacked Heron is only eighteen inches tall. It has dark green plumage above and pale green underparts. Its bill is black and yellow-green, and its head has a black cap. Eyes change from yellow to orange when breeding and the leg color changes from yellow to red-orange. It is usually silent, but sometimes utters a sharp *kyah!* when flushed. Other vocalizations include croaks and harsh sneezing sounds.

The Greenbacked Heron forages with great patience. It stands motionless for long periods with its

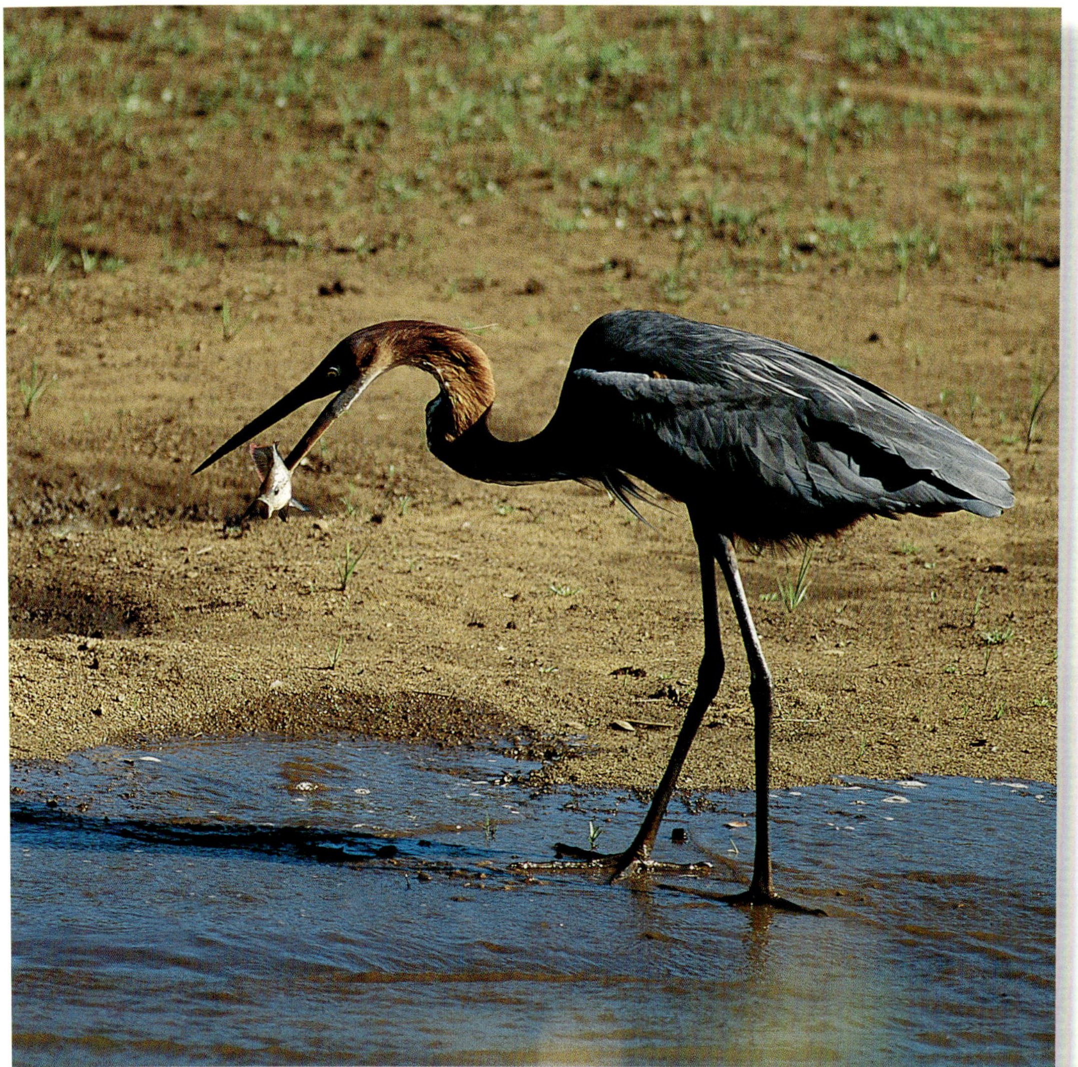

*It was rather unusual for me to see a fish that has been speared by a Goliath Heron. He will normally catch a fish by grabbing it in his beak.*

neck hunched into its shoulders and its head positioned horizontally, or angled downward. When it spots prey, it strikes rapidly – sometimes submerging itself in the water, to catch fish frogs or insects.

The Greenbacked Heron's nest is a flimsy platform only twelve inches wide, built on the end of a tree branch or in a bush. Both parents care for two to five young, which grow quickly and depart within thirty-five days.

The Grey Heron is a larger bird at forty inches tall. As named, it is mostly grey, but it has a white head and black patches on its wings and breast. A black eye stripe ends on the nape with a wispy plume. Its bill is yellow, its legs and feet are yellow and brown. In addition to wading in shallows, the Grey Heron can swim in deeper water. Pairs create a racket on their nests, raising their crests and croaking noisily – *kraaunk-kraaunk-kraaunk.*

FACING PAGE
*The brooding patch located on a male Saddlebilled Stork is characteristic of his tending a nest.*

The Goliath Heron is well named, for at five feet tall, it is the world's largest. This stately bird is slate grey above with chestnut colored underparts, neck and head. It has yellow eyes with a greenish eye ring, a very large bill that is black above, black and white below, and long black legs and toes.

The Goliath Heron chooses to live near the larger inland waterways and estuaries. Unlike other herons, it hunts in deeper water where it spears the larger fish, frogs and reptiles. Sometimes it will also eat carrion. When foraging, it walks slowly or will stand still for long periods of time, waiting for prey to appear. In the air, the Goliath Heron flies gracefully, with slow even wing beats.

At breeding time. it can be found in pairs or groups, nesting in colonies. A great bird naturally needs a great nest. In a tree or bush, on a rock, cliff or island, the Goliath Heron builds a platform nest up to four and a half feet wide. The female lays two to five pale blue-green eggs.

That such a large bird is shy and solitary seems odd, but it is. Should it care to make itself known, the Goliath Heron may grunt or call a deep raucous – *kowoorrk-kowoorrk-woork-work-work* – or utter a hippo-like sound – *mmmmmm-haw-haw-haw-haw.*

Sharing the inland waterways with the waders are waterfowl, which are herbivorous birds. One such bird is the Egyptian Goose. It is a large goose, at twenty- four inches tall, and, it is a colorful goose. Its eyes vary from orange to red-brown to red. The pink bill is edged in dark brown and maroon. The body is brown above and greyish below. It has a dark brown eye patch, neck ring and breast spot. Under the tail is bright yellow-ochre, and the brown and white wings are edged with green. This handsome bird stands on legs and web-toed feet of pinkish red. The Egyptian Goose has a vocabulary of sounds as colorful as its physique. With neck outstretched, the male hoarsely grunts – *hur-hur-hur-hur,* while the female wheezes – *hisssssss,* like the air brakes on a bus. While taking to the air, they rapidly honk – *ka-ka-ka-ka-ka.*

In early morning and late afternoon, they fly to graze in pans, marshes and estuaries, where they eat grass, leaves, seeds and seedlings, grain, aquatic rhizomes and tubers. At midday, these gregarious birds gather in flocks and leisurely swim or float. At night they roost in trees or gather on shorelines.

Egyptian Geese nest in a variety of places. Within dense vegetation they may line a hollow with grass and down, or utilize holes in cliffs, caves or trees. They reuse abandoned crow, raptor or hammerkop nests as well. In whatever nest they choose, the female lays five to seven cream-colored eggs. The downy, precocial goslings that hatch leave the nest only six hours later. In less than two months

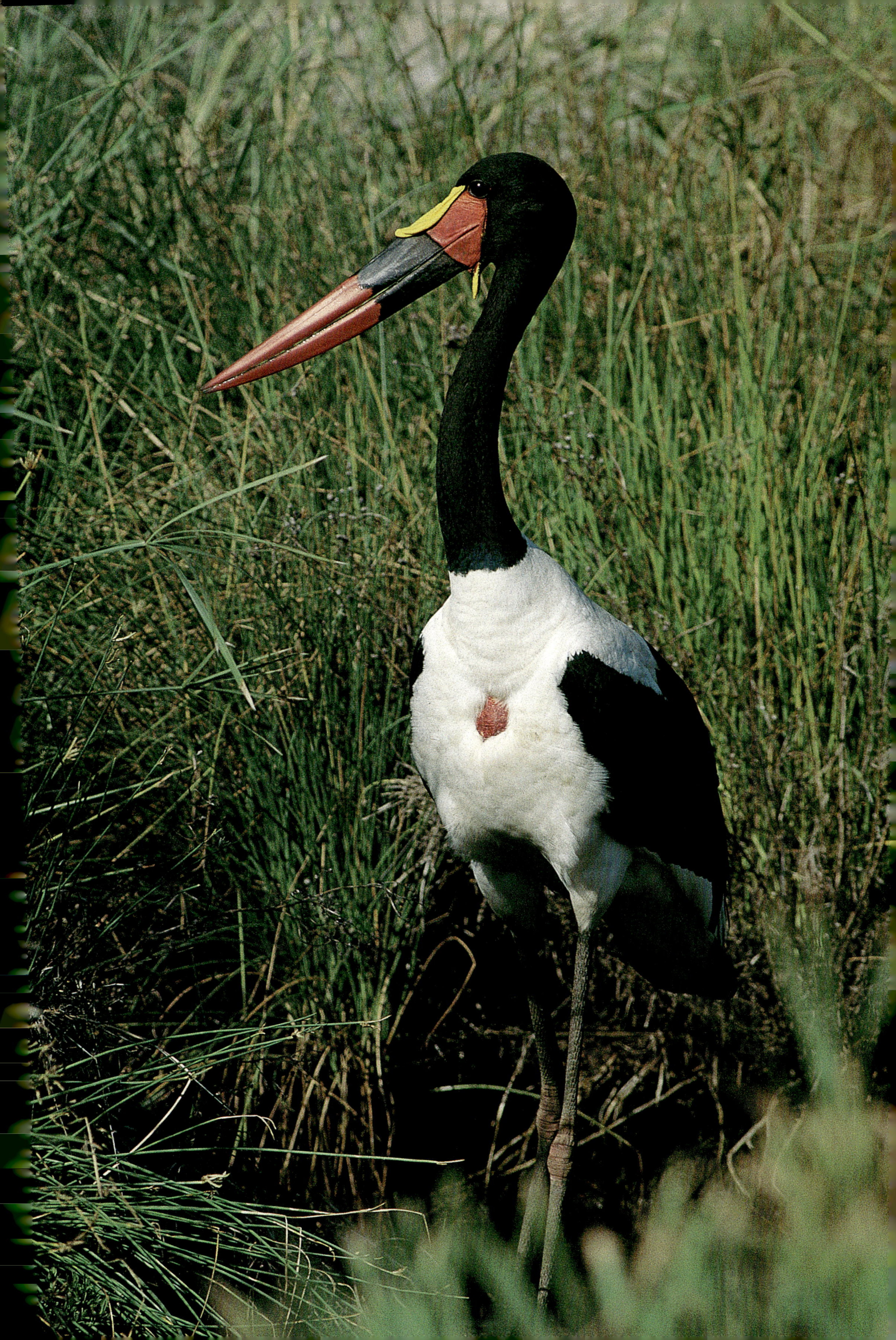

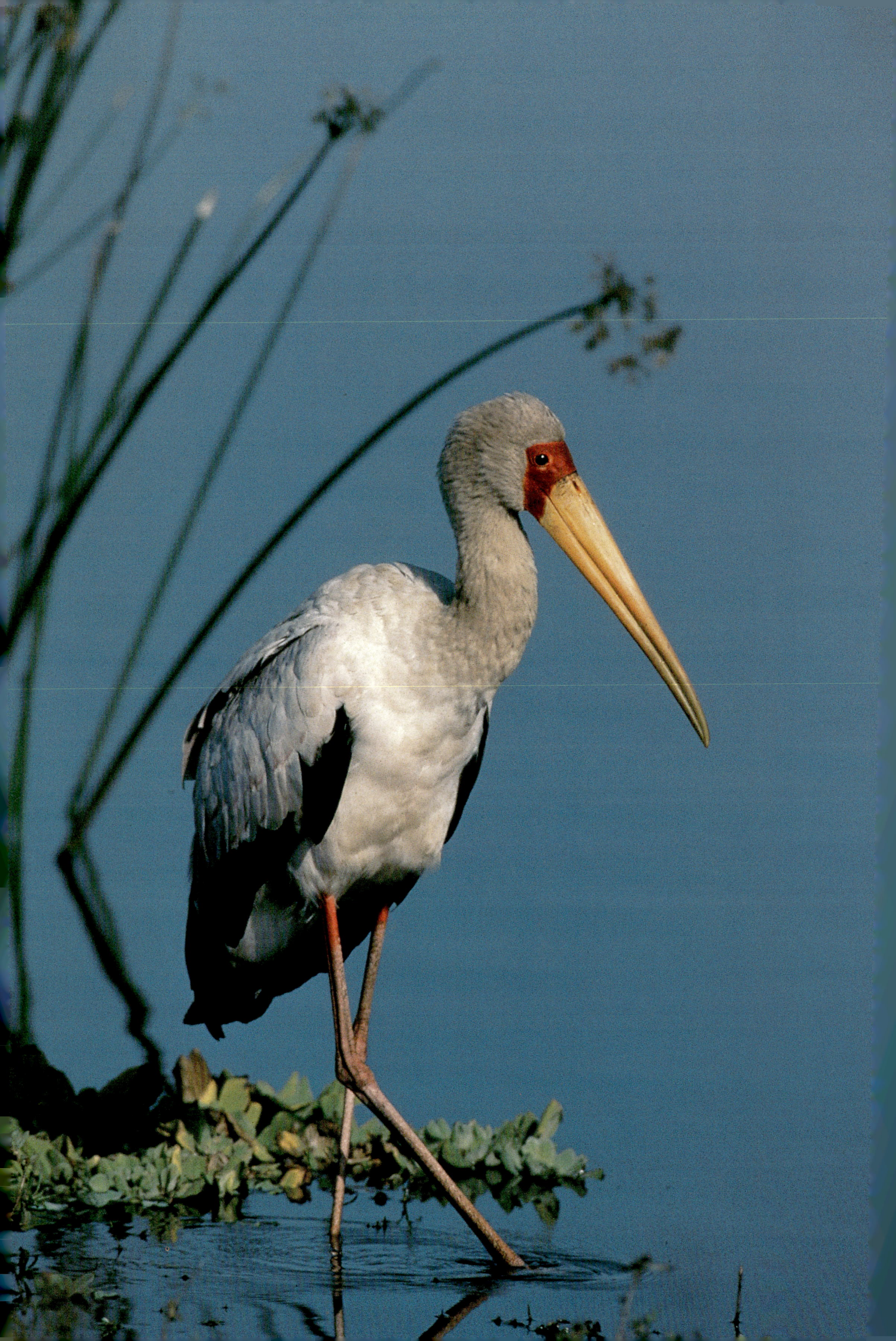

they are fledged and foraging like adults.

Some of the largest wading birds are the storks. There are eight species in Africa. Like the herons, they have long legs, necks and bills. Their wings are long and broad and most have black and white plumage. Unlike herons, they fly with their necks outstretched and perform aerobatics such as diving, plummeting and flipping in flight. They also soar to great heights. Storks are less vocal than other waders, more apt to clap or rattle their bills than to use their voices.

FACING PAGE
*Yellowbilled Storks are one of the more impressive, large birds in the Kruger Park.*

Stork bills are sensitive and storks often feed by touch. Touch receptors within the bill respond by causing the bill to snap shut when touched by prey in muddy or turbid water, in densely vegetated pools or at night. Stork society is complex and they are known for their physical displays. Some greet one another with up-down head movements and bill clattering. Lateral head shaking on a nest looks like a signal for 'no, you cannot join me.' Some displays are species specific.

The Saddlebilled Stork stands five feet tall. It is black and white and has black legs with red knees. Red patches ring the eyes, and red or yellow wattles hang below the chin. On the chest is a red patch. The bill is red and black with a yellow 'saddle' for which it is named. The Saddlebilled Stork is a shy, wary bird, usually solitary and silent. It forages by walking slowly in shallow water, stirring the mud to rouse its prey. When it catches a frog or fish, it first tosses it in the air, then catches and swallows it whole. If a fish is very large, the stork will first nip off the spine before eating. A Saddlebilled Stork nest is a platform of sticks six feet in diameter, made of sticks and lined with reeds, sedges and earth. It is built on top of a bush or tree near water, by both male and female.

The Yellowbilled Stork is a smaller bird at three feet tall. As named, it has a golden-yellow bill. Its face and legs are red and become more intensely so when breeding. Plumage is pinkish white with black wings and tail. Unlike the Saddlebilled, this stork is gregarious and it roosts communally on sand banks or in trees with other wading birds. By day, it stands on shore or walks slowly, foraging with slightly open bill and head immersed – hoping to feel the touch of prey that will cause its bill to snap shut – catching lunch. Even when flying it exhibits behavior suited for hunting, for it raises one wing to block the sun's glare while looking for fish. On a typical stork-style nest the Yellowbilled Stork atypically whines, squeaks, and screams as well as clatters its bill.

The Black Stork is actually brownish-black with a greenish sheen and white underparts. The bill, legs and feet are red, as are bare-skin rings around the eyes. It stands four feet tall. The Black Stork is a solitary bird with feeding habits much like the Yellowbilled Stork. Black Stork nesting habits concur with typical stork behavior, although they tend to breed in mountainous areas. They soar well, roost in trees and sometimes vocalize with whistles, croaks and bill clattering. Curiously, the Black Stork's chicks are downy white.

The Marabou Stork is a contradiction. Classified with the stork family, it acts more like the vultures with which it associates. For the most part it looks like a stork, with its black and white plumage, long legs and bill. But, the Marabou's pinkish-red head and neck are naked – like a vulture's. It also has a naked red throat patch and other skin pouches on its neck. The association with vultures occurs around carcasses of large mammals. The Marabou Stork is a carrion eater for the most part. But, it also hunts rodents, insects and birds. At times it behaves like a stork, when it feeds in water, wading and stirring the mud for fish or amphibians.

By day the Marabou stands still or assumes a curious position, squatting on bent tarsal joints, feet projecting forward. When it decides to fly, it takes off running with a lowered head. Once airborne it may soar for hours at great heights. Unlike its stork cousins, the Marabou flies with its neck tucked and folded.

FACING PAGE
*A pairs of Black Storks were roosting on this tree limb; they will also roost in cliffs.*

At nesting time, the Marabou male collects sticks for the female, which builds a three foot wide platform nest in seven to ten days. They nest in the dry season when more carrion is available to feed their young. Two or three young are produced. They are dependent on the parents for several months. When fledged, the young have a wooly head covering that looks like a bad wig. On the nest, Marabou Storks vocalize with a variety of whines, whistles, squeals and – moos! .... Don't they know they're storks?

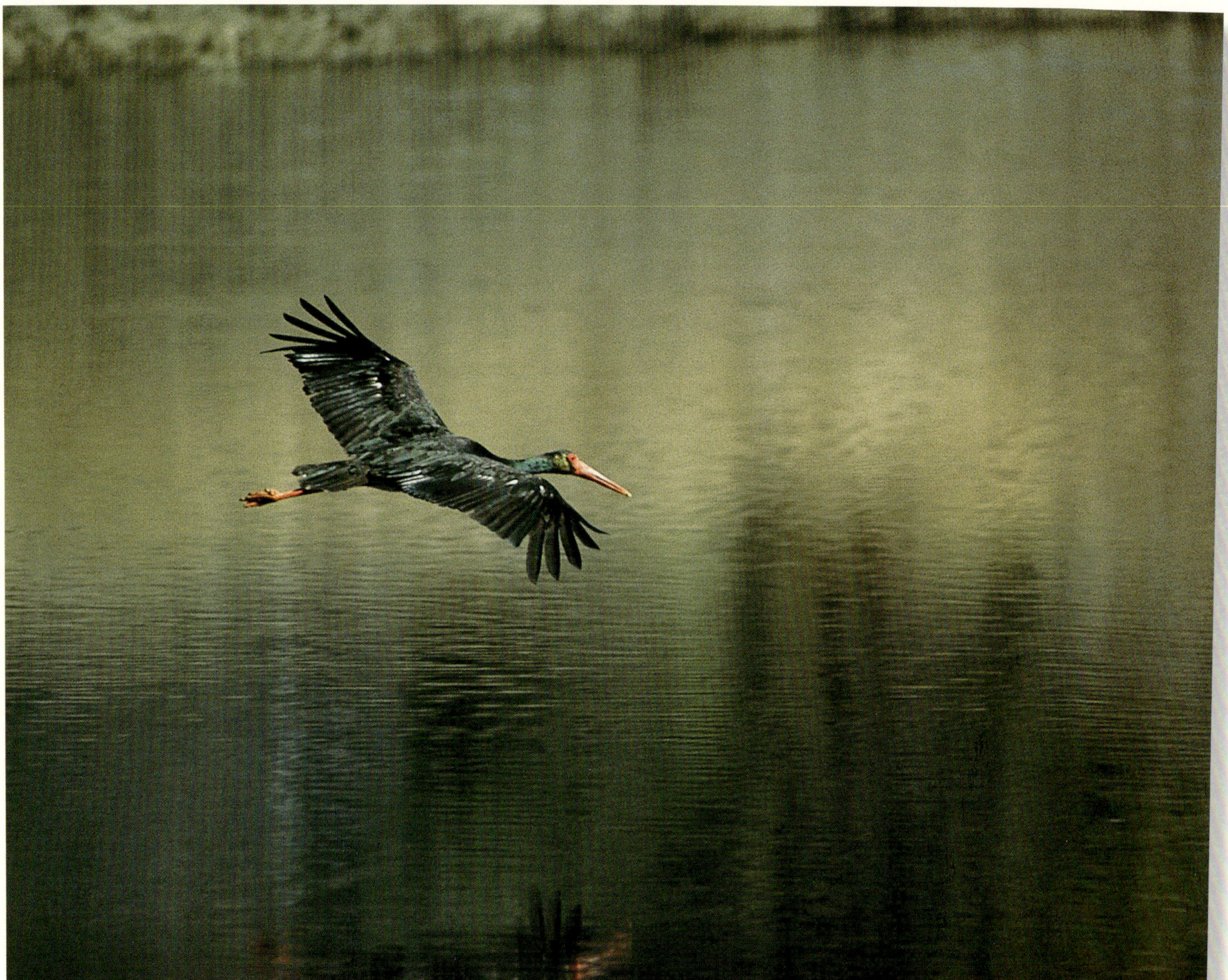

*A Black Stork is flying across one of many dams in the Kruger Park.*

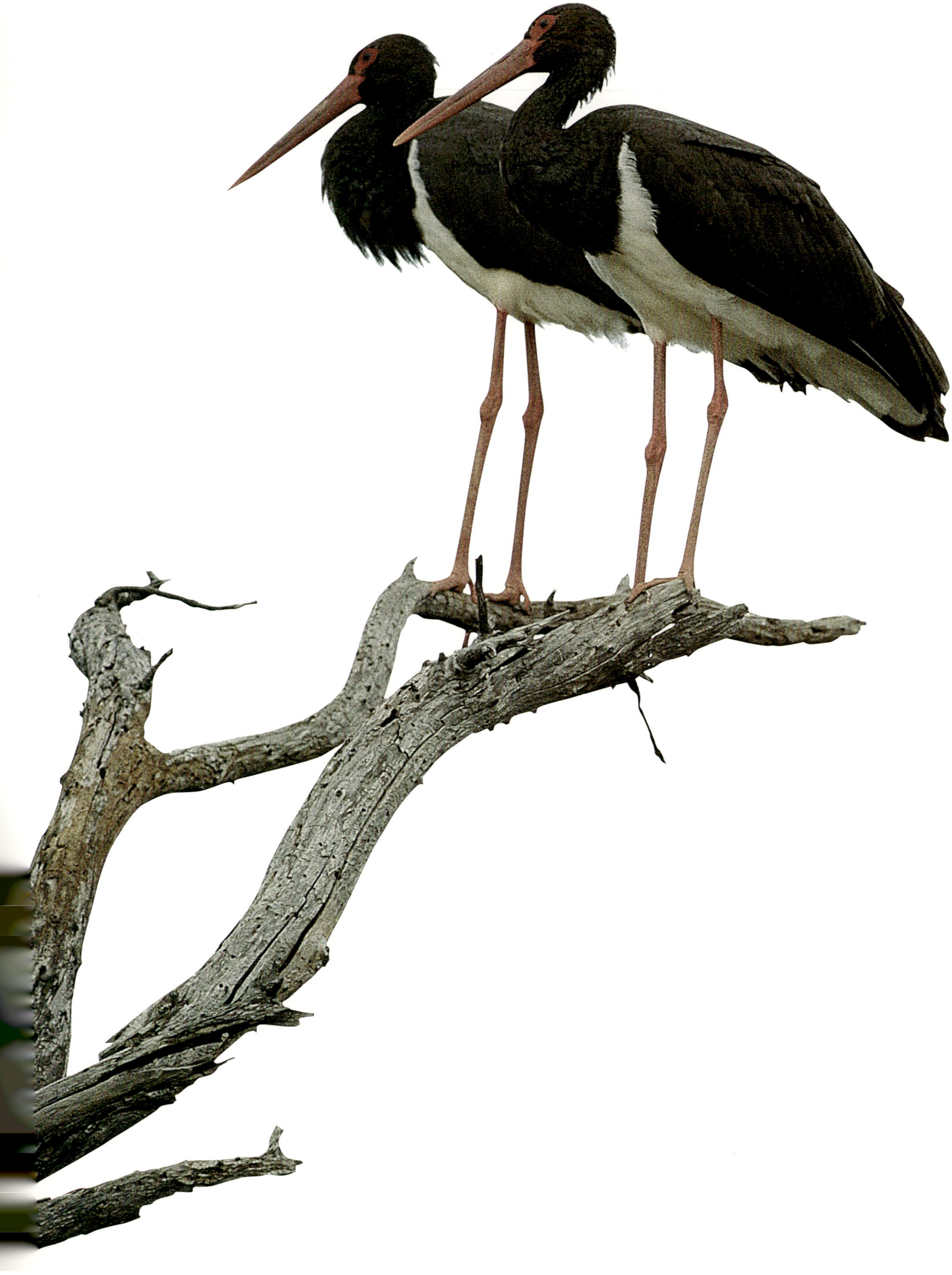

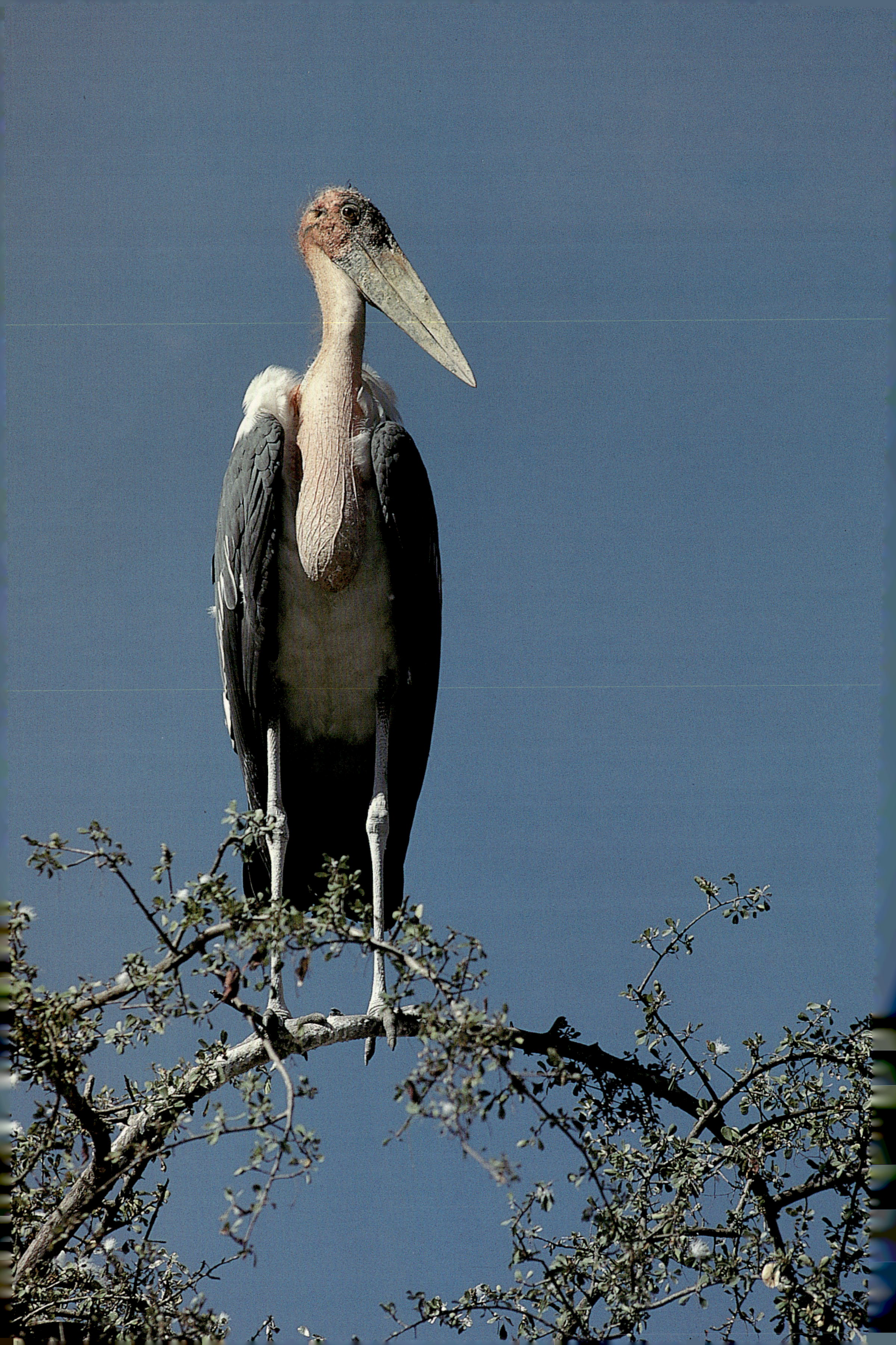

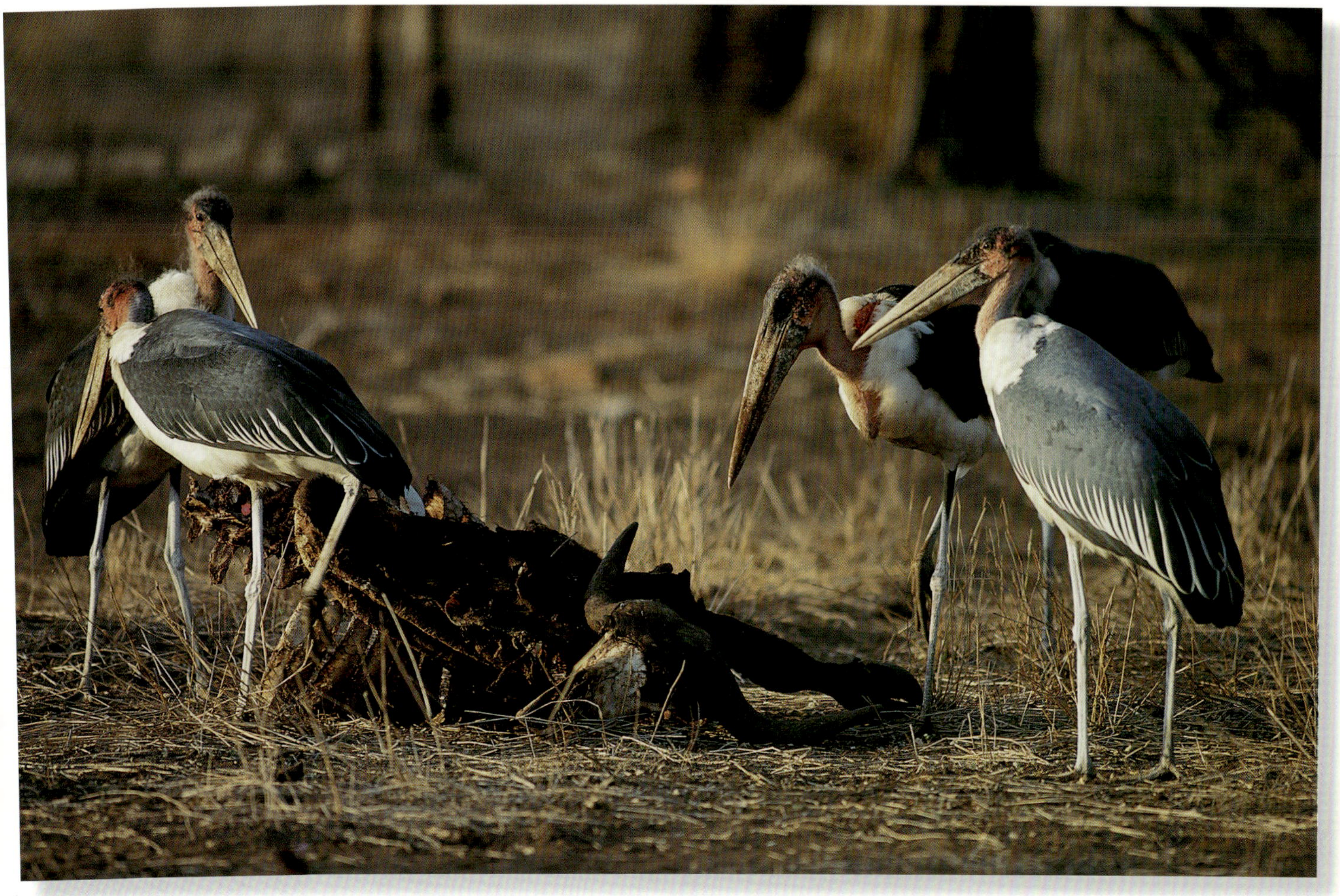

FACING PAGE

*A Marabou Stork is considered "homely" by a good many observers. Maybe it would be better to say that he's just not quite as handsome as some of the other storks.*

ABOVE

*It is not unusual to see Marabou Storks around a carcass; in fact it is quite common.*

CHAPTER FIFTEEN

# Grey Ghosts of Africa

*Elephants*

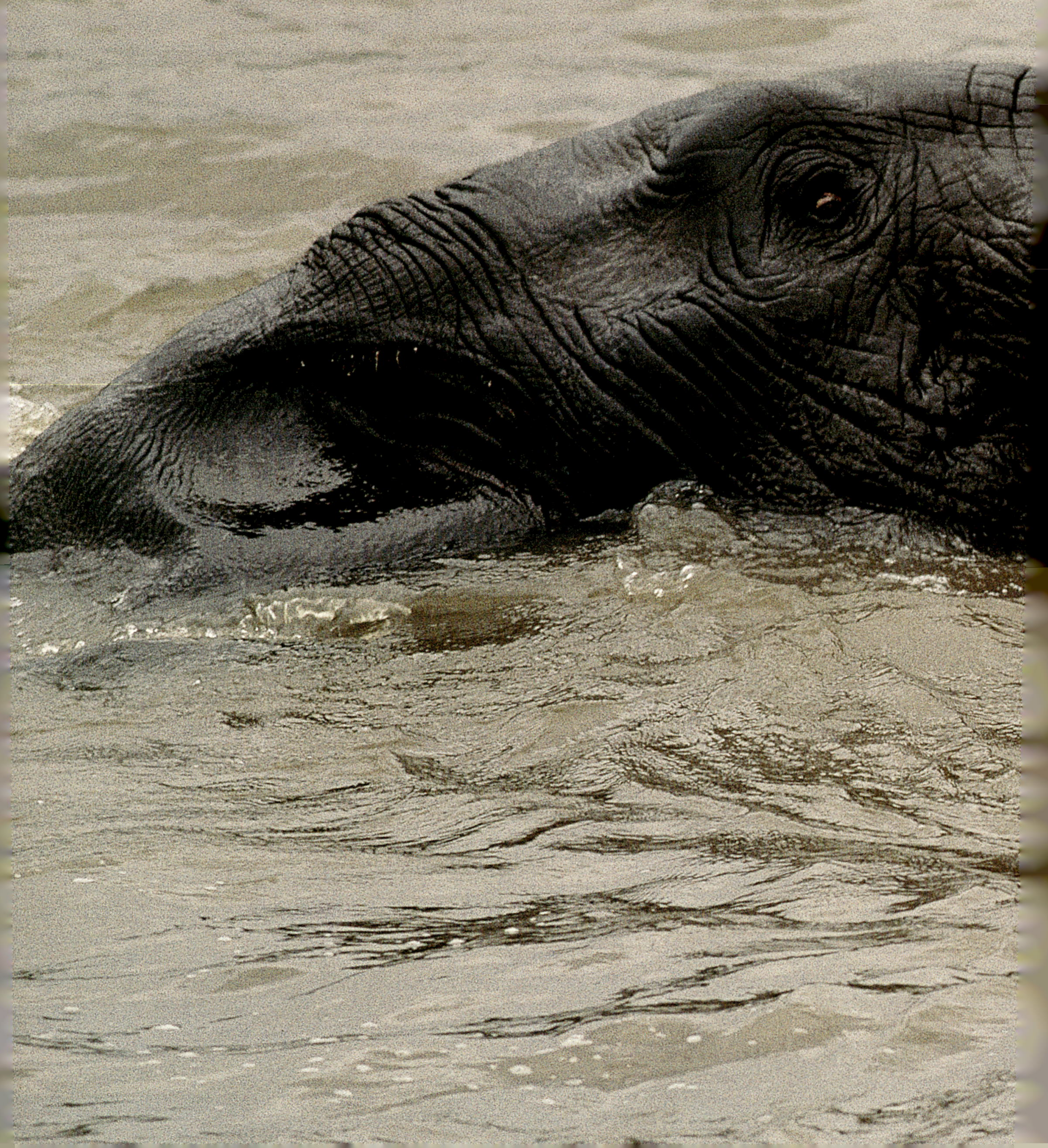

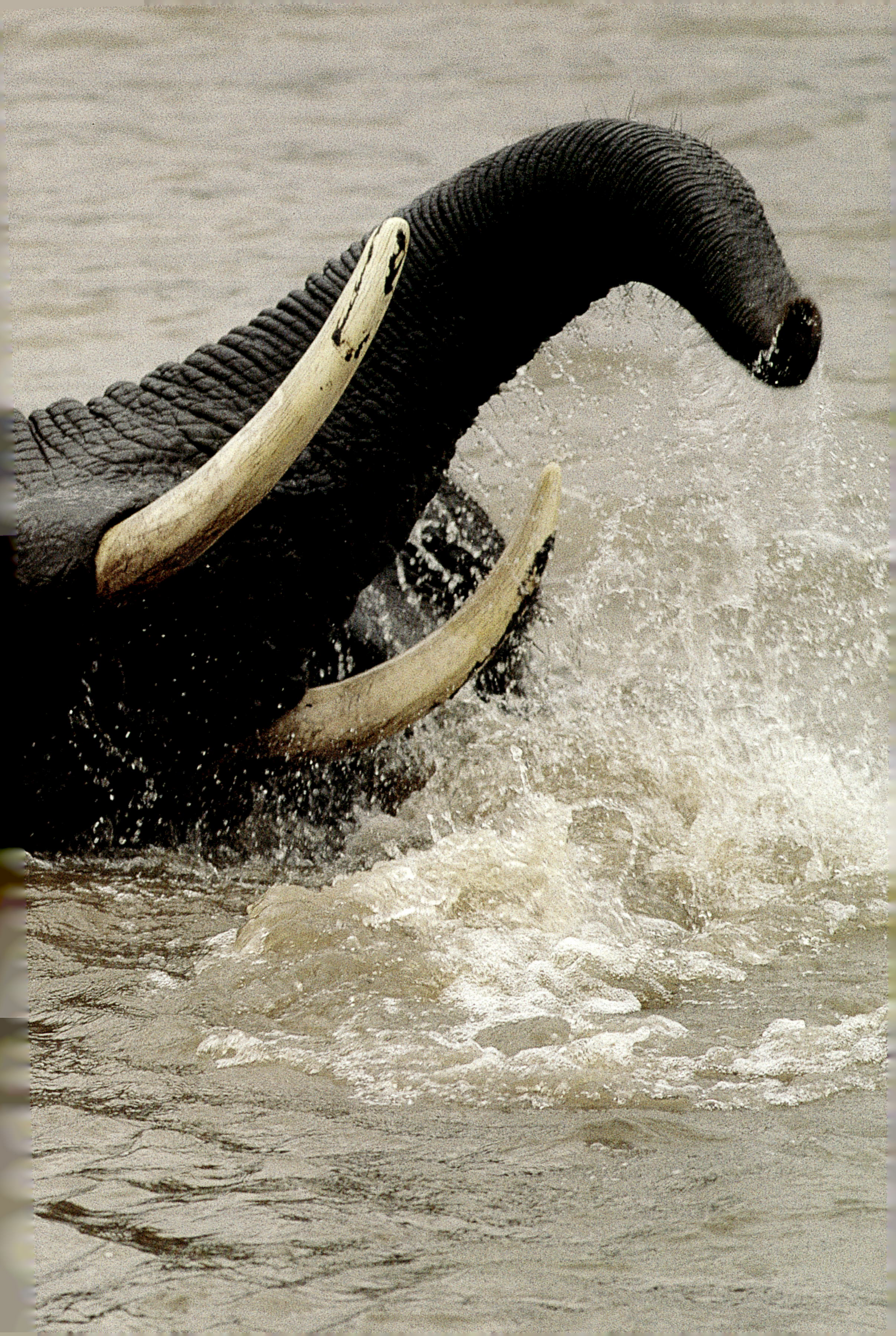

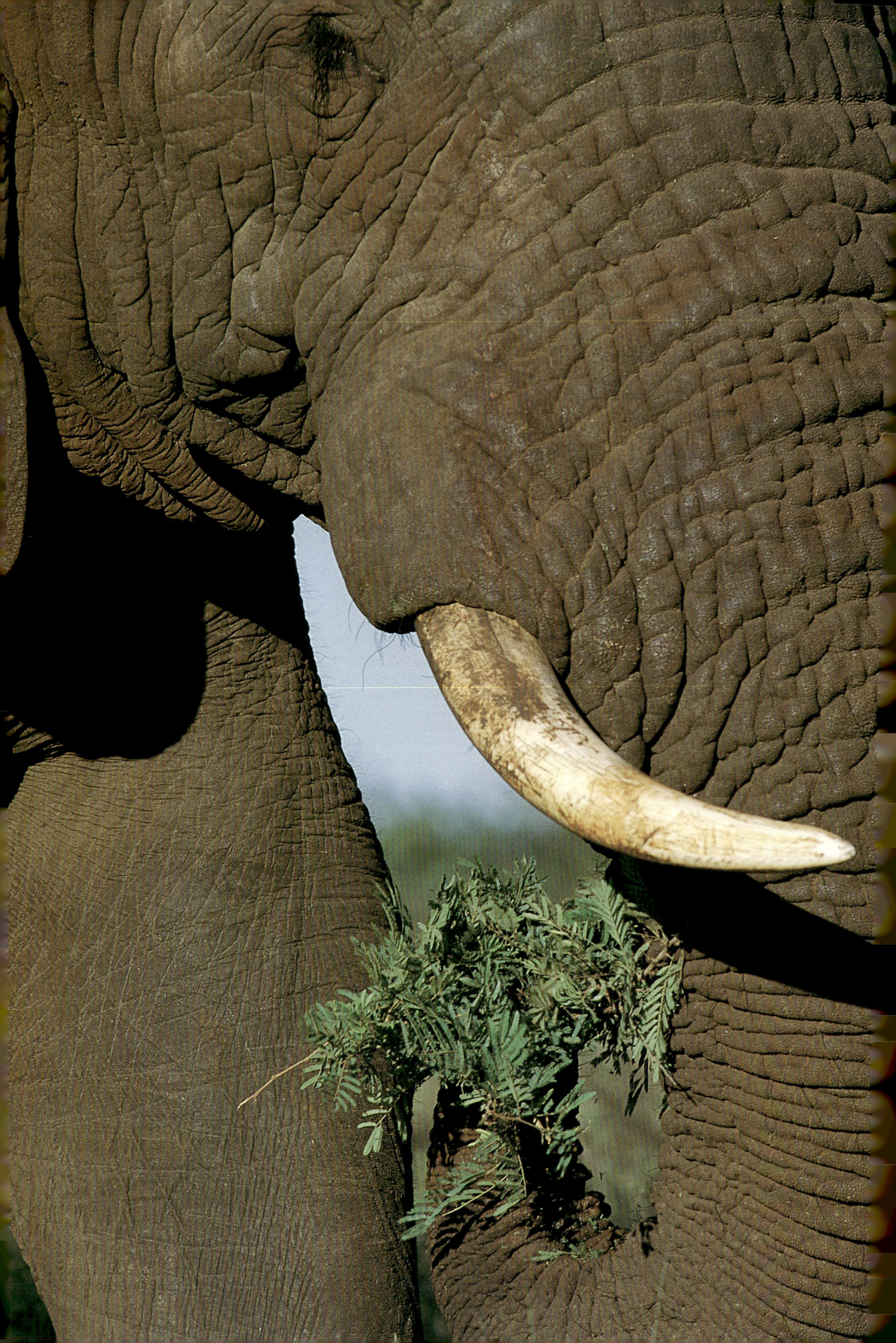

When earth was younger, giant beasts ruled the land. They reigned through many an age. Now, long after time and nature's way have sent most of those giants to oblivion, we know them only from what they left behind – bones, teeth and footprints – trapped in the strata of the eons. Discovery of these mammoth remains fascinates us. Their size ignites curiosity and we study every fact and clue. Our imaginations concoct stories of what life may have been like in that prehistoric time when majestic giants roamed the earth. That fascination is sustained whenever we humans encounter elephants, mighty vestiges of an age of giants, and we can link our curiosity and imaginings with reality.

PRECEDING PAGES
*A playful bath is a welcome diversion for elephants.*

FACING PAGE
*Elephants will feed for the better part of the day, during which a large bull will consume about five-hundred pounds of fodder.*

Wherever elephants have lived, throughout recorded history, stories and myths have arisen. The mystique of the elephant has become part of culture and religion. It is a symbol of strength, wisdom, dignity, long memory and long life. The awe with which we regard the elephant is inspired by not only its size, but by its unique features; massive ivory tusks and a long versatile trunk.

The elephant is indeed enormous. It is the largest living land mammal on earth. An adult male of the African bush elephant reaches an average of ten to eleven feet tall at the shoulder and weighs eleven to thirteen-thousand pounds. A female reaches eight to nine feet and can weigh six-thousand six hundred to seven-thousand seven hundred pounds. A record size bull stood over thirteen feet tall and weighed over twenty-two thousand pounds.

The elephant's thick legs and saucer-shaped feet are quite adequate for supporting its weight and distributing it so that, when walking, the elephant's footprints are shallower than one might imagine. The toes are imbedded in a cushion of fibrous material and fat. When the elephant's foot hits the ground, it spreads, and when lifted, it shrinks – making for a whisper-soft footfall. The front feet are larger than the back to bear the weight of the massive head, trunk and tusks.

The elephant's large head supports the heavy tusks and long muscular trunk, but is itself rather light due to a bone structure full of air cells and cavities. The male has a domed forehead while the female's is angular. The head, tusks and trunk account for twelve to twenty percent of the body weight. The impressive tusks are actually elongated upper incisors made of dentine, cartilage and calcium – together called ivory. They first appear when the elephant reaches two years old and grow continuously throughout its life. By age sixty, the tusks of a bull may reach over seven feet long and weigh one-hundred thirty-two pounds each. One very old bull had a record size tusk measuring ten feet long and weighing two-hundred eighty-seven pounds. Many elephants have one tusk shorter than the other because they favor it for digging salts or roots from the ground, wearing it down in the process. A cows tusks are smaller and usually weigh under fifty pounds. Some elephants have no tusks at all. Those individuals are often ill-tempered.

The elephant is born with six sets of teeth. It uses one set at a time, which includes two upper and two lower teeth. When they wear out from eating fibrous foods, they are replaced by another set that pushes forward. By age six, an elephant is already using its third set of teeth. The fourth set appears around age fifteen; the fifth at twenty-eight and the sixth, and final set, around forty-seven years of age. When the last set wears out, usually around sixty, the aged elephant has trouble eating and may starve to death.

Another feature for which the elephant is well known is its skin. An alternate name, *pachyderm*, meaning 'thick skin', aptly describes the two inch thick hide of the elephant. The animals true skin color is dark brownish-grey, but it may often appear white or reddish due to a dust or mud covering. It is hairless except for sparse bristles. Baby elephants have more hair. The skin has no pores for sweating. With such a large skin surface area the elephant would dehydrate quickly in its hot habitat if perspiration occurred. Instead, it regulates body temperature in several other ways. The elephant's large ears function like a car radiator. Densely packed with blood vessels, they release heat into the atmosphere as the animal flaps them. The elephant will also keep cool with water, mud, or simply by standing in shade. The richly textured skin looks rugged, but it is actually quite sensitive. It needs protection from parasites and disease. Frequent baths in water, or wallowing in mud do the trick.

FACING PAGE
*A bull elephant was feasting on coarse Lala palms. The crunching noise while he ate these fronds made them sound delicious.*

When the elephant bathes, it uses its long trunk to suck up water and spray the cooling liquid over its back. The trunk is a marvelous all-in-one tool used for bathing, squirting, smelling, communicating and grasping. Evolved over sixty million years, it is a fusion of the nose and upper lip with two finger-like nodes at the tip. The trunk contains thousands of muscles that allow it to be both sensitive enough to pick berries and mighty enough to rip up trees. The trunk extends from the head to the ground enabling the animal to graze, for it is unable to bend down and eat with its mouth. The trunk also serves as a handy snorkel when the elephant swims or walks under water.

On the other end of this marvelous beast hangs a rope-like tail, four feet long and fringed with long, black, bristly hair. At times, the tail is useful as well. Young elephants hold onto their mother's tail with their trunks as they walk.

In ancient times, elephants roamed the entire continent of Africa, excluding the Sahara Desert. They are now confined to National Parks and only a few wild areas. The smaller, forest elephants live in equatorial forest areas. The larger bush or savanna elephants, which are pictured here, inhabit grassy open woodland. The habitat must provide for all their needs; food, water and shade. An average home range covers two-hundred ninety square miles, but a range may be as small as five square miles in a lush, well-watered area, or over thirteen-hundred fifty square miles in an arid region. Elephants will travel long distances to areas where rain has recently fallen, knowing that new vegetation will soon sprout. They travel paths well worn by their ancestors. They walk in single file at a pace of three or four miles per hour. Sometimes, the speed quickens to six or eight miles per hour. They must keep three feet on the ground at all times while they walk. Many of the '*elephant roads*' have been widened and paved into present-day highways.

Whether traveling to find sustenance or staying in a small home range, elephants spend most of their time feeding. They sleep sporadically, for they must move to forage and only have time for naps. By both day and night these herbivores graze and browse to fulfill their constant food requirements. A bull will eat about five-hundred pounds of food per day, and a cow consumes about three-hundred fifty pounds. In the rainy season, fresh grass is the staple food. The elephant curls the end of its trunk around a bunch of grass and pulls it up by the roots. After gently beating the bundle on a leg to remove the clinging soil, the trunk is

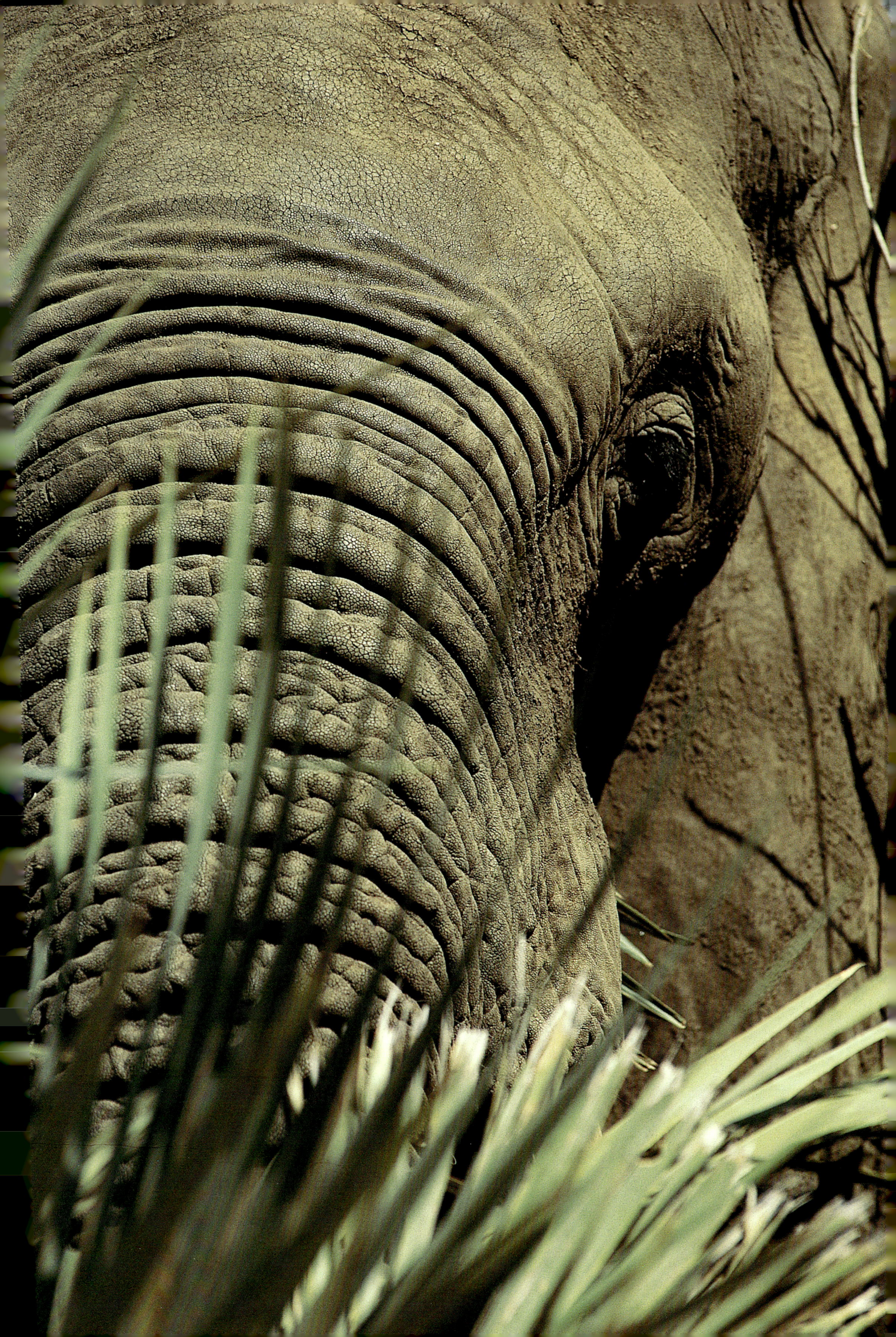

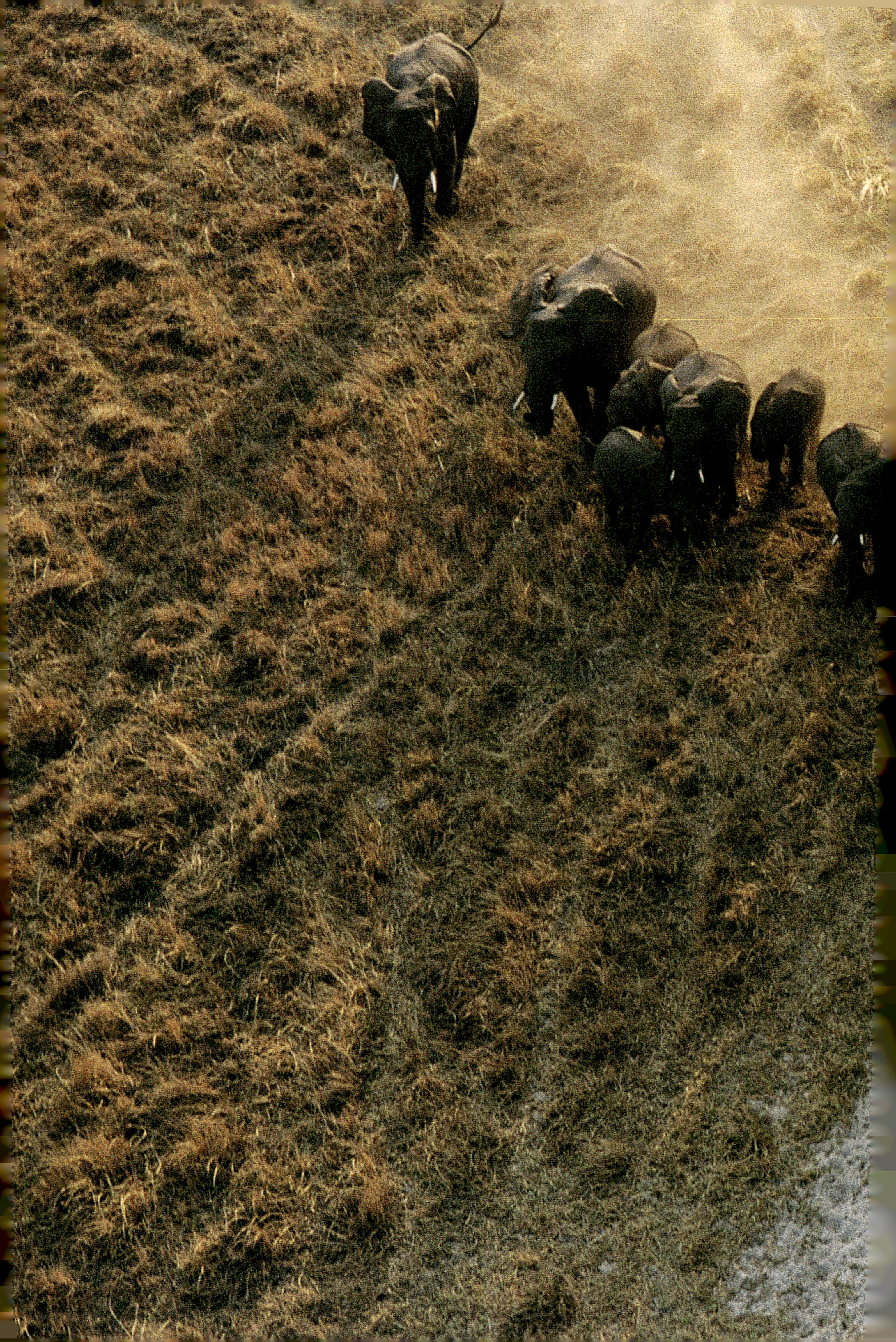

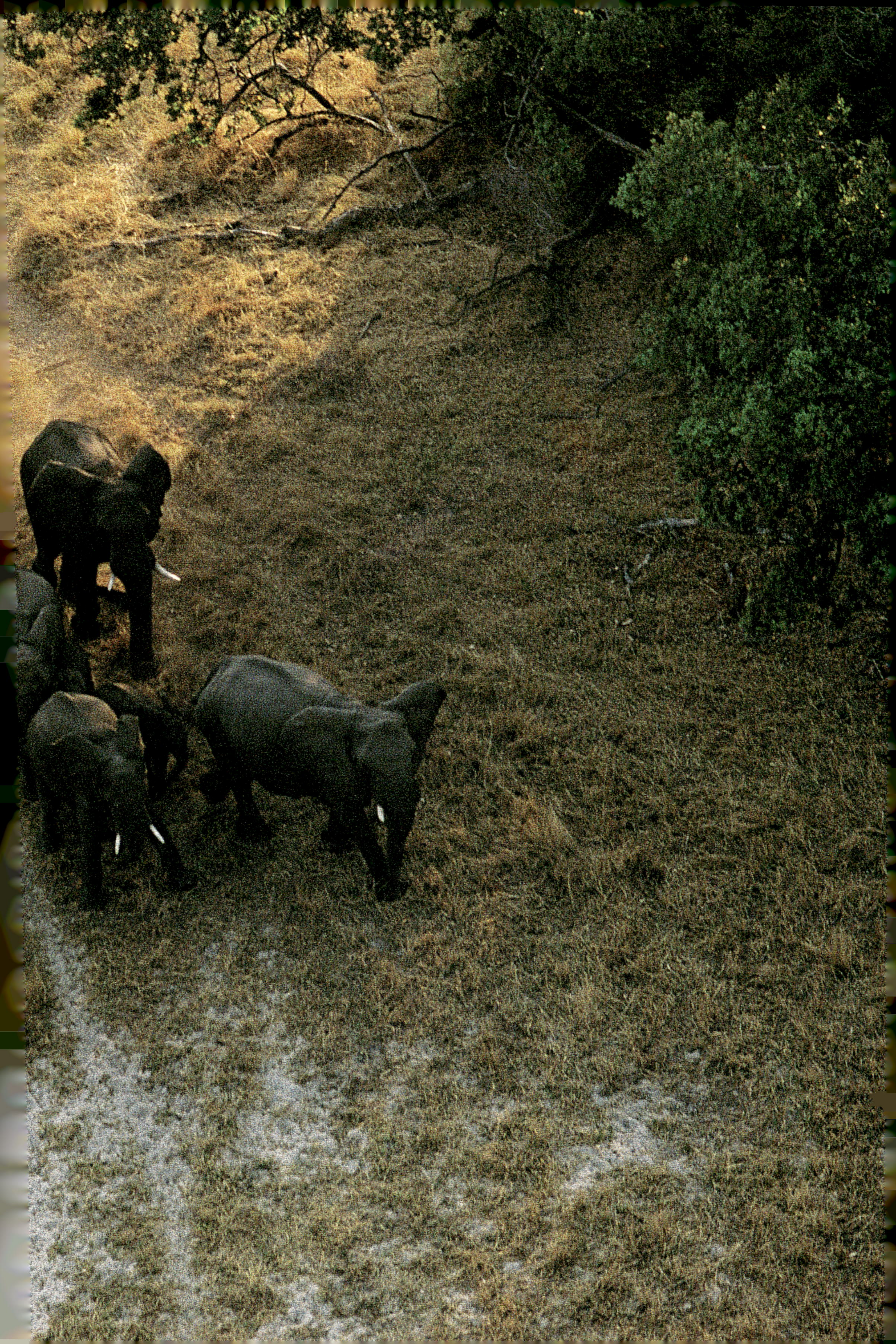

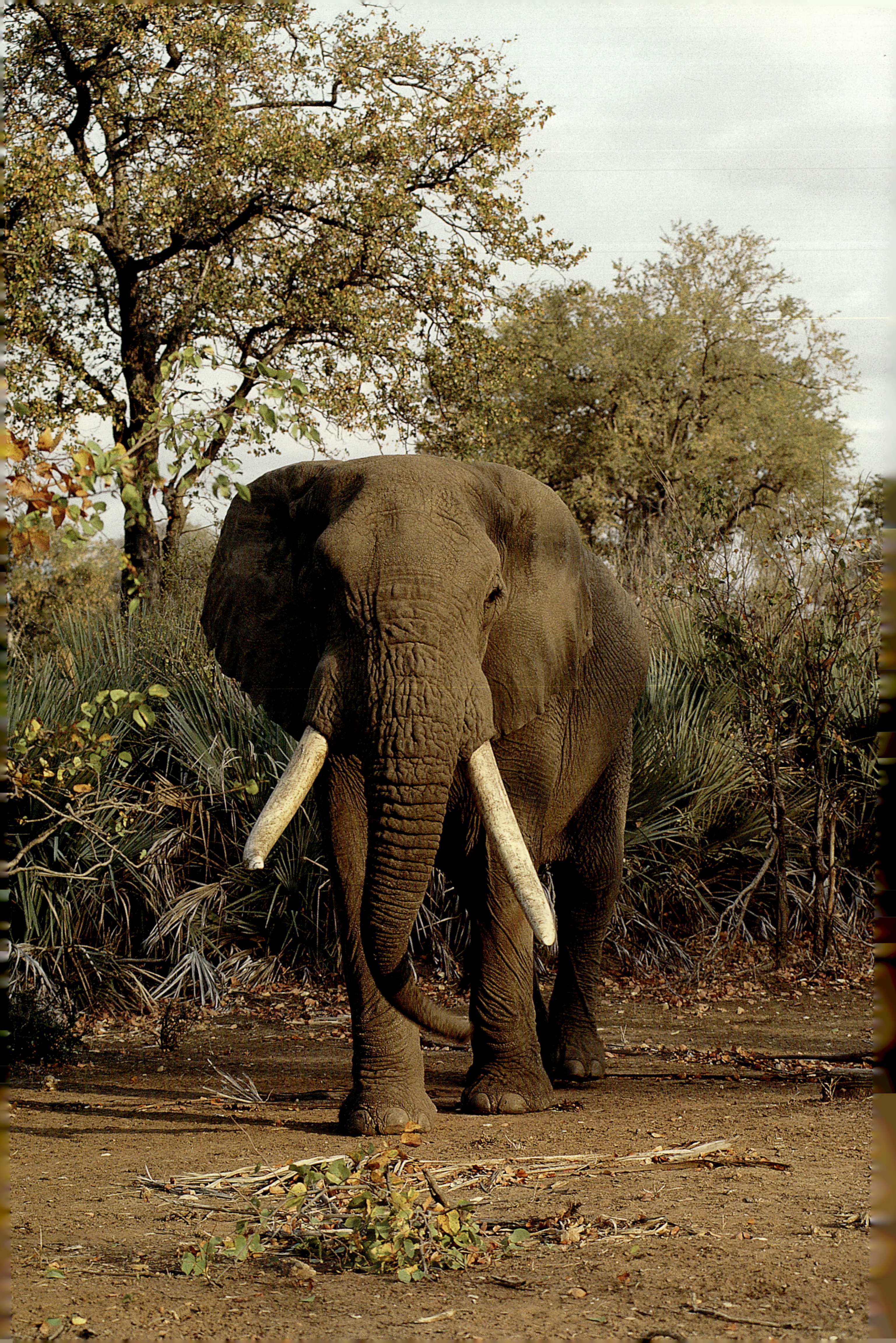

curled up to the mouth and the morsel is eaten. When available, flowers, fruit, leaves and roots are also eaten. Elephants feed from the ground up to twenty feet into the trees – sometimes standing up on hind legs and stretching further with their trunks. They can reach higher than a giraffe. In the dry season they resort to eating twigs, branches and tree bark. They must spend most of their time feeding because only half of what they eat is digested. Hungry elephants tear down trees and extract their inner cores, sometimes breaking a tusk in the process. A large herd can be very destructive, second only to man in altering an environment. But the destruction has a bright side. Other animals benefit by gaining new food sources when the land is opened up for new growth, or by attaining new areas for cover.

PRECEDING PAGES

*A small herd of elephants was led by its matriarch across a savanna in Botswana.*

FACING PAGE

*This is an example of a master and a slave tusk. The right tusk is substantially shorter than the left because he uses it as a working tool.*

Elephants need a lot of water. They drink every day when possible, but can go for three or four days without it. A large bull will drink nineteen to twenty-four gallons of water at a time and up to fifty gallons a day. The trunk becomes a vessel for drinking. He fills the trunk and squirts the water down his throat. Young elephants kneel and drink with their mouths, for it takes much time and practice to master the trunk technique. When water is scarce, elephants will dig holes in a dry riverbed and wait for underground water to seep in. The wells benefit other animals also.

Elephants love water for more than drinking. They spend hours in water, bathing and sometimes swimming. They squirt themselves and each other with their trunks. Calves especially enjoy playing in water – trunk wrestling, shoving one another and teasing other animals. Mud wallowing is a pleasant pastime as well. The mud cools and protects them from parasites. All the water and mud activity these giant pachyderms enjoy expands water holes and pans, once again benefiting other wildlife. Dust baths often follow a time in the water; the dust also serving to repel parasites. In times of drought, they may regurgitate water from their stomachs to cool their skin or the skin of their calves. Calves are well cared for in elephant society.

The family structure of these great beasts is closely knit. A family group consists of nine to eleven related cows and calves. It is led by the oldest and wisest cow, called the matriarch. She may be over fifty years old and past breeding age, but she has much to offer with her long life experience. The saying '*elephants never forget*' applies to the matriarch, for she remembers how, where, and when to find food and water. She makes the decisions for the herd and teaches the younger members. When she dies she is succeeded by her eldest daughter.

When a family group gets too large, it will subdivide. The smaller groups are independent, but remain bonded and will regroup at water holes or in times of drought to share resources. Large herds of two-hundred or more are likely to be related family groups. Male elephants congregate in bachelor herds where they constantly challenge each other for dominance. Those that are the oldest, largest and have the longest tusks usually win out. Very old bulls live singly or in pairs and never rejoin the herd – often ending their days in swampy areas where the vegetation is more tender for their well worn sixth set of teeth.

The curious sight of a group of elephants with their trunks held up in the air does not mean they are snobs. It indicates that they are sniffing the air. The posture is a common one, for the elephant's sense of smell is acute. They use it to smell each other over distances or to detect the presence of danger.

Elephants have lovely green or hazel eyes – but vision is not their best sense. They do, however, respond to visual communication through body language. Posture changes, as well as ear, trunk and tail positions can signal aggression, submission, courtship or the presence of predators. An elephant standing alert, with head and tail raised, ears spread and trunk poised for sniffing, indicates some sort of disturbance. It usually means, also, that it has heard something. The elephant's hearing is excellent.

They communicate with quite a varied vocabulary of sounds. Fifteen different vocalizations have been recorded. The most common sound is a growl – a deep rumbling that carries for half a mile. It is used both as a warning and to maintain contact with the herd when visual contact is obscured by foliage or darkness. The growl sound can dip below the human threshold of hearing into a low frequency level called *'infra sound.'* Herd members keep track of each other over long distances with such sounds. It is possible to detect an elephant signalling with infra sound by seeing its forehead vibrating or even feeling airborne vibrations. Of course to gain that experience, one would be dangerously and foolishly close. Elephants also roar, trumpet, scream, squeal, and snort. The famous trumpeting sound they make is used when they are excited, surprised, widely separated from the herd or when about to attack. All the sounds are amplified by the marvelous trunk, which acts as a resonating chamber. The trunk is used also as an organ of touch. Elephants greet by touching one another's mouth with the trunk's tip – a sort of kiss. And, they intertwine trunks, and caress one another with them – especially calves.

Family is everything to elephants. The rainy season is the time for both calving and mating. Females begin to breed at age nine or ten. Males reach sexual maturity at twenty-five, but do not successfully compete for breeding until age thirty-five. The oldest and largest bulls do most of the mating. Female oestrus is stimulated by a change in diet. High protein grasses brought by the rains bring on ovulation.

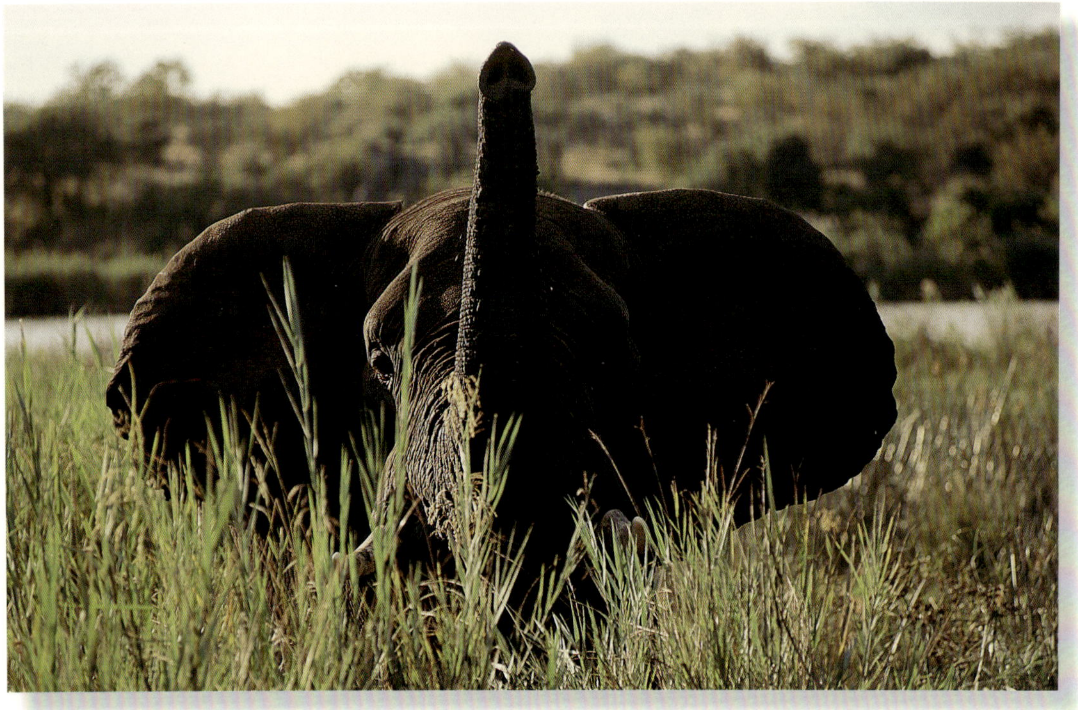

*An elephant's senses of smell and hearing are quite acute.*
*Just after I took this photograph the wind shifted and he picked up my scent and moved off.*

At the same time males ready for mating develop a condition called '*musth*', which indicates higher testosterone levels. At this time glands on the bull's temples secrete a tear-like discharge that stains the face. The bull becomes aggressive, unpredictable and somewhat uncontrollable. Musth may last for five months in males over thirty-five. Musth males compete for females in season but hierarchy is usually well established and fighting is not often serious. Mock charging, tusk and trunk wrestling, and head-to-head shoving, rarely escalate to more serious ramming or goring, that causes injury.

The largest and strongest bulls win the privilege to mate, but the females do the choosing. They prefer males in full musth. A cow may signal to a favorite bull with an infra sound love call. He will join her and stay close for three days while mating takes place. Cows may also mate with more than one bull.

Elephants have the longest gestation period of all mammals – twenty-two months. This insures that births also take place in the rainy season. When a cow is ready to calve, she leaves the herd to find a shady, secluded area, preferably near water. The matriarch, and one or two close relatives accompany her. She will clear the ground and squat to give birth. The calve's short fall to the ground breaks the umbilical cord and the mother clears away the birth sac. The birthing party shows great excitement when the calf arrives and helps the mother in attending to the pinkish newborn. At birth, a calf weighs between two-hundred and two-hundred fifty pounds, and is small enough to stand under its mother. The calves first response after standing is to locate nourishment. The mothers breasts are between her front legs, an unusual placement for a four-legged animal. The calf quickly finds the breast, moves its trunk aside and suckles. Calves suckle for two to three years before weaning, although they are taught to find solid food within two months. A mother elephant dotes on her calf. She frequently touches and caresses it, gently bathes and dusts it and pushes it beneath her for protection from the sun or danger. When traveling she guides the calf by holding its tail with her trunk as it walks ahead, or the calf may follow, clinging to her tail. Thus begins a long and loving relationship between mother and calf.

When daughters are grown, they remain with their mother's herd. Mother-daughter bonds last for decades. Sons leave the herd at twelve to fourteen years old and go off to live alone or to join a bachelor herd where they learn the ways of a bull elephant. Young bulls will not return to the herd until several years later, when they are ready to mate.

Elephants have few natural enemies. Small calves must be protected from lions and hyenas. Humans have historically been the greatest enemy to elephants - hunting them both for meat and for their valuable ivory. When an enemy is sensed, elephants become quiet and alert. Usually very noisy when foraging, an entire herd will fall dead silent… raise their trunks… and warily sniff the air.

Response to danger involves both defensive and offensive action. Cows will stand close together in a circle, shielding the calves in the center. The matriarch faces the danger first, with an alert posture, then with threat displays; such as head tossing, ear slapping, tusk swishing and kicking up dust. She may coil up her trunk and abruptly uncoil it with a trumpet blast – like a party noisemaker.

Usually, these displays are enough to scare off an intruder, but sometimes, a charge ensues. The elephant will tuck its trunk below its jaw and charge, trumpeting and screaming, with tusks aimed at the enemy. A running elephant can attain a speed of twenty-five miles per hour, faster than a man can run.

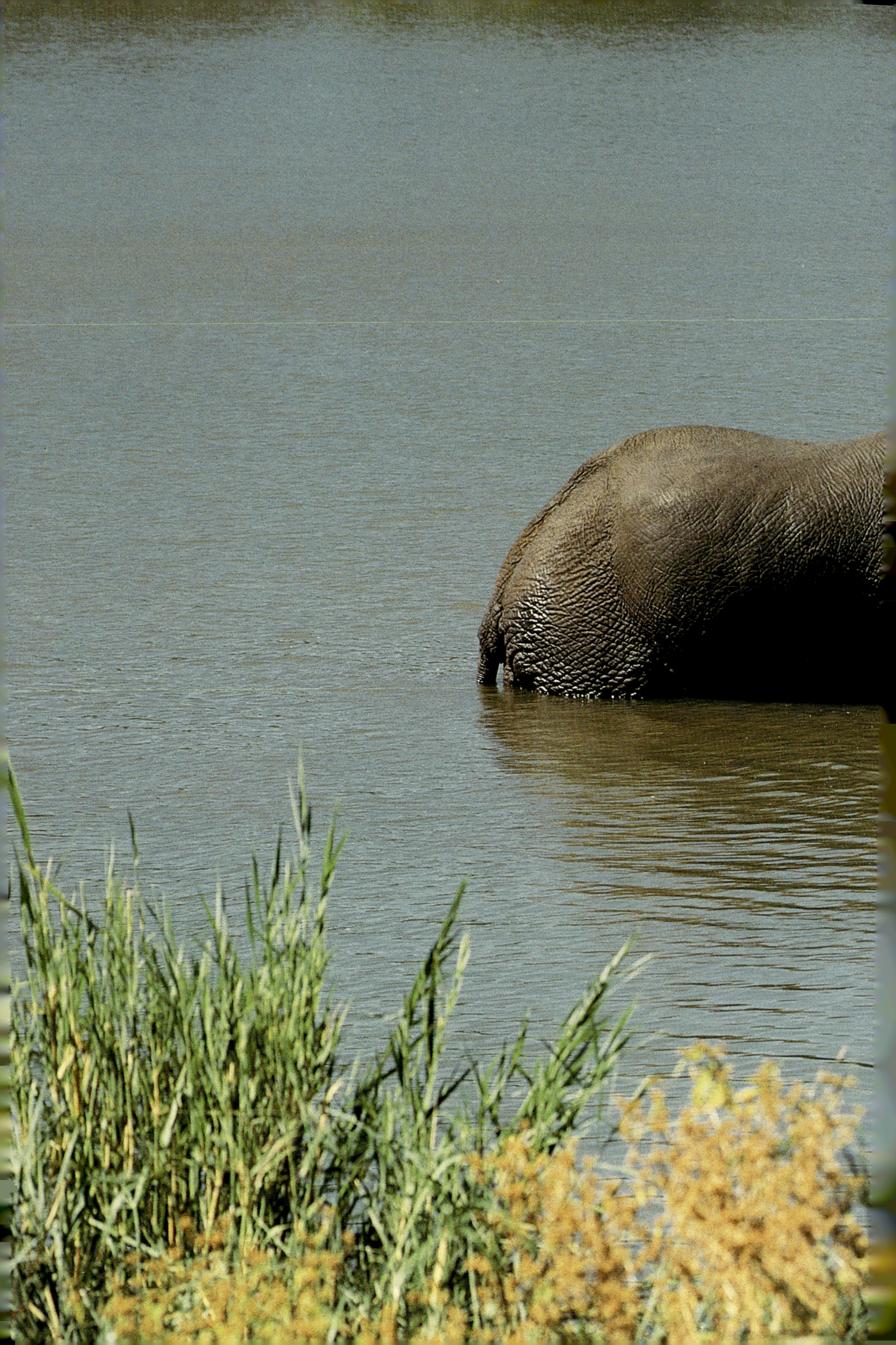

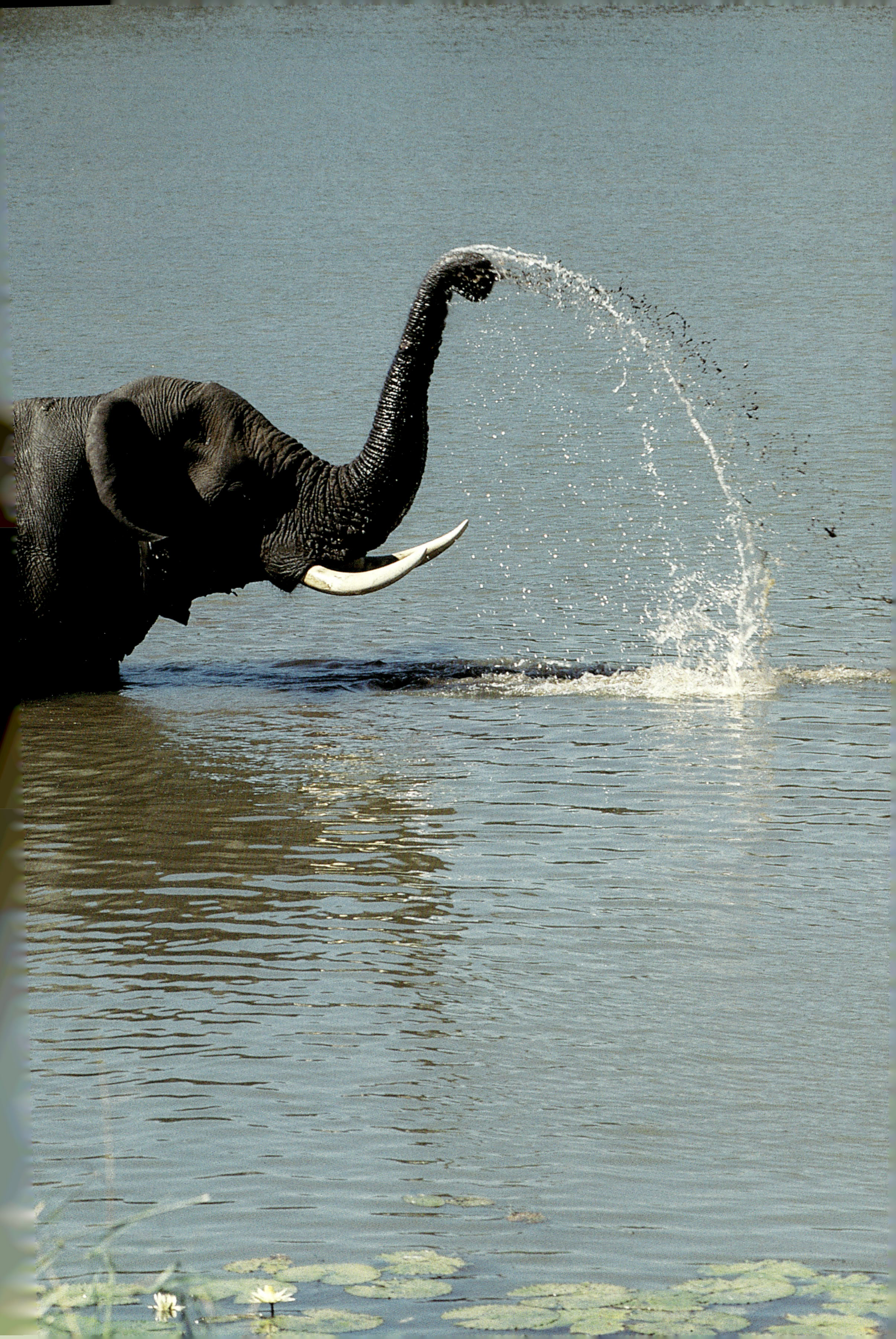

If one of the herd is injured in an attack, the others will stay to assist, protect and defend the injured member. That loyalty is admirable, but has been unfairly exploited by ivory hunters.

PRECEDING PAGES
*It was a hot day and I had wished that I could have enjoyed the cool water also.*

Centuries of ivory hunting decimated elephant populations in the past. Today, conservation measures and a ban on the ivory trade has lessened their decline. Deforestation, roads and human encroachment are greater threats for wild populations. An elephant's life expectancy is sixty years. A few live longer, but many die much younger.

It has been widely reported that elephants migrate to special areas to die. These 'elephant grave-

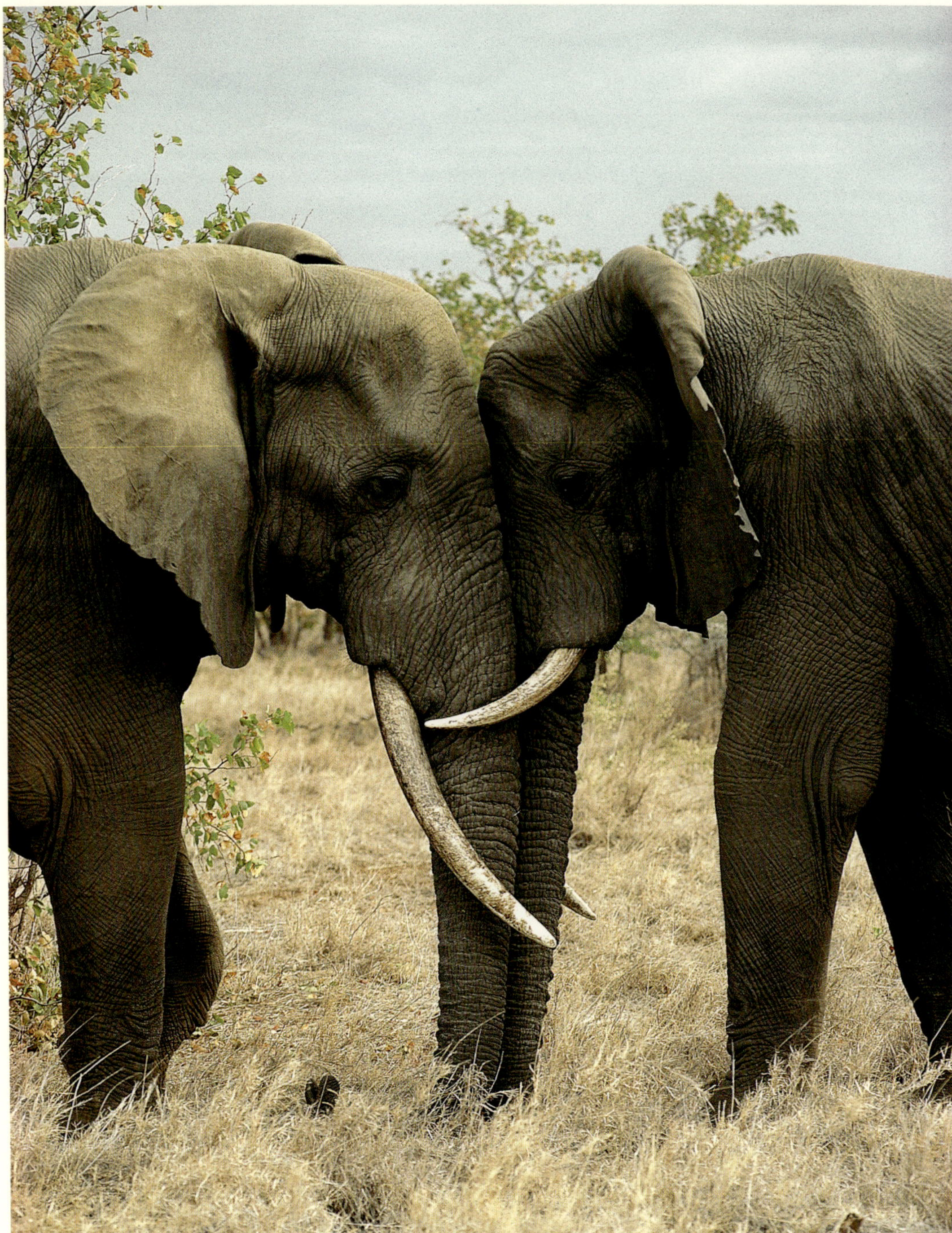

*These two bulls were measuring each other's strength. The bull on the left had a decided advantage in weight and tusk length.*

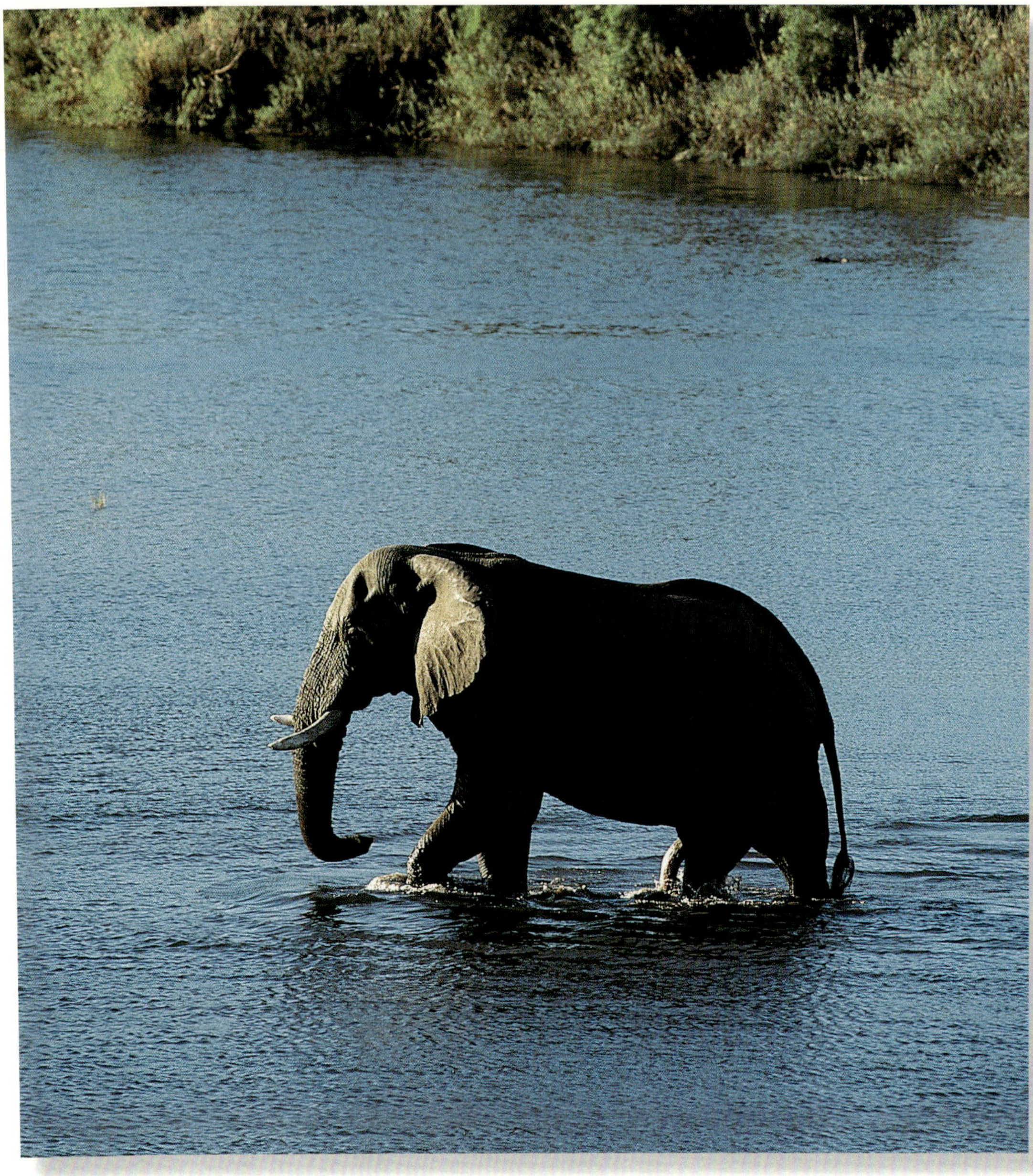

*A large bull is crossing the Letaba River in the Kruger National Park.*

yards' with large collections of bones may be accounted for in other ways. They may be sites where many died at the hands of ivory hunters, or the last stand at a diminished water hole during a drought. Flood waters may have washed bones together on a high point of a flood plain. Whatever the practical cause might be, many still choose to believe that these highly social, caring animals would certainly revere their dead in special graveyards.

We too hold these sublime creatures in high esteem – as we come to know them and feel empathy for their plight. Our hearts go out to these mammoth beings that both link us to Earth's history and grace the present. Only caring humans can provide for their future – assuring that the grey ghosts of Africa remain, living and breathing, and not simply as ghosts of the past.

FOLLOWING PAGES

*Etosha Park in Namibia is an arid region where temperatures can soar beyond one-hundred degrees Fahrenheit during the day. The water hole offers welcome relief for this herd.*

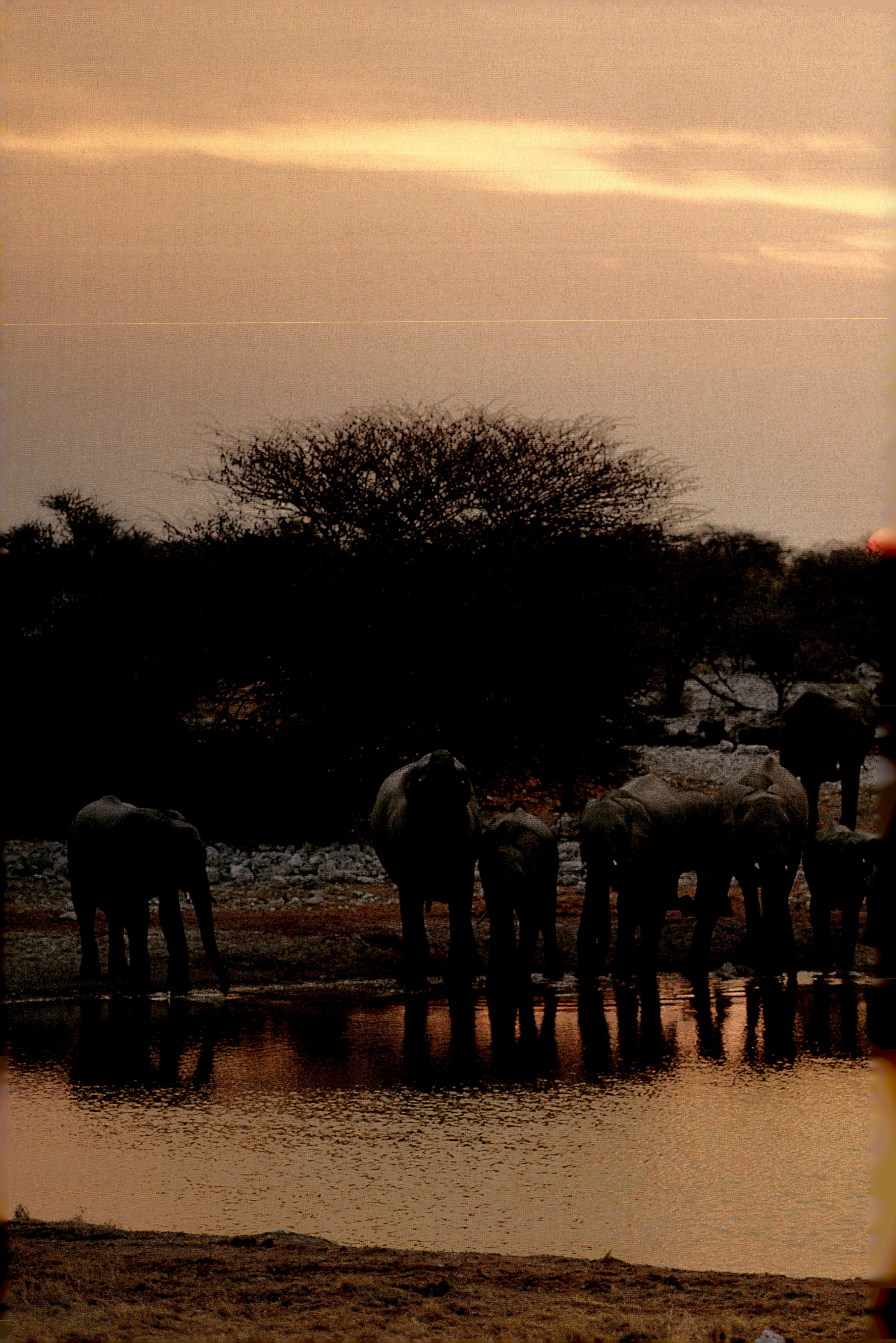

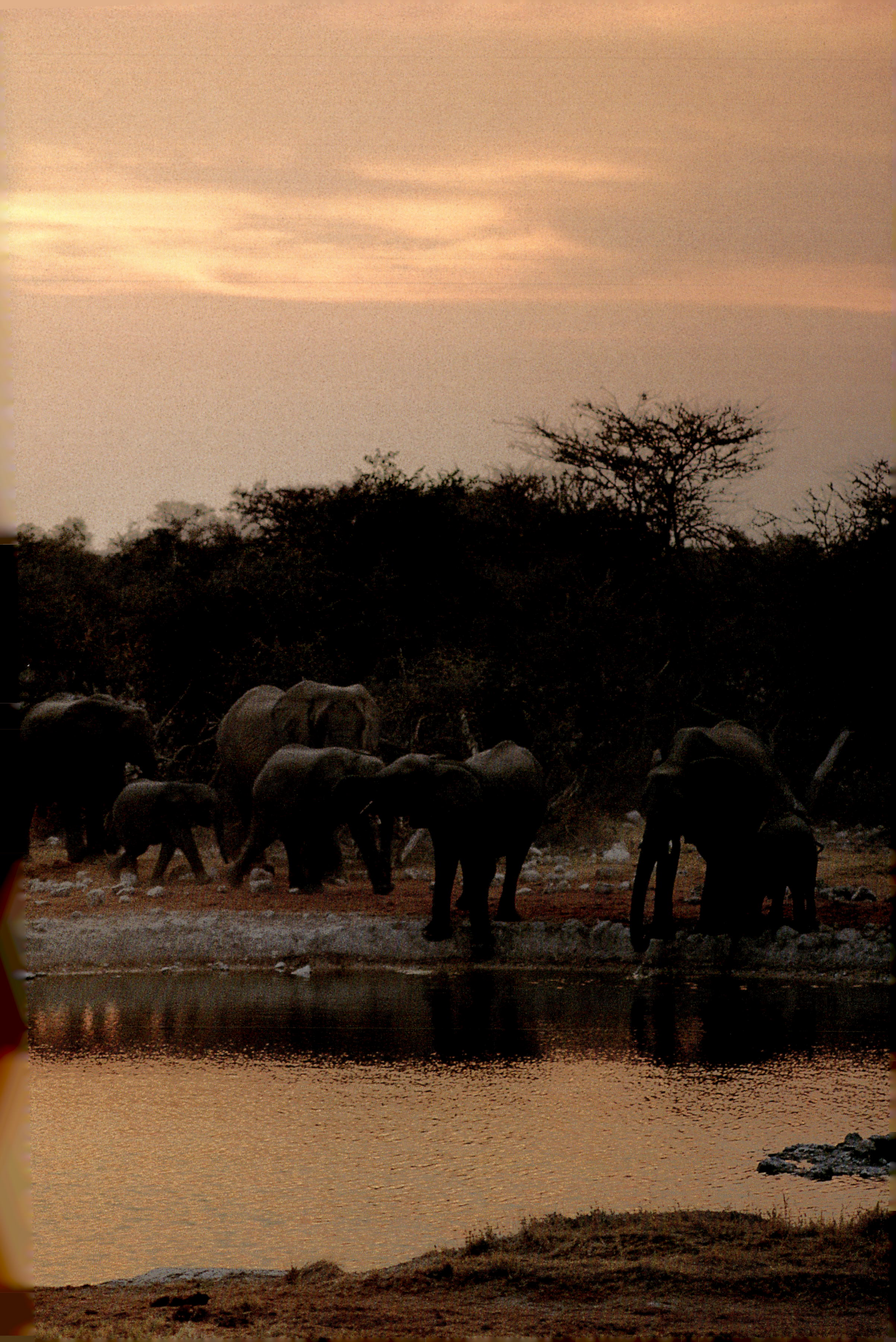

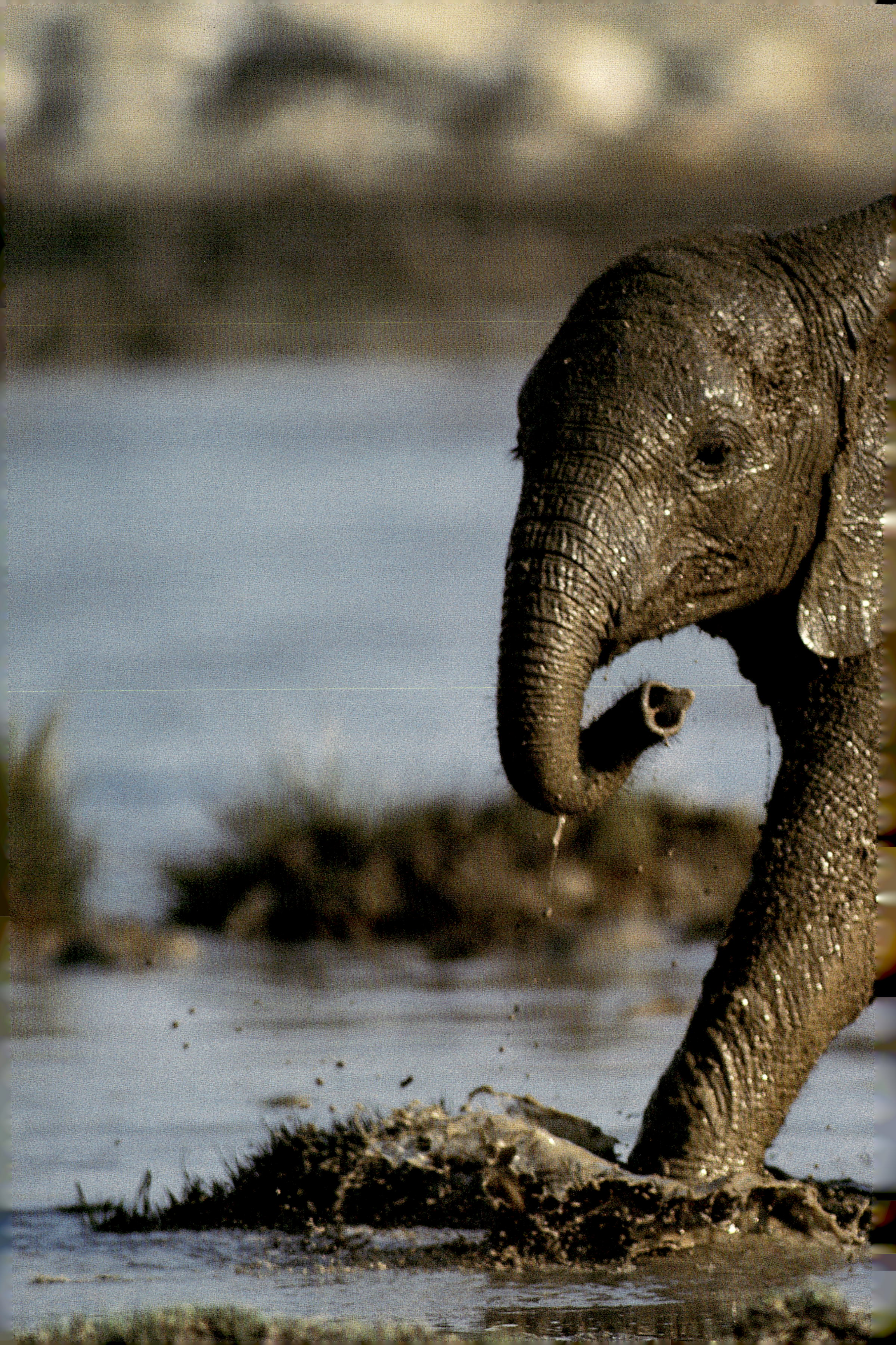

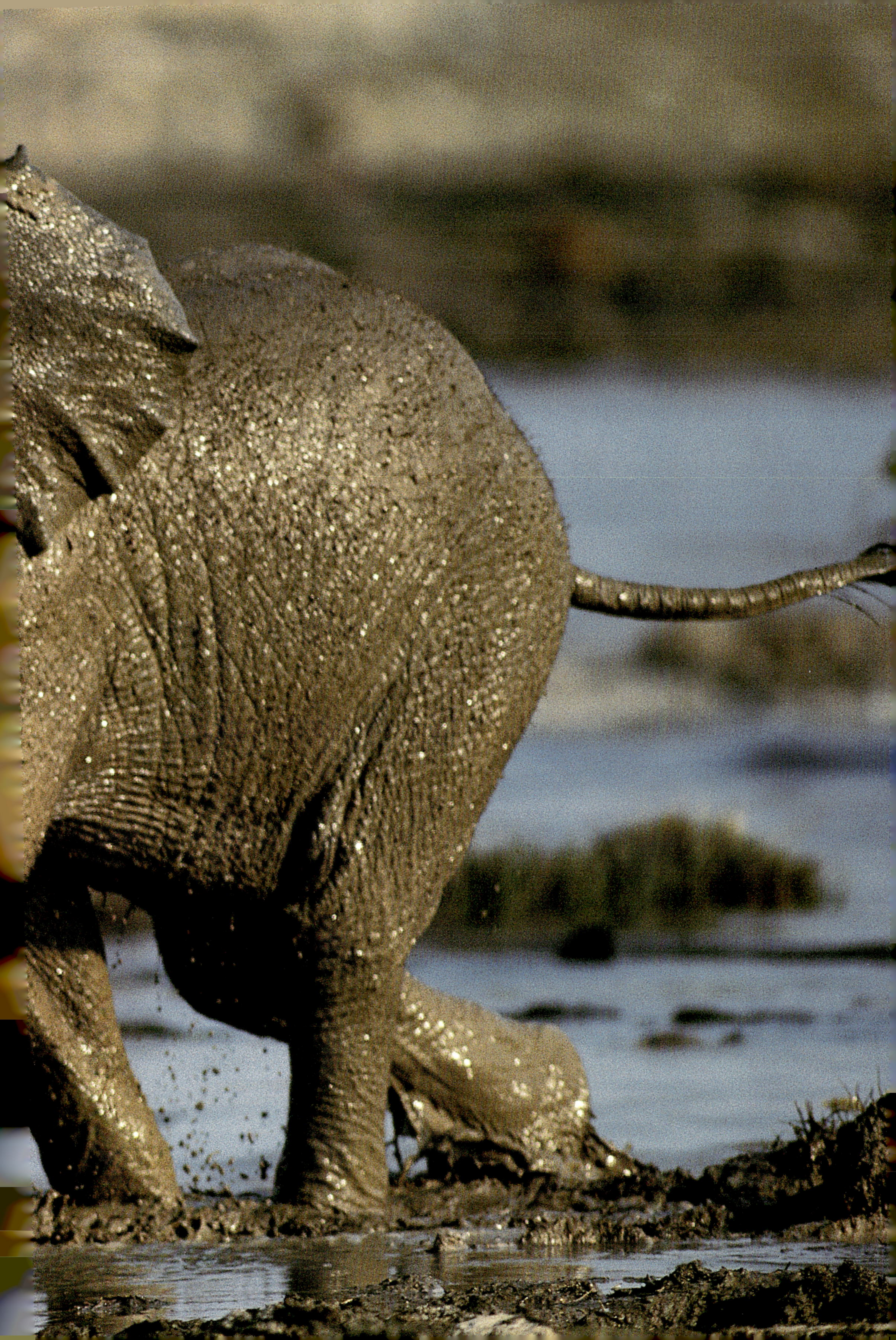

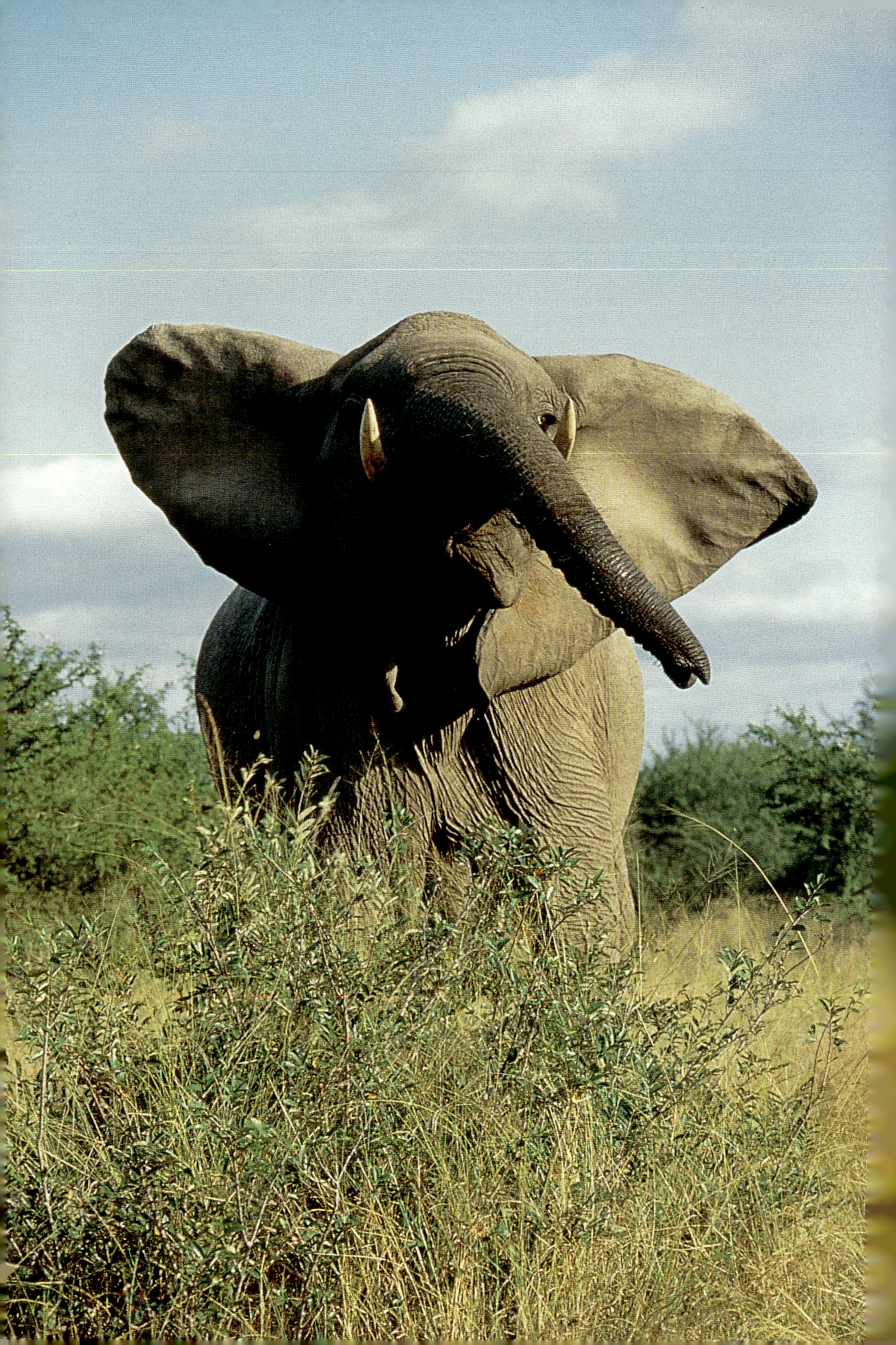

***How Elephant Got His Trunk*** (Traditional Southern African Tale)

*At the dawn of time, when Creator brought forth all the animals from the roots of the Baobab tree, Elephant was made to look quite different than he does today. Although he was large and grey, his nose was more like that of a pig, and he had to eat with his mouth. Because of his size, he had to eat constantly to satisfy his hunger. Stooping to eat was uncomfortable, and kneeling to drink was even more so. Elephant's life was little more that eating and drinking and his life was otherwise unfulfilled. One day as Elephant knelt to drink at the river, Crocodile, who was very hungry at the time, spotted him from his sunning rock. Crocodile slipped silently into the water and swiftly swam toward the unsuspecting Elephant, hoping for a big meal. Before Elephant could move, Crocodile had grabbed his piggy snout and started to drag Elephant into the water. But Elephant was strong and resisted, pulling with all his might in the opposite direction. Each animal pulled and pulled in desperation, and with each tug, Elephant's nose stretched and stretched and stretched. The tug-o-war went on and the hours passed into night. Finally, Elephant's strength won out and Crocodile released his grip, exhausted – and still hungry! Elephant sat down, exhausted as well and slowly came to the realization that he was different than before the struggle. His nose now hung low to the ground and he felt helpless as to what to do. Nothing would change it back, so Elephant had to learn to live with it and adapt to new ways of eating and drinking. It took much practice, but after a while it clearly became an advantage to have a trunk. Elephant could reach in all directions to find food and easily bring it to his mouth with little effort. He no longer had to stoop to drink, and he found many new uses for his trunk. He could pick branches to scratch his back. He could spray himself and use it to breathe when crossing the deep river by holding his trunk above water. To this day, each of Elephant's descendants is happy to have a long trunk, but every youngster must patiently learn to use it. Until they learn, young elephants can be seen nursing with their mouths and kneeling to drink from the river as Elephant once did, when time began.*

PRECEDING PAGES

*Happiness is a mud bath.*

FACING PAGE

*This angry cow became quite cantankerous, so my "chief scout" and I decided that a cautious retreat was in order. A calf that was hidden behind the shrubbery was the reason for her belligerence.*

FOLLOWING PAGES

*The end of a long, hot day in Namibia was happily balanced with the arrival of this herd at the water hole.*

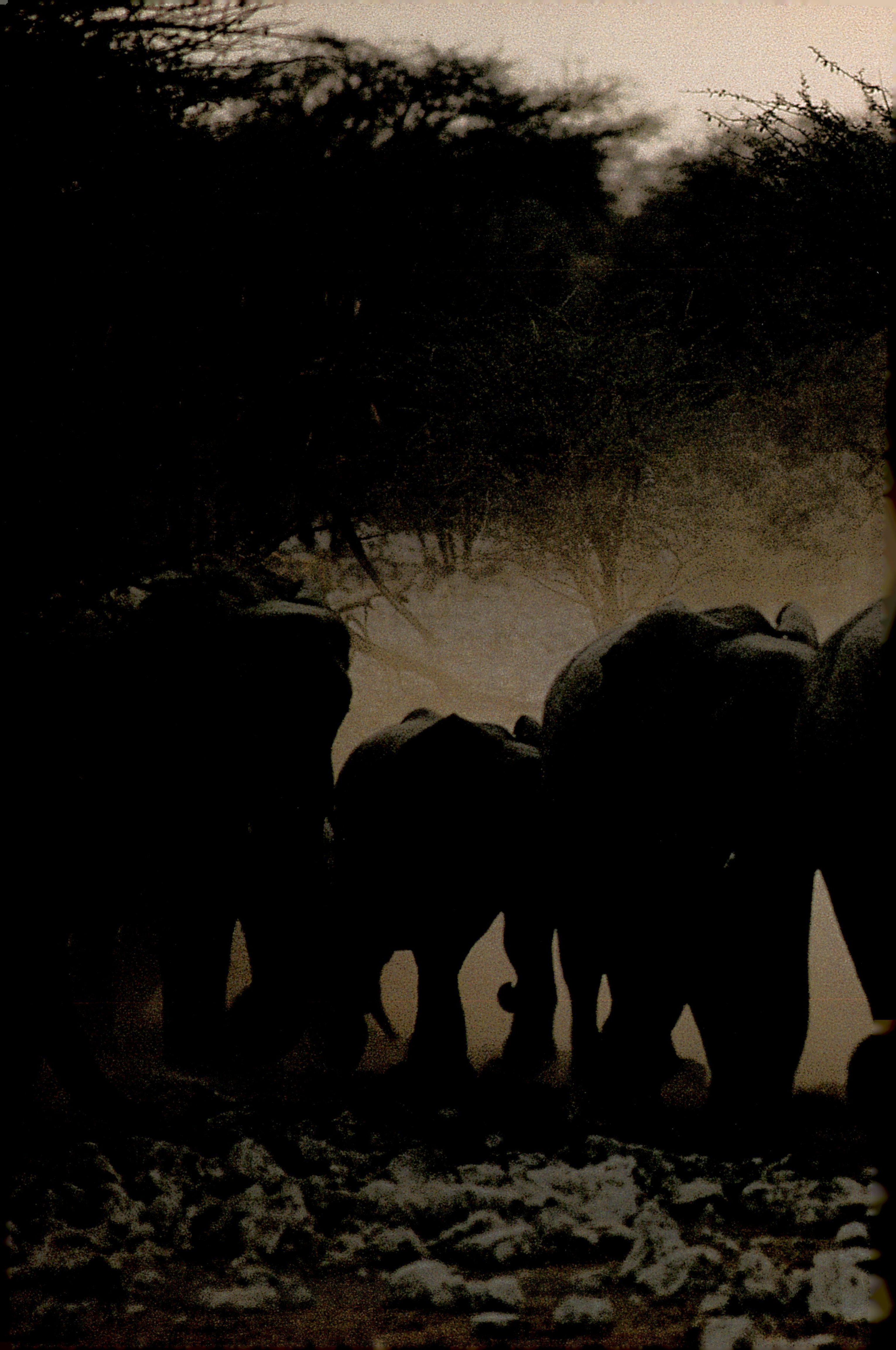

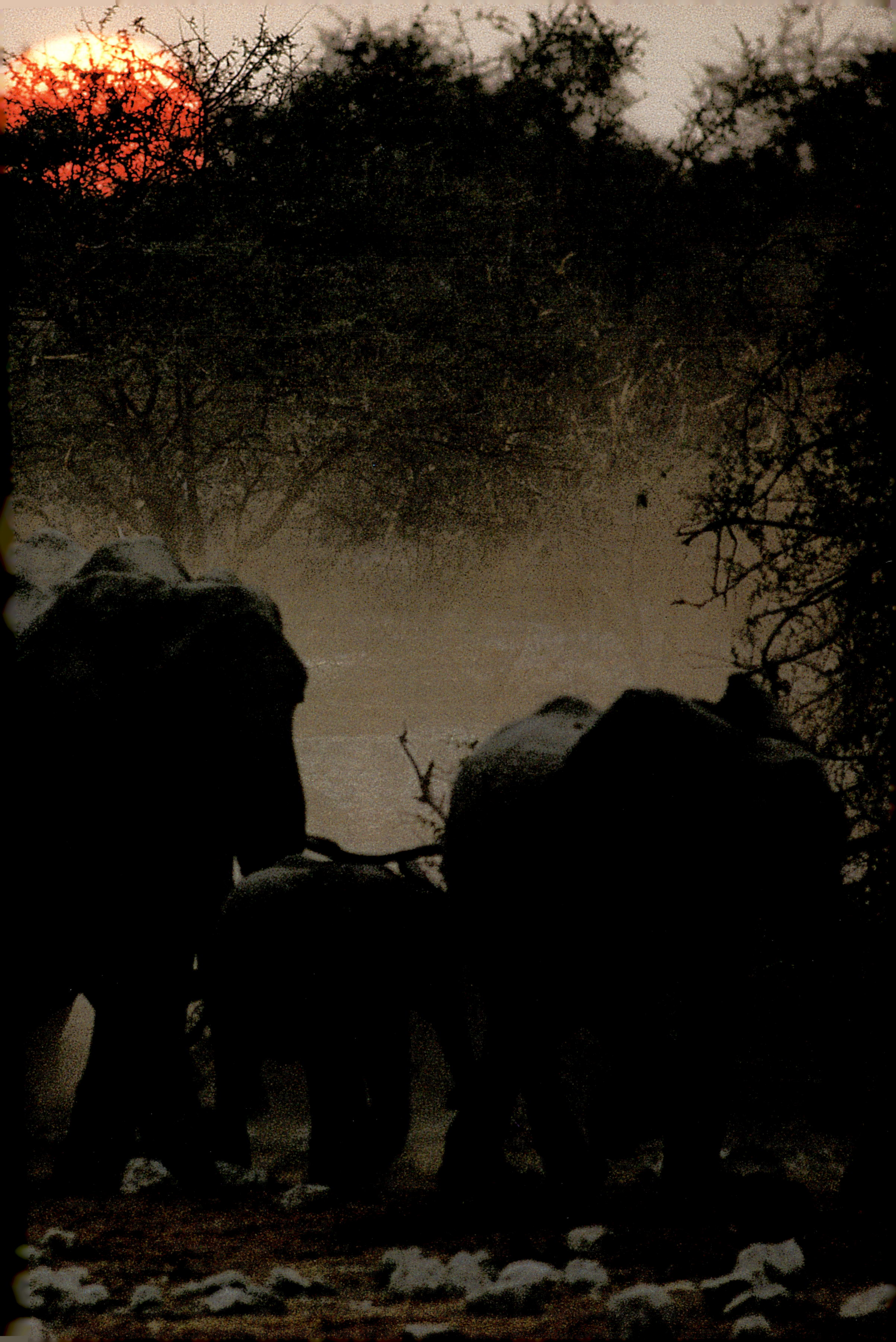

# EPILOGUE

The photos in this book are the result of numerous extended trips to Africa and I would enjoy sharing with you some of the ambiance and everyday occurrences that make up our days "on safari."

Many times the first impression people get of our travels is a mental image of us trekking through a hot jungle loaded with cameras and lenses. First of all there are no jungles in the areas where we traveled and secondly, we work diligently to NOT load ourselves with too much gear. Our visits to the bushveld and desert are really quite civilized and comfortable.

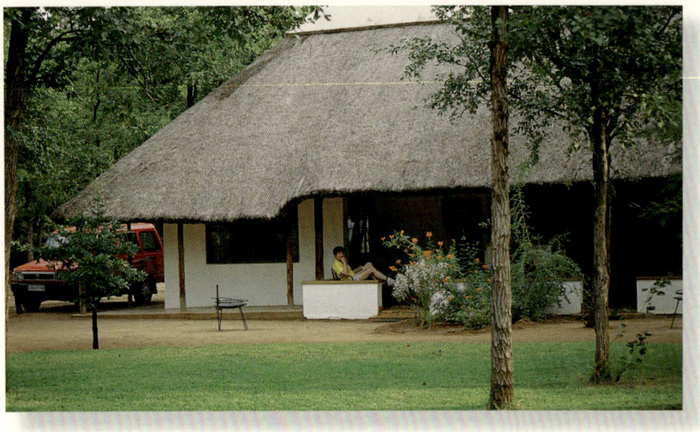

*Not all camps are tented or rustic. My chief scout, Jan, and I have spent many nights in this "hut" that is large and comfortable.*

The camps in which we stay and the huts or rondovels we use have everything we need. We may shop, dine at a restaurant or cafeteria, cook our own meals, take a hot or cold shower and turn on the air conditioner if need be. We have done all of the above. Camp life also affords the diversions of seeing hyenas at night just outside the fence, or in the Okavango Delta, the occasional elephant who meanders calmly through our back yard.

When we leave camp it is always in a vehicle. Any South African National Park containing predators requires visitors to remain in their vehicle at all times except in designated areas. The only time I was exempt from this rule was in the company of a park ranger. They, of course, are armed, although in all my experiences with them they never came close to firing a rifle. These are knowledgeable, conscientious men who know the habits of these animals and also know the yearnings and desires of photographers.

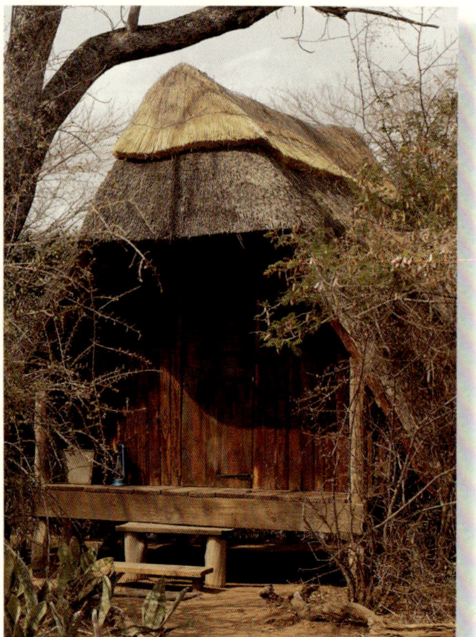

*After a hard day of hiking and photographing crocodiles and hippos, this rustic hut looked like a 5-star hotel to us.*

In addition to an exciting and rewarding day trip with rangers, overnight excursions are also offered. My good friend, Johann van der Merwe, led such a trip when I arrived at the Kruger Park with another good friend, Arden "Spook" Skelton. Camping with Johann meant no cafeteria, no indoor plumbing, no fences and certainly no air conditioner. A small wooden hut was available with simple cots and lanterns. "Spook" was quick to become the "coffee chef" and he made a great pot of coffee. Johann led us on a trek and delighted us with photo opportunities of Nile crocodiles. It was a fantastic camping experience.

However, as with any "thrilling" occupation there are hours of tedium, punctuated with moments of excitement. Patience is an ever-present virtue when working with animals. Waiting for them to act or re-act during long, hot hours is fundamental. However the opportunities to get a pleasing photograph are unquestionably worth every minute.

My hope is that you will find two or three (maybe more) elements in this book of special interest to you. The animals and their environment are certainly special to me and concern for their future should be an important consideration to all of us. A quite poignant statement was made by Baba Diom, a Senegalese diplomat, during a World Congress on the Environment.

*"In the end we will save only what we love.*

*We will love only what we understand.*

*We will understand only what we are taught."*

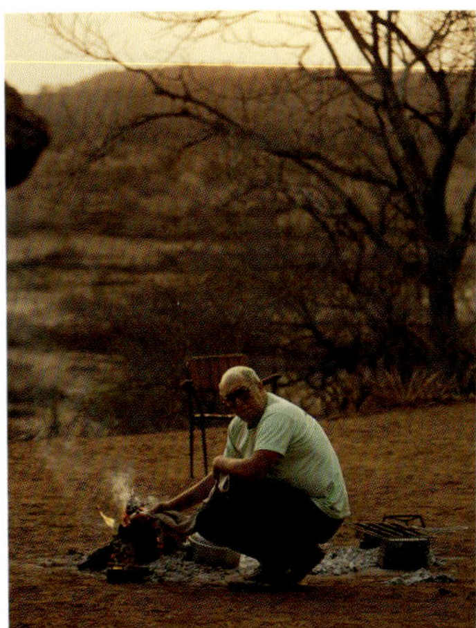

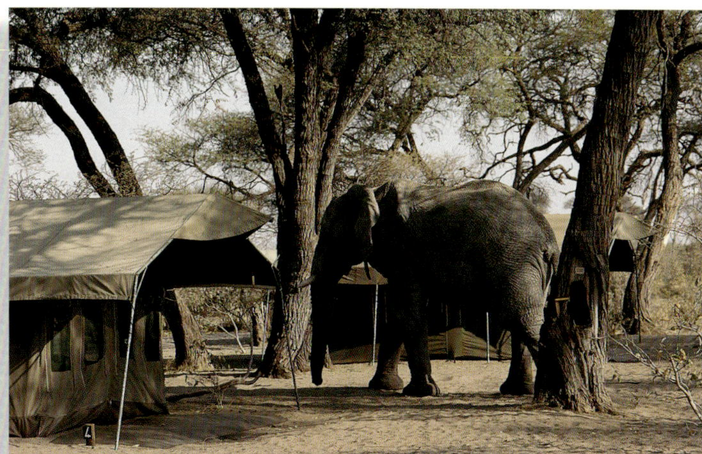

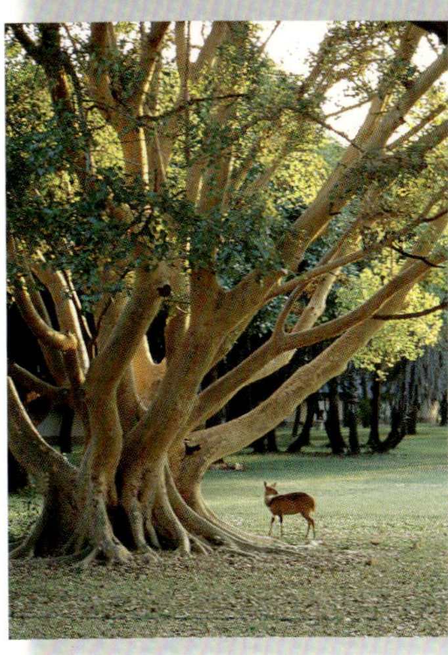

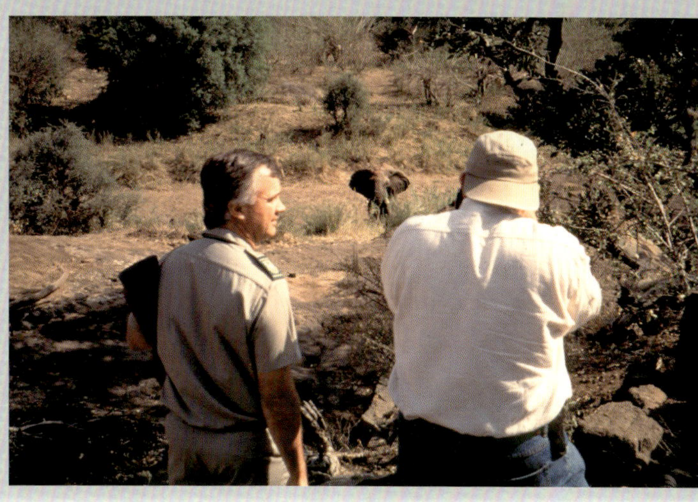

(PHOTO: ARDEN SKELTON)

ABOVE LEFT

*On two occasions a fellow photographer and close friend, Arden "Spook" Skelton, accompanied me to Africa. We spent some time at the Olifants trail camp where we were able to trek through the area on foot under the close supervision and protection of Johann van der Merwe who was, at the time, a trail's officer. Spook and I were shown some grand sights by Johann and we often reminisce over our experiences.*

ABOVE RIGHT

*When I returned to camp in Botswana's Okavango Delta early one evening I thought for a moment that I was going to have a roommate. The season was very dry and some elephants had the habit of looking for water in our camp. Fortunately, the bull passed through the tent area with no mishaps.*

BELOW LEFT

*One morning this bushbuck wandered into the Letaba rest camp to taste the sweet grasses.*

BELOW RIGHT

*If I were facing the camera you would easily see the joy and excitement on my face. My friend Johann knows how much I love elephants.*

# ANIMAL TAXONOMY-*MAMMALS, REPTILES & BIRDS*

## MAMMALS

| | |
|---|---|
| Steenbok | *Raphicerus campestris* |
| Klipspringer | *Oreotragus oreotragus* |
| Springbok | *Antidorcas marsupialis* |
| Waterbuck | *Kobus ellipsiprymnus* |
| Sable Antelope | *Hippotragus niger* |
| Gemsbok | *Oryx gazella* |
| Blue Wildebeest | *Connochaetes taurinus* |
| Impala | *Aepyceros melampus* |
| Bushbuck | *Tragelaphus scriptus* |
| Nyala | *Tragelaphus angasii* |
| Greater Kudu | *Tragelaphus strepsiceros* |
| African Buffalo | *Syncerus caffer* |
| Giraffe | *Giraffa camelopardalis* |
| Common Warthog | *Phacochoerus africanus* |
| Hippopotamus | *Hippopotamus amphibius* |
| White Rhinoceros | *Ceratotherium simum* |
| Burchell's Zebra | *Equus burchelli* |
| African Elephant | *Loxodonta africana* |
| Dwarf Mongoose | *Helogale parvula* |
| Suricate or Meerkat | *Suricata suricatta* |
| Spotted Hyena | *Crocuta crocuta* |
| African Wildcat | *Felis sylvestris* |
| Leopard | *Panthera pardus* |
| Lion | *Panthera leo* |
| Cheetah | *Acinonyx jubatus* |
| Black-backed Jackal | *Canis mesomelas* |
| Wild Dog | *Lycaon pictus* |
| Chacma Baboon | *Papio ursinus* |

## REPTILES

| | |
|---|---|
| Nile Crocodile | *Crocodylus niloticus* |
| Nile Monitor Lizard | *Varanus niloticus niloticus* |

## BIRDS

| | |
|---|---|
| Goliath Heron | *Ardea goliath* |
| Grey Heron | *Ardea cinerea* |
| Cattle Egret | *Bubulcus ibis* |
| Greenbacked Heron | *Butorides striatus* |
| Black Stork | *Ciconia nigra* |
| Yellowbilled Stork | *Mycteria ibis* |
| Marabou Stork | *Leptoptilos crumeniferus* |
| Saddlebilled Stork | *Ephippiorhynchus senegalensis* |
| Hammerkop | *Scopus umbretta* |
| Egyptian Goose | *Alopochen aegyptiacus* |
| Lappetfaced Vulture | *Torgos tracheliotus* |
| Whitebacked Vulture | *Gyps africanus* |
| Hooded Vulture | *Necrosyrtes monachus* |
| African Fish Eagle | *Haliaeetus vocifer* |
| Bateleur | *Terathopius ecaudatus* |
| Tawny Eagle | *Aquila rapax* |
| Martial Eagle | *Polemaetus bellicosus* |
| Gymnogene | *Polyboroides typus* |
| Dark Chanting Goshawk | *Melierax metabates* |
| Little Sparrowhawk | *Accipiter minullus* |
| Helmet Guineafowl | *Numida meleagris* |
| Ground Hornbill | *Bucorvus leadbeateri* |
| Ostrich | *Struthio camelus* |
| Secretarybird | *Sagittarius serpentarius* |
| Kori Bustard | *Ardeotis kori* |
| Redcrested Korhaan | *Eupodotis ruficrista* |
| Giant Eagle Owl | *Bubo lacteus* |
| African Scops Owl | *Otus senegalensis* |
| Giant Kingfisher | *Ceryle maxima* |
| Pied Kingfisher | *Ceryle rudis* |
| Pygmy Kingfisher | *Ispidina picta* |
| Whitefronted Bee-eater | *Merops bullockoides* |
| Lilacbreasted Roller | *Coracias caudata* |
| Southern Yellowbilled Hornbill | *Tockus flavirostris* |
| Crested Barbet | *Trachyphonus vaillantii* |
| Glossy Starling | *Lamprotornis nitens* |
| Redbilled Oxpecker | *Buphagus erythrorhynchus* |
| Lesser Masked Weaver | *Ploceus intermediu* |

# REFERENCES


Allen, Thomas B. (1997). *ANIMALS OF AFRICA.* Washington, DC.: Beau Arts Edition

Estes, R.D. (1992). *THE BEHAVIOR GUIDE TO AFRICAN MAMMALS.* Berkeley & Los Angeles, CA.: University of California Press

Estes, R.D. (1999). *THE SAFARI COMPANION.* White River Junction, VT.: Chelsea Green

Fourie, P.F. (1984). *KRUGER NATIONAL PARK, QUESTIONS AND ANSWERS.* Nelspruit, RSA.: South African Country Life

Greaves, Nick (1996). *WHEN ELEPHANT WAS KING.* Southern Book Publishers

Greaves, Nick (1989). *WHEN HIPPO WAS HAIRY.* Swaziland and Durban: Bok Books International

Kindgon, J. (1997). *THE BEHAVIOR GUIDE TO AFRICAN MAMMALS.* Berkeley and Los Angeles, CA.: University of California Press

Maclean, Gordon Lindsay (1985). *ROBERTS' BIRDS OF SOUTHERN AFRICA.* Cape Town, RSA.: The Trustees of the John Voeloker Bird Book Fund

Mundy, P., Butchart, D., Ledger, J., Piper, S., (1992). *THE VULTURES OF AFRICA.* Randburg, RSA.: Acorn Books and Russel Friedman Books

Newman, Kenneth (1983). *NEWMAN'S BIRDS OF SOUTHERN AFRICA.* Johannesburg, RSA.: Macmillan

Palgrave, Keith Coates (1992). *TREES OF SOUTHERN AFRICA.* Nelspruit, RSA.: Struik Publishers

Pienaar, U. DE V., Joubert, S.C.J., Hall-Martin, A., De Graaff, G., Rautenbach, I.L. (1987). *FIELD GUIDE TO THE MAMMALS OF THE KRUGER NATIONAL PARK.* Cape Town, RSA.: C. Struik Publishers

Proctor, N.S., and Lynch, P.J. (1983). *MANUAL OF ORNITHOLOGY.* New Haven and London: Yale University

Sinclair, Ian: Hockey, Phil: Tarboton, Warwick (1993). *SASOL BIRDS OF SOUTHERN AFRICA.* Cape Town, RSA.: Struik Publishers

Skinner, J.D. and Smithers, R.H.N. (1990). *THE MAMMALS OF THE SOUTHERN AFRICAN SUBREGION.* Pretoria, RSA.: University of Pretoria

Tingay, Paul (1995). WILDEST AFRICA. New York, N.Y.: St. Martin's Press

Wyk, Piet van (1988). *FIELD GUIDE TO THE TREES OF THE KRUGER NATIONAL PARK.* Cape Town, RSA.: Struik Publishers

Whitfield, Dr. Philip, General Consultant (1998). The Simon & Shuster *ENCYCLOPEDIA OF ANIMALS.* New York, N.Y.: Simon & Shuster Editions

*ZOO BOOKS.* San Diego, CA.: Wildlife Education Ltd. (various issues)

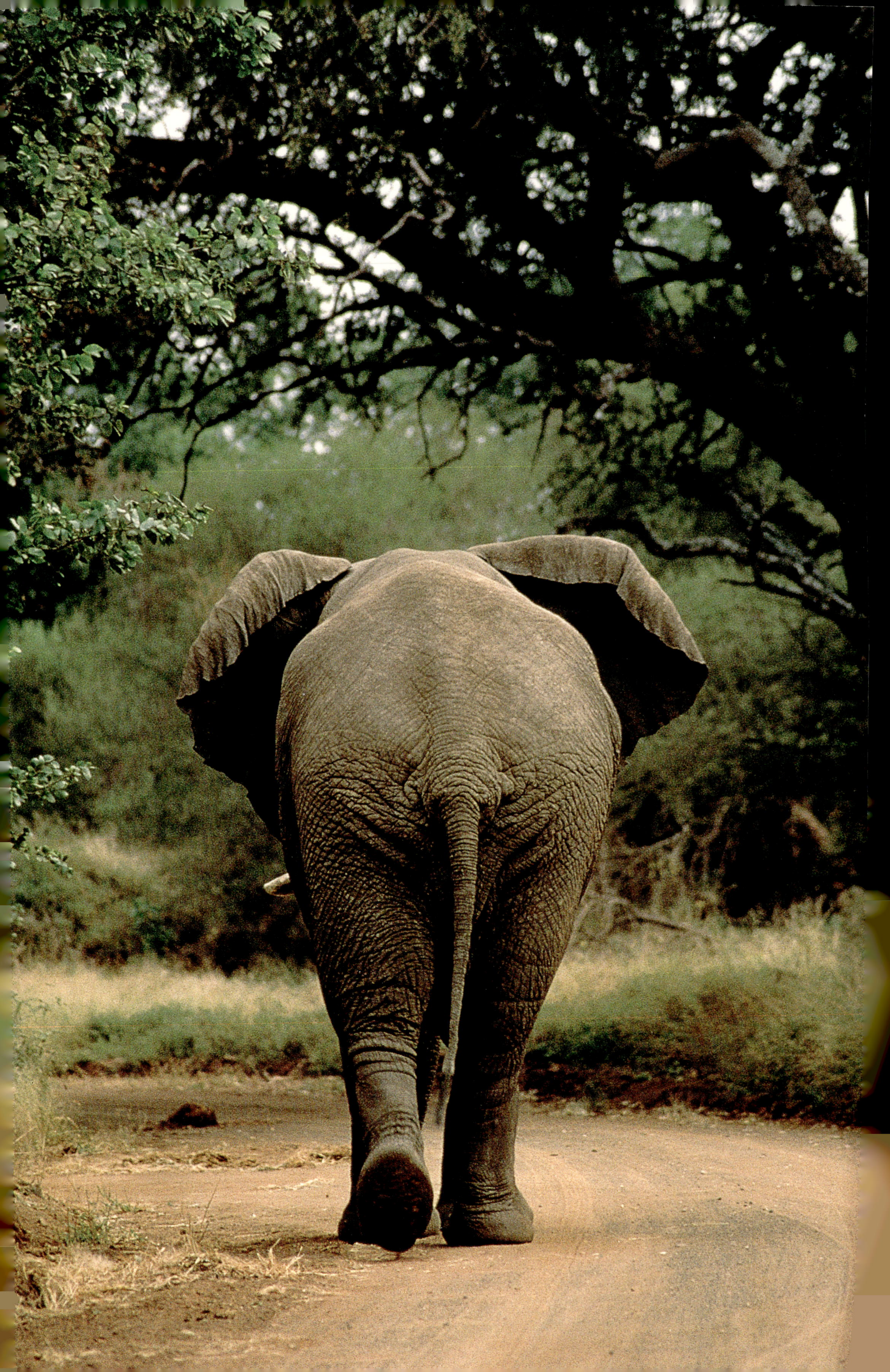